AF581878

# Environmental Sustainability in Maritime Infrastructures

# Environmental Sustainability in Maritime Infrastructures

Editors

**M. Dolores Esteban**
**José-Santos López-Gutiérrez**
**Vicente Negro**
**José Marcos Ortega**

MDPI • Basel • Beijing • Wuhan • Barcelona • Belgrade • Manchester • Tokyo • Cluj • Tianjin

*Editors*
M. Dolores Esteban
Departamento de Ingeniería Civil, Urbanismo y Aeroespacial
Universidad Europea
Madrid
Spain

José-Santos López-Gutiérrez
Departamento de Ingeniería Civil, Hidráulica, Energía y Medio Ambiente
Universidad Politécnica de Madrid
Madrid
Spain

Vicente Negro
Departamento de Ingeniería Civil, Hidráulica, Energía y Medio Ambiente
Universidad Politécnica de Madrid
Madrid
Spain

José Marcos Ortega
Departamento de Ingeniería Civil
Universidad de Alicante
Alicante
Spain

*Editorial Office*
MDPI
St. Alban-Anlage 66
4052 Basel, Switzerland

This is a reprint of articles from the Special Issue published online in the open access journal *International Journal of Environmental Research and Public Health* (ISSN 1660-4601) (available at: www.mdpi.com/journal/ijerph/special_issues/Maritime_Infrastructures).

For citation purposes, cite each article independently as indicated on the article page online and as indicated below:

| LastName, A.A.; LastName, B.B.; LastName, C.C. Article Title. *Journal Name* **Year**, *Volume Number*, Page Range. |
|---|

**ISBN 978-3-0365-2260-9 (Hbk)**
**ISBN 978-3-0365-2259-3 (PDF)**

# Contents

# About the Editors

**M. Dolores Esteban**

María Dolores Esteban (Ph.D.) is Associate Professor in Coastal Engineering at the Universidad Politécnica de Madrid and in Coastal Engineering and Renewable Energies at the Universidad Europea. She has more than 15 years of experience in the fields of maritime engineering and renewable energies as an engineer in a private company and as a lecturer and researcher at the university. She has more than 50 technical papers in SCI indexed journals and numerous oral presentations in international conferences. Her main fields of research are coastal and maritime engineering, harbor and ports, beaches, loads actions and effects in maritime structures, renewable energies like offshore wind, wave, and currents. She is very active in R&D projects, being an important asset to her research group the Environmental, Coast and Ocean Research Laboratory (ECOREL) at the Universidad Politécnica de Madrid.

**José-Santos López-Gutiérrez**

José-Santos López-Gutiérrez (Ph.D.) is Associate Professor in Coastal Engineering at the Universidad Politécnica de Madrid. He has more than 25 years of experience in the fields of maritime engineering and renewable energies as a lecturer and researcher at the university. He has supervised 5 Ph.D. dissertations. He has more than 60 technical papers published in SCI indexed journals and more than 60 oral presentations in national and international conferences. He is responsible of the research line on marine energies within the group of researchers of the Environmental, Coast and Ocean Research Laboratory (ECOREL) at Universidad Politécnica de Madrid. Apart from research fields related to marine energies, such as load and effect actions on marine structures for offshore wind installations, and wave and current exploitation, other research fields in which he is involved include coastal and maritime engineering, ports, and beaches.

**Vicente Negro**

Vicente Negro (Ph.D.) is a Full Professor in Coastal Engineering He is the head of the Environmental, Coast and Ocean Research Laboratory (ECOREL) at Universidad Politécnica de Madrid. He has supervised more than 10 Ph.D. dissertations. He has more than 70 technical papers in SCI indexed journals and more than 40 scientific and dissemination articles at national level. He has more than 40 contributions in national and international conferences with peer arbitration. He is an expert in marine sciences in the Program of the Directorate General XII of Research in Brussels, Drafting Commission of the ROM (Puertos del Estado), being an Editorial Member of the ROM0.3/91, Environmental Actions I: Waves; ROM 0.0/2002, General procedure and calculation bases in the maritime and port works project; ROM 1.0/09, Recommendations for the design of Breakwaters and ROM 2.0/11, Recommendations for the project and execution of Docking and Mooring Works.

**José Marcos Ortega**

José Marcos Ortega (Ph.D.) is Associate Professor in Civil Engineering Department at University of Alicante (Spain). He belongs to the research group "Durability of materials and constructions in Engineering and Architecture" of the same university. His research topics are focused on sustainability and durability of construction and building materials, performance of cement-based materials exposed to harmful conditions, particularly marine environments, and non-destructive techniques for characterizing cementitious materials. He is the co-author of around 40 scientific manuscripts in SCI indexed journals, 5 book chapters and more than 30 communications to national and international conferences. He has participated in several competitive R&D projects funded by national and regional agencies, being the researcher responsible in three of them. He has supervised one Ph.D. dissertation. He has received several awards, highlighting the "Young Investigator Award in Environmental and Sustainable fields 2018" given by *Applied Sciences* journal.

# Preface to "Environmental Sustainability in Maritime Infrastructures"

In 2019, we started working together with the editorial team of the *International Journal of Environmental Research and Public Health* (MDPI editorial) to act as guest editors of a Special Issue related to sustainability in the sea. This opportunity was welcomed with great enthusiasm by the Guest Editorial Team. We started working on the project, with the great support of the Main Editor who has guided us to achieve a Special Issue that has exceeded our initial expectations. This Special Issue is entitled "Environmental Sustainability in Maritime Infrastructures". Oceans and coastal areas are essential in our lives from several different points of view: social, economic, and health. Given the importance of these areas for human life, not only for the present but also for the future, it is necessary to plan future infrastructures, and maintain and adapt to the changes the existing ones. All of this taking into account the sustainability of our planet. A very significant percentage of the world's population lives permanently or enjoys their vacation periods in coastal zones, which makes them very sensitive areas, with a very high economic value and as a focus of adverse effects on public health and ecosystems. Therefore, it is considered very relevant and of great interest to launch this Special Issue to cover any aspects related to the vulnerability of coastal systems and their inhabitants (water pollution, coastal flooding, climate change, overpopulation, urban planning, waste water, plastics at sea, effects on ecosystems, etc.), as well as the use of ocean resources (fisheries, energy, tourism areas, etc.). This Special Issue achieved published papers. The papers have very good quality, being their title "Economic aspects of a concrete floating offshore wind platform in the Atlantic Arc of Europe", "A floating offshore renewable energy economic software", "Transparency of financial reporting on greenhouse gas emission allowances: The influence of regulation", "Life cycle assessment on wave and tidal energy systems: a review of current methodological practice", "Carbon Footprint of a Port infrastructure from a Life Cycle approach", "Subsystem Hazard Analysis on an Offshore Waste Disposal Facility", "Wave energy assessment at Valencia Gulf and comparison of energy production of most suitable wave energy converters", "A review on environmental and social impacts of thermal gradient and tidal currents energy conversion and application to the case of Chiapas, México", "Community-Based Portable Reef to Promote Mangrove Vegetation Growth: A Bridging between Ecological and Engineering Principles", "A GIS-Based Artificial Neural Network Model for Flood Susceptibility Assessment", "Evaluation of Formal and Informal Spatial Coastal Area Planning Process in Baltic Sea Region", "Socio-economic context and community resilience among the people involved in fish drying practices in the South-East coast of Bangladesh", and "Matrix of Architectural Solutions for the Conflict between Transport Infrastructures, Landscape and Urban Habitat along the Mediterranean Coastline: The Case of the Maresme Region in Barcelona, Spain". It is a pleasure for the Guest Editorial Team to have put all these interesting manuscripts together. We thank the authors of the articles that have allowed this Special Issue to have such a high level.

**M. Dolores Esteban, José-Santos López-Gutiérrez, Vicente Negro, José Marcos Ortega**
*Editors*

International Journal of
*Environmental Research and Public Health*

MDPI

*Article*

# A GIS-Based Artificial Neural Network Model for Flood Susceptibility Assessment

**Nanda Khoirunisa, Cheng-Yu Ku * and Chih-Yu Liu ***

Department of Harbor and River Engineering, National Taiwan Ocean University, Keelung City 20224, Taiwan; 10786013@mail.ntou.edu.tw
* Correspondence: chkst26@mail.ntou.edu.tw (C.-Y.K.); 20452003@email.ntou.edu.tw (C.-Y.L.)

**Abstract:** This article presents a geographic information system (GIS)-based artificial neural network (GANN) model for flood susceptibility assessment of Keelung City, Taiwan. Various factors, including elevation, slope angle, slope aspect, flow accumulation, flow direction, topographic wetness index (TWI), drainage density, rainfall, and normalized difference vegetation index, were generated using a digital elevation model and LANDSAT 8 imagery. Historical flood data from 2015 to 2019, including 307 flood events, were adopted for a comparison of flood susceptibility. Using these factors, the GANN model, based on the back-propagation neural network (BPNN), was employed to provide flood susceptibility. The validation results indicate that a satisfactory result, with a correlation coefficient of 0.814, was obtained. A comparison of the GANN model with those from the SOBEK model was conducted. The comparative results demonstrated that the proposed method can provide good accuracy in predicting flood susceptibility. The results of flood susceptibility are categorized into five classes: Very low, low, moderate, high, and very high, with coverage areas of 60.5%, 27.4%, 8.6%, 2.5%, and 1%, respectively. The results demonstrate that nearly 3.5% of the study area, including the core district of the city and an exceedingly populated area including the financial center of the city, can be categorized as high to very high flood susceptibility zones.

**Keywords:** geographic information system; back-propagation neural network; rainfall; historical flood; prediction

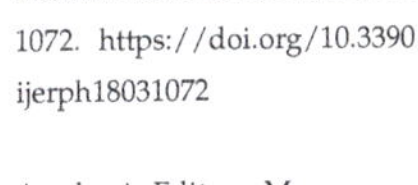

**Citation:** Khoirunisa, N.; Ku, C.-Y.; Liu, C.-Y. A GIS-Based Artificial Neural Network Model for Flood Susceptibility Assessment. *Int. J. Environ. Res. Public Health* **2021**, *18*, 1072. https://doi.org/10.3390/ijerph18031072

Academic Editors: M. Dolores Esteban, José-Santos López-Gutiérrez, Vicente Negro and José Marcos Ortega

Received: 12 December 2020
Accepted: 21 January 2021
Published: 26 January 2021



## 1. Introduction

Coastal areas are vulnerable to climate change, particularly sea-level rises and floods related to extreme rainfall [1,2]. Taiwan is an island which is prone to flood disasters triggered by heavy rainfall and typhoons every year. In Taiwan, extreme weather conditions, such as heavy precipitation and typhoons generated by climate change, strengthen the phenomenon of flood disasters [3]. Keelung City, one of the coastal cities in Northern Taiwan, has become highly urbanized and densely populated in recent years [4]. Flooding events have frequently occurred in the past because typhoons and rainstorms typically sweep over the upstream basins between May and October; this trend is expected to increase in the future [5–7].

Various approaches have been proposed to evaluate flood disaster risk based on the susceptibility of the system and hydrology [8–10]. The artificial neural network (ANN) is one of the most implemented machine-learning techniques in engineering risk assessment [11]. The ANN model is a network of machine learning that is based on the human brain [12]. Nowadays, ANNs and computational intelligence (CI) methods are often used for flood disaster modeling [13]. Machine-learning technologies have been applied for analysis of flood susceptibility assessment, including logistic regression, radial basis function (RBF) neural network, and support vector machine (SVM) [14–16]. For logistic regression, this algorithm is easier to implement and interpret, and more efficient to train [14]. However, it may lead to overfitting. For the RBF neural network, it performs more robustly and tolerantly than conventional neural networks, especially when dealing with noisy data [15]. For SVMs, it is more effective in high-dimensional spaces [16]. Despite the success of the above machine-learning technologies as effective numerical tools for engineering risk assessment, there is still growing interest in the development of a more accurate predictive risk model. The analysis of the spatial distribution of flood disaster risk plays an important role, especially regarding disasters occurring along coasts and rivers [17–19]. Spatial analysis is applied to define the relationship between flood factors in hazards, vulnerability, and risk through the map, without focusing on complex hydrological modeling [20]. For flood analysis, studies using geographic information system (GIS) technologies, remote sensing, and numerical models, and adopting an artificial neural network approach, have been widely used around the world [21–26].

This article presents a GIS-based artificial neural network (GANN) model for the flood susceptibility assessment of Keelung City, Taiwan. Various factors, including elevation, slope angle, slope aspect, flow accumulation, flow direction, topographic wetness index (TWI), drainage density, rainfall, and normalized difference vegetation index, were generated using a digital elevation model and LANDSAT 8 imagery. Historical flood data from 2015 to 2019, including 307 flood events, were adopted for a comparison of flood susceptibility. Using these factors, the GANN model, based on the back-propagation neural network (BPNN), was employed to assess flood susceptibility. The main contribution of this work is that the proposed method, based on ANN and GIS, may improve the ability to establish further precise flood models, and present the results in a spatial environment. The advantages of the GIS spatial analysis capability were integrated into the artificial neural network model. This work is organized as follows. In Section 2, the methodology is introduced. Results are presented in Section 3. Discussions are presented in Section 4, and key findings of this pioneering work are summarized in this section. Conclusions are made in Section 5.

## 2. Materials and Methods

### 2.1. Description of the Study Area

The research area was Keelung City, which is located in the northeastern part of Taiwan. The city area covers 132.7589 km$^2$, and is divided into 7 districts and 157 villages, as shown in Figure 1. The city is also known as the rainy port for its high frequency of rain, with a yearly rainfall average upwards of 3700 mm. Keelung City is one of the major coastal cities in Northern Taiwan which has become highly urbanized and densely populated in the last few years [4].

### 2.2. Preparation of Data and Geospatial Layer

Table 1 depicts the source data of the factors. The geomorphologic area of Keelung City and the relevant factors are shown in Figure 2. Factors including elevation, slope angle, slope aspect, flow accumulation, flow direction, and TWI were generated from the digital elevation model (DEM), with a resolution of 20 m. The LANDSAT 8 imagery, with a resolution of 30 m, from the United States Geological Survey, was used to generate the normalized difference vegetation index. Detailed descriptions of the factors are as follows.

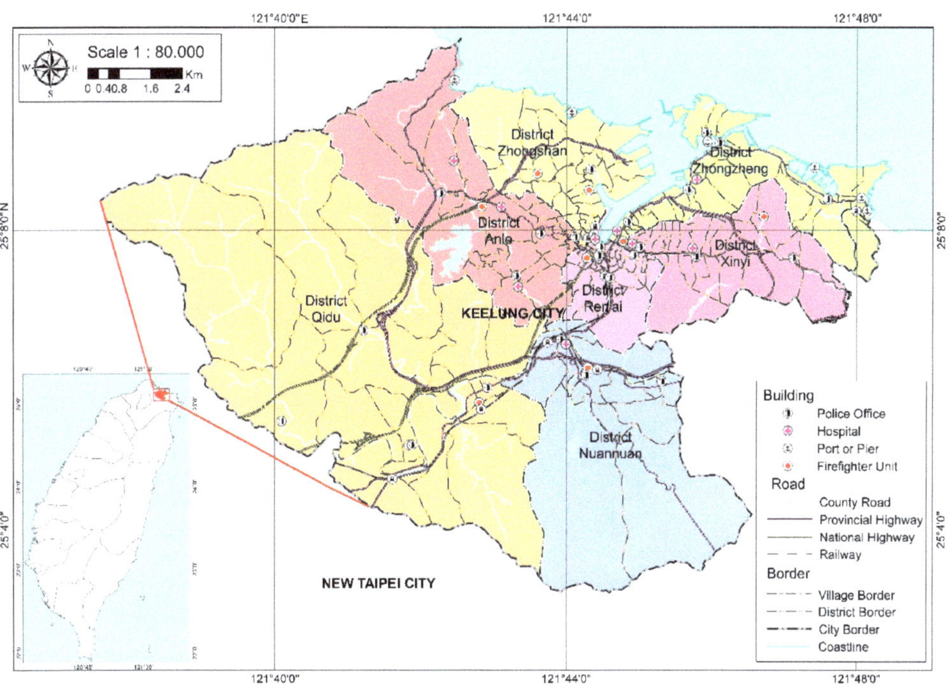


**Figure 1.** Administration map of Keelung City, Taiwan.

**Table 1.** Source data of the factors and outliers. DEM: Digital elevation model.

| Source Data | Factors | Resolution (m) | Maximum Outliers |
|---|---|---|---|
| DEM | Elevation | 20 | 704 |
| DEM | Slope | 20 | 60.40 |
| DEM | Slope aspect | 20 | 358.95 |
| DEM | Flow direction | 20 | 128 |
| DEM | Flow accumulation | 20 | 247,638 |
| DEM | Topographic wetness index (TWI) | 20 | 19.74 |
| Stream river | Drainage density | 20 | 4.07 |
| Rainfall data from the Central Weather Bureau (CWB) of Taiwan | Rainfall interpolation | 96 | 102.65 |
| LANDSAT 8 imagery | Normalized difference vegetation index | 30 | 0.41 |

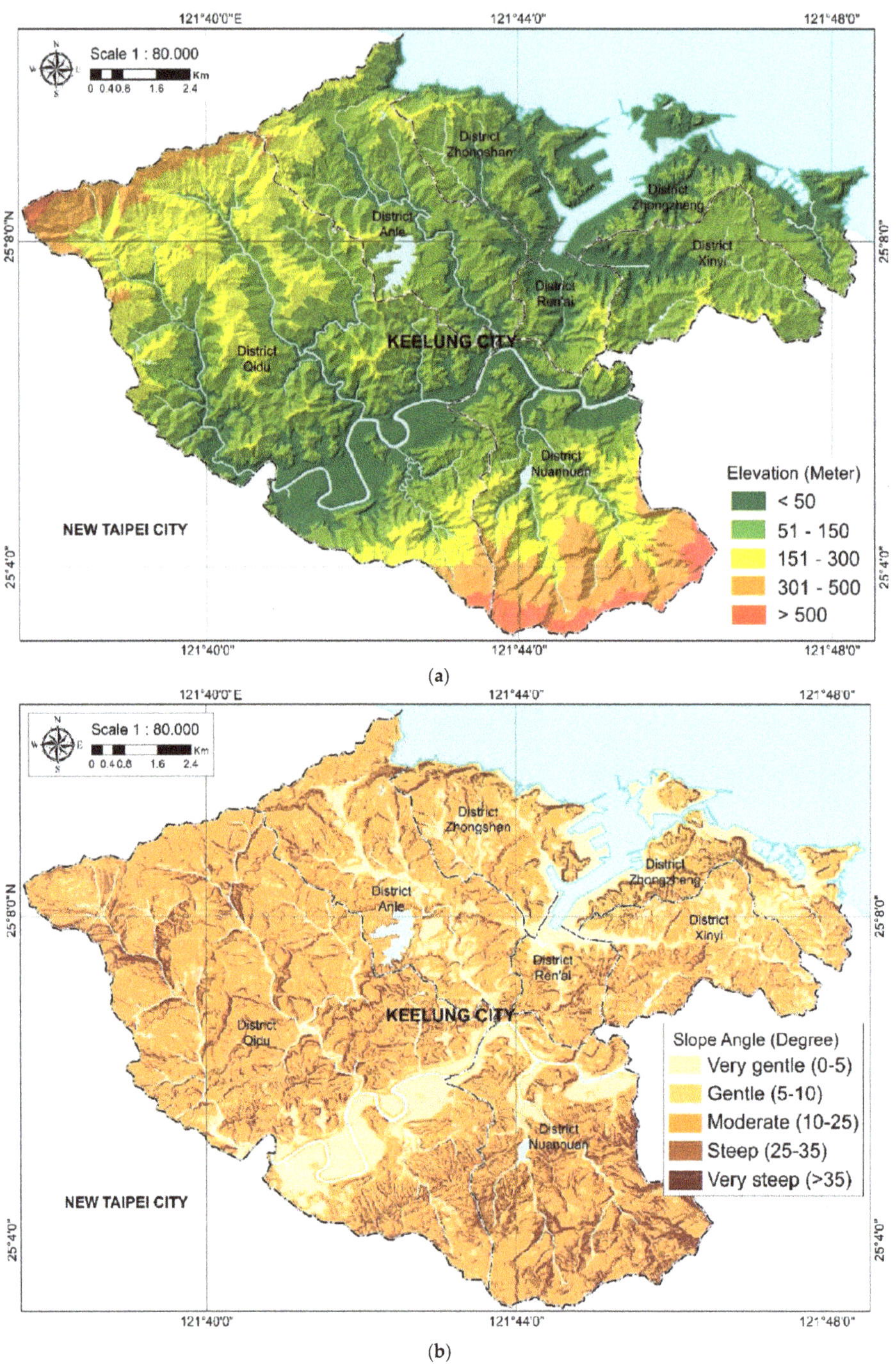


(a)

(b)

**Figure 2.** *Cont.*

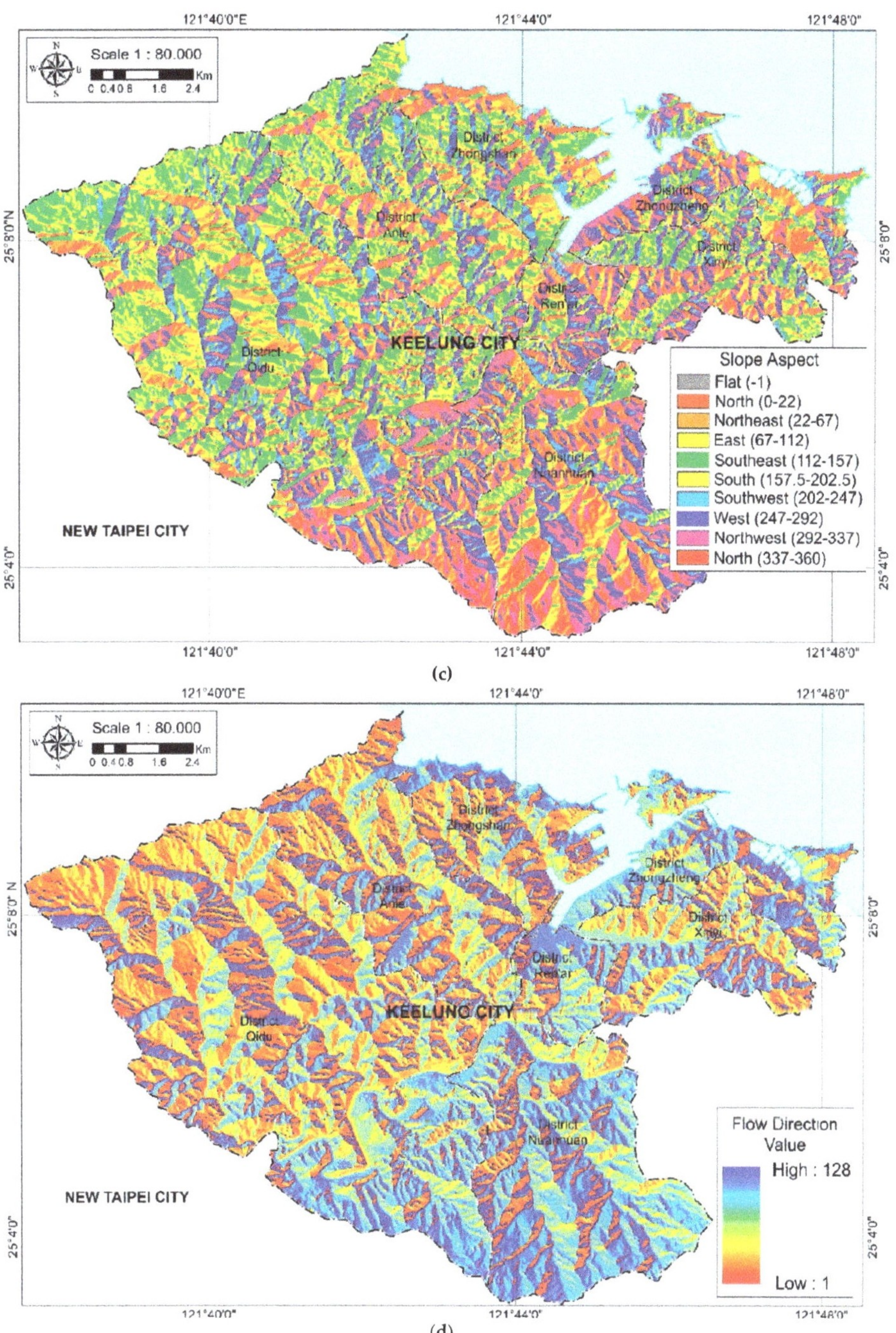


(c)



(d)

**Figure 2.** *Cont.*

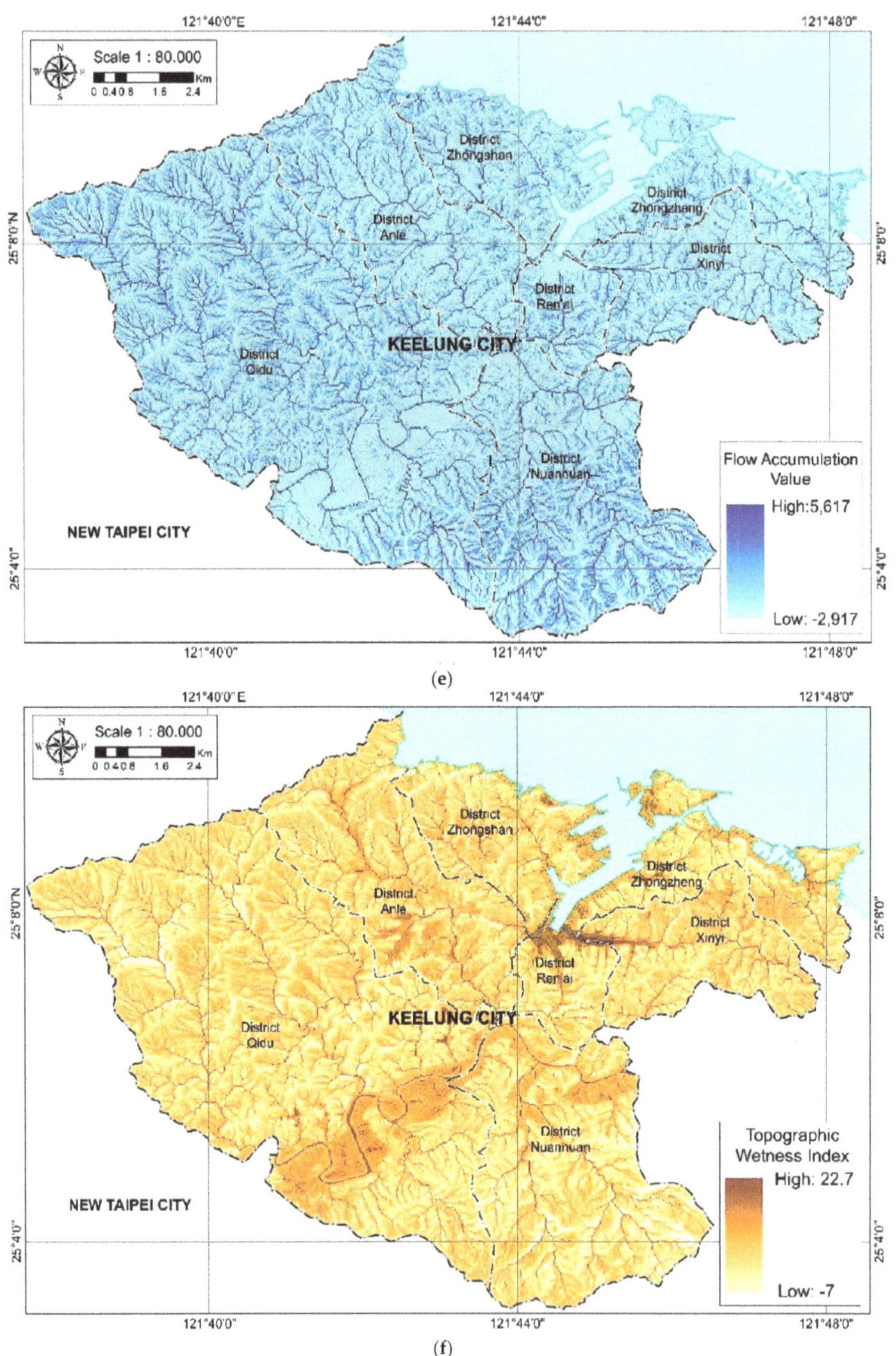


(e)

(f)

**Figure 2.** *Cont.*

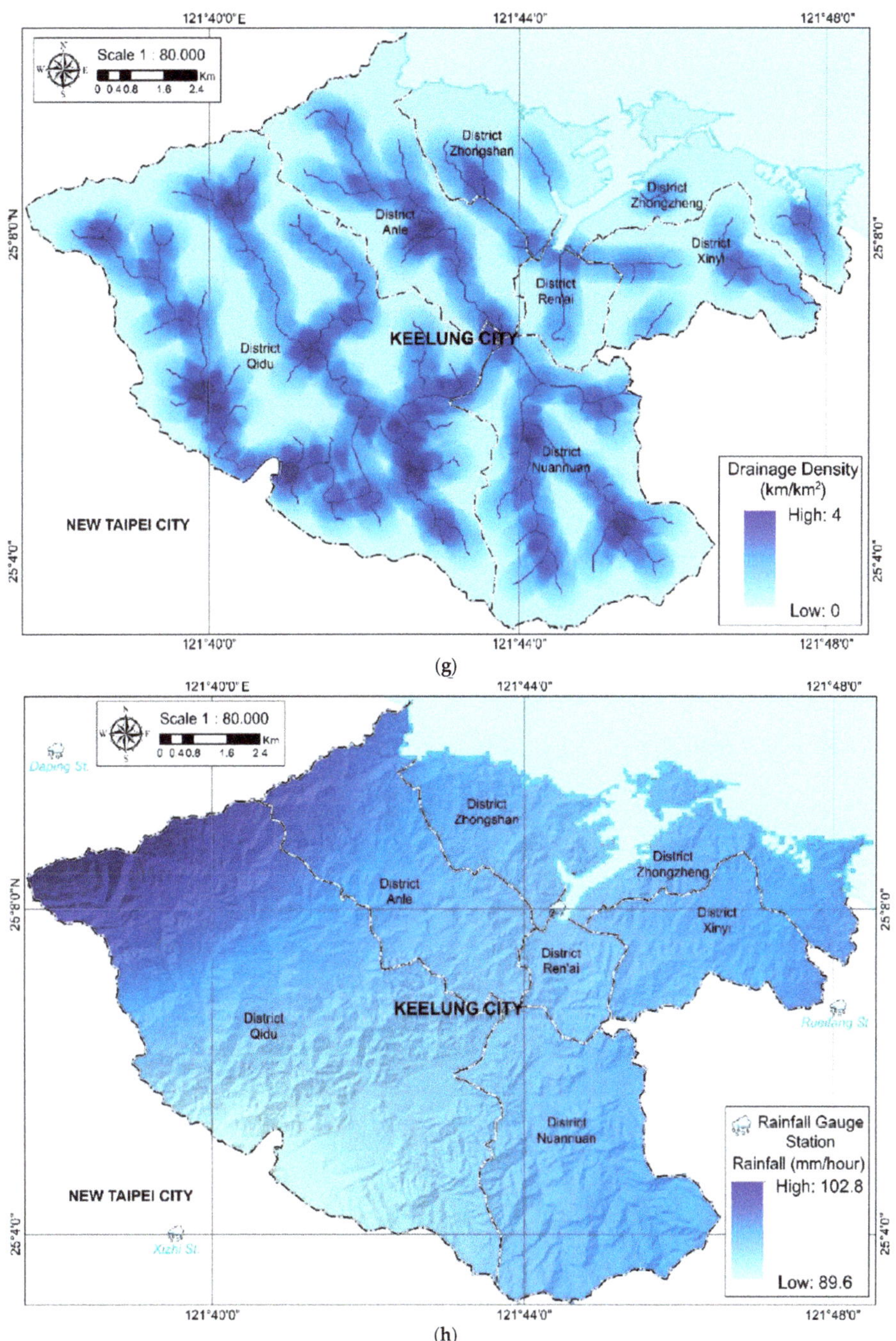


(g)



(h)

**Figure 2.** *Cont.*

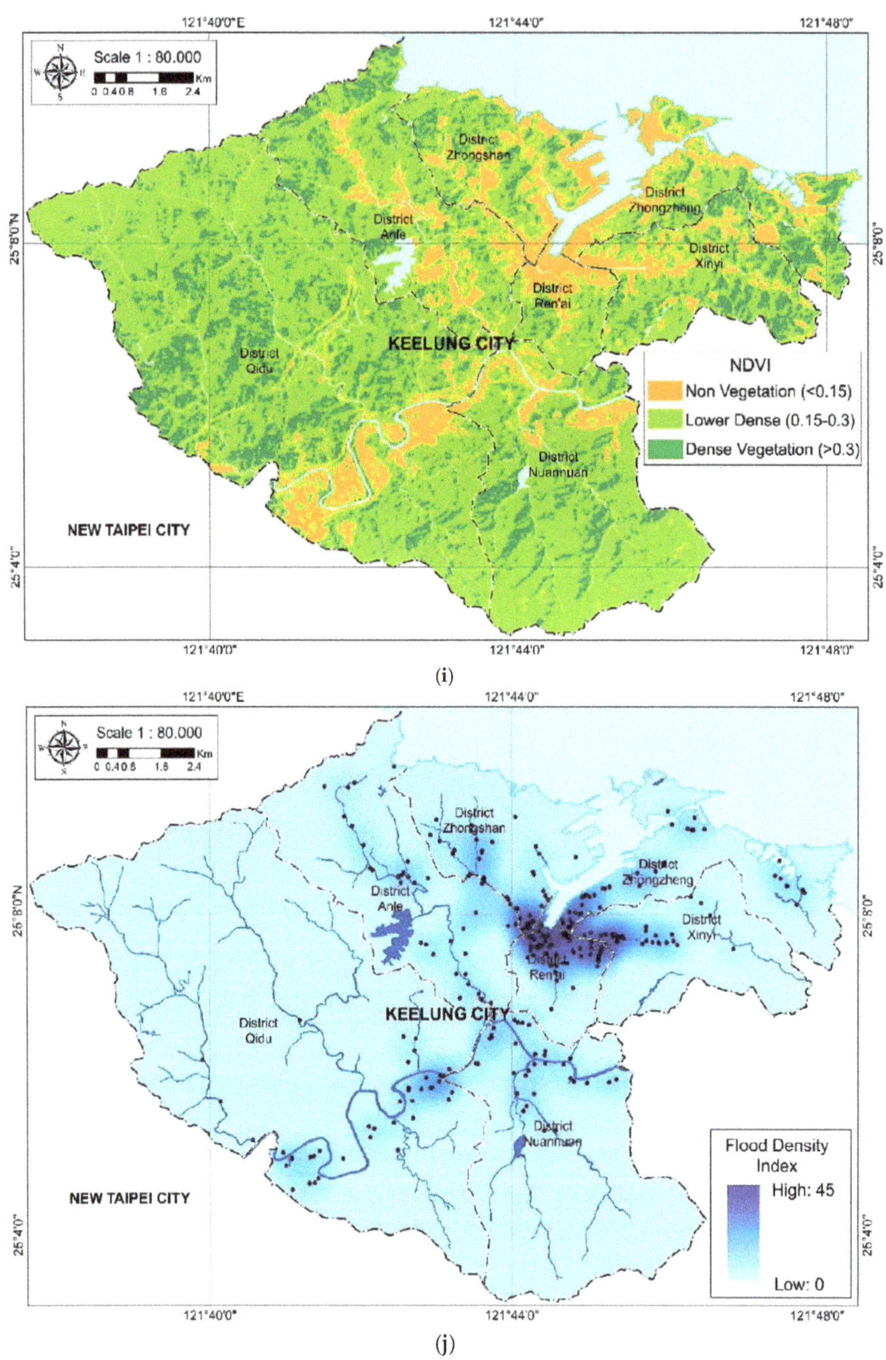


**Figure 2.** Geomorphologic area of Keelung City and relevant factors: (**a**) Elevation; (**b**) slope; (**c**) slope aspect; (**d**) flow direction; (**e**) flow accumulation; (**f**) TWI; (**g**) drainage density; (**h**) rainfall interpolation; (**i**) normalized difference vegetation index (NDVI); and (**j**) historical flood data.

### 2.2.1. Elevation

The elevation map is a representation of altitude, with ranges between 0 and 783 m. Elevation is another frequently used parameter and one of the key factors controlling the floods of an area [21,22]. Flood disasters tend to occur in low altitude areas, compared with landslides that have a tendency to happen at higher altitudes [23]. Generally, water continually flows from higher elevations to lower elevation areas.

### 2.2.2. Slope

This topographic factor is fundamental in hydrological studies. The relationship with rainfall is likely that the slope is directly influenced by the infiltration of rainfall [24]. The slope of an area and the surface flow velocity could have a positive correlation. The slope is also very closely connected to the flow of runoff directly toward downstream; a higher magnitude of slope in an area might accelerate precipitation-related runoff. The surface runoff increases significantly as the gradient increases; consequently, the infiltration decreases. As an outcome of this, regions with a sudden decrease in the slope have a higher probability of flooding as a massive volume of water becomes stationary, which causes a severe flood disaster situation.

### 2.2.3. Slope Aspect

This aspect recognizes the downslope direction, and is also thought of as the slope orientation [24]. The aspect of the slope presents the steepness of the surface, and is represented by three groups based on color brightness or saturation. The pixel values in the output aspect–slope raster represent a combination of aspect and slope. The aspect is one of the significant factors in producing flood susceptibility maps [25].

### 2.2.4. Flow Direction

The hydrologic characteristics of a surface are the capacity to establish the flow direction of each raster cell. The flow direction is a grid whose value indicates the direction of flow for every cell to its steepest downslope neighbor in the DEM [26].

### 2.2.5. Flow Accumulation

The flow accumulation tool generates accumulated flow as the accumulated weight of all cells flowing into each downslope cell in the output raster. The results of flow accumulation can be used to create a stream network by applying a threshold value to select cells with a high accumulated flow.

### 2.2.6. Topographic Wetness Index

The TWI is a physical representation of flood inundation areas, which is an important component of a river catchment. The TWI of a catchment indicates two types of measurements: Flat lands and hydrographic positions. The TWI is commonly used to quantify topographic control of hydrological processes. It is expressed as

$$TWI = \ln\left(\frac{\alpha}{\tan\beta}\right), \quad (1)$$

where $\alpha$ is the cumulative upslope contribution area draining through a point (per unit contour length), and $\tan\beta$ is the slope angle at the point. It affects the spatial distribution of soil moisture, and the groundwater flow often follows the surface topography. In this study, TWI is considered another contributing factor. Areas with a high wetness index occur where there is a combination of low slope and high flow accumulation and, therefore, may indicate locations that are at greater flood risk [27].

### 2.2.7. Drainage Density

The calculation of the drainage system raster was done by using the line density method, with rivers (polyline) as the main input data. The unit of calculated density was the length per unit of area. A higher probability of flooding is strongly associated with higher drainage density, as it represents greater surface runoff. The drainage density map of Keelung City was calculated from the drainage network map using line density tool in the ArcGIS, and ranges from 0 to 4 km/km$^2$.

#### 2.2.8. Rainfall Interpolation

Rainfall-induced flooding is associated with tropical storms, hurricanes, tropical depressions, and west trade winds that directly strike the windward side of the highlands. A large number of previous studies in the literature have established a relationship between the rainfall and the flood occurrence of an area [28–30]. Preparation of a rainfall map in this study used the pixel-based highest hourly rainfall data of several decades, spread around four rain gauge stations around Keelung City, with the Kriging interpolation method [31]. The rainfall data were collected from Keelung station, Xizhi station, Ruifang station, and Daping station through the Central Weather Bureau (CWB) of Taiwan.

#### 2.2.9. Normalized Difference Vegetation Index (NDVI)

The NDVI is one of the most extensively adopted vegetation indexes using satellite imagery, and for monitoring of global vegetation cover [32]. The NDVI, developed by Rouse in 1973, is used to monitor vegetation health, and to compare outputs across sensors with slightly different specifications [33]. The NDVI equation is defined as

$$\text{NDVI} = \frac{NIR - RED}{NIR + RED}, \tag{2}$$

where *NIR* is the reflection in the near-infrared spectrum and *RED* is the reflection in the red range of the spectrum. *NIR* and *RED* represent near-infrared (λ ~ 0.8 μm) regions of the spectrum and surface reflectance averaged over the visible spectrum (λ ~ 0.6 μm), respectively. The NDVI data source for this layer was the LANDSAT 8 OLI/TIRS C1 Level 1, to match the multispectral bands captured on 13 March 2018 and the area coverage of the study area on path 117 and row 43. The satellite imagery was rendered as NDVI, and colorized for use in the visualization analysis. Specifically, the source of the NDVI layer is the metadata of the imagery, which shows land cloud cover of only 0.8% and scene cloud cover of 1.91%. The NDVI response indicates the effective flood extent as an influential factor for the strength and capacity of an area against flood hazards.

#### 2.2.10. Historical Flood Density

Information on the historical floods in Keelung City was collected from the Emergency Management Information Cloud (EMIC) of Taiwan from 2015 to 2019, and comprised 307 events. The number of events each year is shown in Figure 3. The flood history, also known as the disaster experience, assumes that such areas have higher adaption ability, but also a high probability of flood occurrence in the future [3,34]. The density of point features in the raster cell unit is generated using the kernel density to fit a smoothly tapered surface to each point of the flood history. The kernel density tool in ArcGIS calculates the density of features in a neighborhood around the features, to find the spatial analysis and previous flood history of the area. By calculating the values of all the kernel surfaces, the density of each output raster cell feature is determined.

### 2.3. *Artificial Neural Network*

The ANN is considered a quantitative black-box approach that tries to simulate the functional human biological nervous system [12]. Moreover, the environment nonlinearity analysis and the forecast can be studied by applying this effective and affordable machine-learning tool. Even though the ANN has also been effectively applied to flood analysis in previous investigations [21,28], the GANN model is still rarely used for flood susceptibility assessment. Furthermore, the ANN has been considered as an alternative to physically-based models due to its simplicity regarding the minimum requirements for collecting detailed data [35]. The schematic and conceptual flowchart of this study is depicted in Figure 4. From Figure 4, three major factors, including topography factors, geology and geomorphology factors, and meteorology factors, were considered as a multi-resource aspect of the database in this study. Various factors, including elevation, slope angle, slope aspect, flow accumulation, flow direction, TWI, drainage density, rainfall, and NDVI, were generated using the digital elevation model, as well as LANDSAT 8 imagery. Historical flood data from 2015 to 2019, including 307 flood events, were adopted for the comparison of flood susceptibility. After collecting the possible factors, the proposed GANN model, based on the BPNN, was employed to assess the flood susceptibility. Finally, the accuracy of proposed GANN model predictions was evaluated by calculating the correlation coefficient.

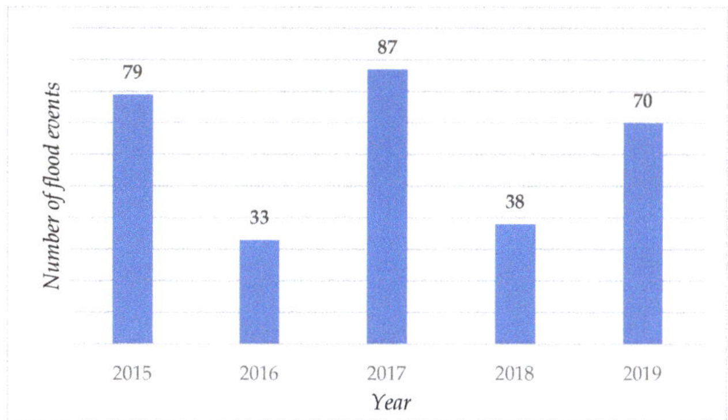


**Figure 3.** Historical flood data in Keelung City, Taiwan.

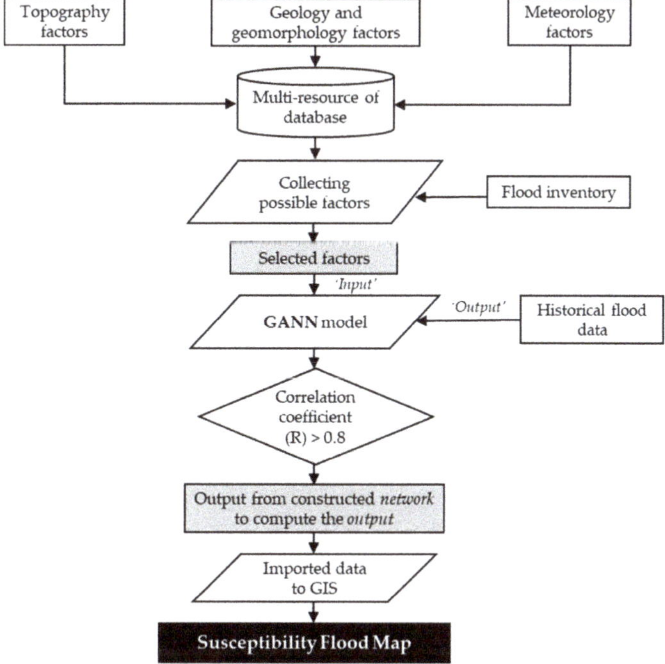


**Figure 4.** Flow chart of the study.

The GANN was employed to assess the flood susceptibility of Keelung City. This research focuses on a BPNN [36,37]. A typical algorithm of artificial neurons comprises three layers: Input, hidden, and output layers. The BPNN algorithm of feedforward shows an essential feature of the training phase, and the progress result is expressed as follows:

$$y_i = F(X_j) = \left(W_{oj} + \sum_{i=1}^{I} W_{ij} x_i\right), \tag{3}$$

$$Z_k = F(Y_k) = \left(W_{ok} + \sum_{j=1}^{J} W_{kj} y_i\right), \tag{4}$$

where $x_i$, $y_i$, and $Z_k$ indicate the input, hidden, output layers, respectively. The bias weights for setting the threshold values are represented as $W_{oj}$ and $W_{ok}$. Meanwhile, $X_j$ and $Y_k$ indicate temporarily calculated results before using the activation function, and $F$ is the activation function applied in the hidden and output layers. The $F$ value ranges from 0 to 1. In this study, we adopted the hyperbolic tangent sigmoid function. A sigmoid function is a mathematical function with a characteristic S-shape curve or sigmoid curve. The sigmoid activation function is a widely used activation function for neural networks [38,39]. The positive input value to the function is transformed into a value between 0.0 and 1.0. Inputs that are much larger than 1.0 are transformed to the value 1.0; similarly, values much smaller than 0.0 are snapped to 0.0. The shape of the function for all possible inputs is an S-shape from zero up through 0.5 to 1.0. Since the probability of anything exists only between the range of 0.0 and 1.0, the sigmoid activation function is adopted as the activation function for the proposed neural networks. Thus, the output $y_i$ and $Z_k$ can be expressed as

$$y_i = F(X_j) = F\left(\frac{1}{1 + e^{-X_j}}\right), \tag{5}$$

$$Z_k = F(Y_k) = F\left(\frac{1}{1 + e^{-Y_k}}\right). \tag{6}$$

For the error back-propagation weight training, the error function can be established as

$$E = \frac{1}{2}\sum_{k=1}^{K} \varepsilon_k^2 = \frac{1}{2}\sum_{k=1}^{K} (t_k - z_k)^2, \tag{7}$$

where $t_k$ and $\varepsilon_k$ are the target value and error in each output node, respectively. The goal is to minimize $E$, the error between the actual output values of the network. The weight adjustment in the link between the hidden and output layers can be expressed as

$$\Delta w_{jk} = \eta \times y_j \times \delta_k, \tag{8}$$

where $\eta$ is the learning rate, with values ranging between 0 and 1. The learning rate values correlate with the speed of convergence of the network of the BPNN. Conversely, a learning rate that is overly large can lead to a widely oscillating network. The new weight herein is updated by the following equation:

$$w_{jk}(n+1) = w_{jk}(n) + \Delta w_{jk}(n), \tag{9}$$

where $n$ is the number of iterations in the network. Similarly, the error gradient in links between the input and hidden layers can be derived from the partial derivative with respect to $w_{ij}$:

$$\frac{\partial E}{\partial w_{ij}} = \left[\sum_{k=1}^{K} \frac{\partial E}{\partial z_k}\frac{\partial z}{\partial Y_k}\frac{\partial_k}{\partial y_j}\right] \times \left(\frac{\partial y_j}{\partial X_j}\right) \times \left(\frac{\partial X_j}{\partial w_{ij}}\right) = -\Delta_j x_i, \tag{10}$$

$$\Delta_j = F'(X_j) \sum_{k=1}^{K} \delta_k w_{jk}. \tag{11}$$

The new weight in the hidden and input links can be regenerated as

$$\Delta w_{ij} = \eta \times x_i \times \Delta_j, \tag{12}$$

$$w_{ij}(n+1) = w_{ij}(1) + \Delta w_{ij}(n). \tag{13}$$

To evaluate the prediction performance of the proposed GANN model, the correlation coefficient (R) is utilized and expressed as

$$\mathrm{R} = \frac{\sum_{i=1}^{n} (t_i - \bar{t})(o_i - \bar{o})}{\sqrt{\sum_{i=1}^{n} (t_i - \bar{t})^2 \sum_{i=1}^{n} (o_i - \bar{o})^2}}, \tag{14}$$

where $t_i$ represents the target value and $o_i$ is the output value. Meanwhile, $\bar{t}$ and $\bar{o}$ are the average values of the target and output values, respectively.

The BPNN algorithm is used in the feedforward GANN, and the structure of the neural network model is displayed in Figure 5. The selected input, containing nine quantitative input variables and one quantitative output variable, includes elevation, slope, TWI, rainfall, NDVI, flow accumulation, drainage density, flow direction, and slope aspect, as shown in Figure 5. The flood history data from 2015 to 2019 in Keelung City were the only output of this model. All the input data in the GANN model were normalized in the range 0–1, with the initial weights automatically assigned to random values.

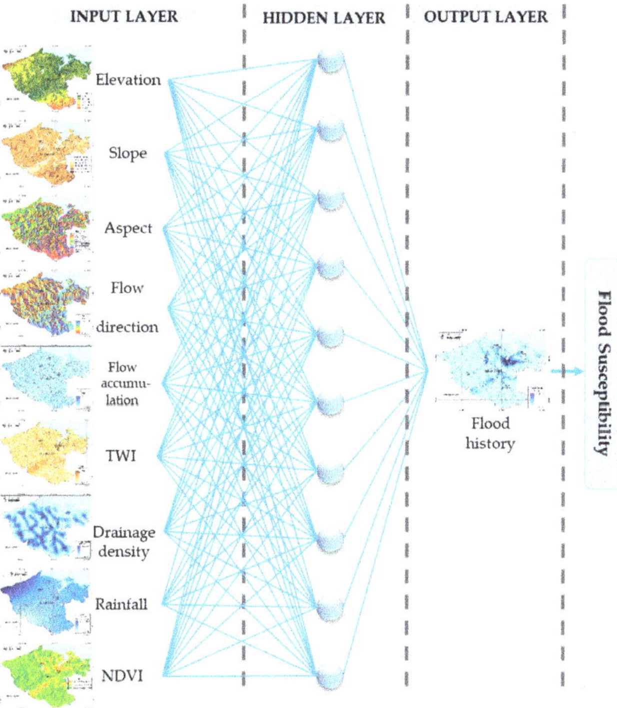


**Figure 5.** Structure of a typical three-layer model for flood susceptibility.

## 3. Results

### 3.1. Results of the Artificial Neural Network Model

The network architecture was designed to determine the flood susceptibility of Keelung City; an output layer of the flood history was required for the architecture of the model. This was used to evaluate the prediction performance of the train validation and test the model using the mean squared normalized error performance function, which measures the network's performance function according to the mean squared error (MSE).

The GANN study needed to classify the variation data in the training process. Kia et al. [21] indicated that 60% should be used for training, 20% for validation, and 20% as testing data, while Aziz et al. [29] and Latt and Wittenberg [40] used a combination of 80% and 20% for training and the testing process, respectively. In this study, the flood data were divided into three groups: 70% for training the network, 15% for validating the model, and 15% for testing the data to completely independently test the network generalization. The chosen training algorithm of this research was Levenberg–Marquardt (LM), which has the fastest training compared to Bayesian regulation back-propagation. This latter method takes longer, but may be better for challenging problems, while scaled conjugate gradient back-propagation is suitable for low-memory situations and was not used here [41]. The LM algorithm was selected as the training function, which combines the Gauss–Newton method and the gradient descent method. The LM algorithm was used to solve non-linear least squares problems and for its fault tolerance and fast convergence ability.

The number of hidden layers, along with neurons inside the model, is frequently defined by trial and error. The number of neurons in the output layers is fixed by the application, and is represented by the class being processed. The GANN model of flood susceptibility used 10 hidden layers, as at that point the model starts to reach the minimum requirement for the correlation coefficient, according to some research [8,28,42].

Figure 6 shows that the correlation coefficient value is very low, with only single nodes, and increases rapidly in five hidden layers, finally gradually stabilizing in 10 hidden layers. This is despite the fact that several researchers used only one hidden layer in their ANN architecture, or a small value of hidden layers from 1 to 7 [29,40]. Some studies show that using more hidden layers obtains the best result, such as Campolo et al. [43], who trained the variables with 20, 25, 30, 35, and 40 hidden layers, and Islam [35] as a comparison, with 15, 20, 25, 30, and 35 hidden layers. There are no strict rules for assigning the number of hidden layers and neurons in the literature [21]. The best design for GANN architecture depends on the problem type under investigation. In this study, the GANN model of flood susceptibility used 9 input layers, 10 hidden layers, and 1 output layer.

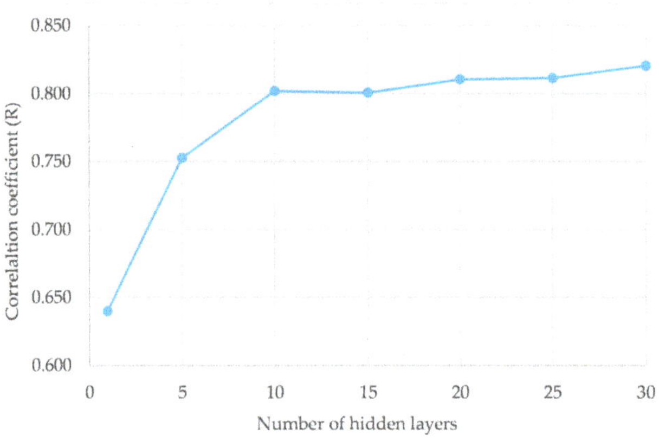


**Figure 6.** Number of hidden layers versus the value of R.

The training process reduced the MSE value from $10^3$ to 9.27 in 161 iterations. However, the best performance, at epoch 155, was 9.7189. The performance declined sharply in the first 20 epochs, and then gradually decreased until epoch 161. Figure 7 presents the linear regression for targets relative to the output of the different sub-divisions after the training process was completed.

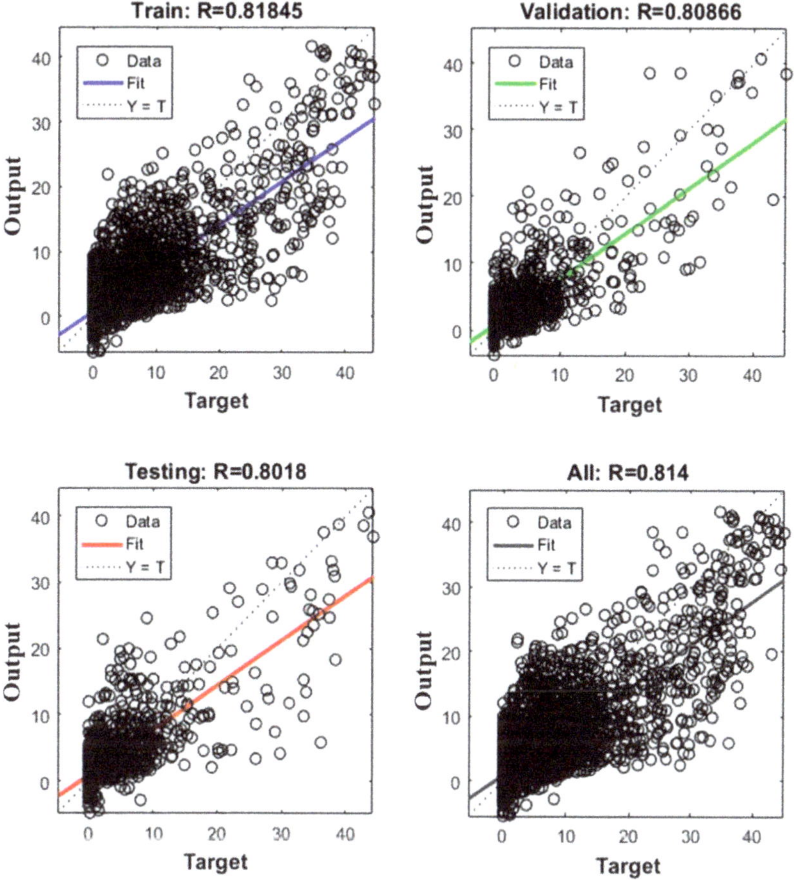


**Figure 7.** Training process for the correlation coefficient.

The efficiency of training the network is represented by the correlation coefficient (R). The cumulative R value was equal to 0.814, which reached the minimum standard of the study requirement (R > 0.8). The training, validation, and testing data sub-division were R = 0.818, 0.808, and 0.801, respectively. The results show that there is a good correlation between the historical flood data and those predicted by the proposed GANN model. According to previous studies [44,45], correlation coefficients whose magnitudes are between 0.7 and 0.9 demonstrate variables which can be considered highly correlated. These results highlight the efficiency of the constructed neural network during the training process and in forecasting flood susceptibility. Our results show that the proposed GANN model could efficiently predict flood susceptibility.

### 3.2. Flood Susceptibility Map

The flood susceptibility map is helpful for disaster planning, in addition to being useful during an actual emergency response to floods. Mapping flood susceptibility could be the first step toward mitigating flooding, because flood susceptibility identifies the most vulnerable locations and provides sufficient lead time for an individual to respond to flooding in an anticipatory rather than reactive manner [46]. The computation of the weight of the factors and artificial neural network modeling was performed in MATLAB; the outputs were exported to GIS for map production and visual interpretation. The flood susceptibility map was analyzed qualitatively using natural breaks (jenks) classification schemes [47,48]. The classification of flood susceptibility mapping in Keelung City could be categorized into five classes [23,27,30] based on the value of GANN prediction of the output: Very low, low, moderate, high, and very high. The flood susceptibility index of the five classes is shown in Table 2, and the flood sustainability of Keelung City is displayed in Figure 8. In Figure 8, the results show that a total of 12.1% of the Keelung City area is classified as flood-prone (very high, high, and moderate), according to the GANN model, with a coverage area of 16.14 km$^2$. Nearly 3.5% of the study area is in high to very high flood susceptibility zones, as shown in Table 2. The highest susceptibility area was only 1% of the total, and should be prioritized for flood management. To prevent urban inundation in the study, the Water Resources Agency, Ministry of Economic Affairs, ROC has developed a two-dimensional flood forecasting system using the Delft-FEWS platform to integrate the SOBEK models and the precipitation data from the Central Weather Bureau [49]. The flood sustainability obtained from the SOBEK model is adopted for comparison. The comparative results can be seen in Figure 9. It is demonstrated that the overall flood-prone areas from both approaches agree with each other. However, it was also found that the flood sustainability from the SOBEK model was much more conservative than our proposed GANN model.

**Table 2.** Classification of the flood susceptibility area in Keelung City.

| Classification | Flood Susceptibility Index | Area (km$^2$) | Percentage (%) |
|---|---|---|---|
| Very low | −3.0–1.7 | 80.30 | 60.5 |
| Low | 1.7–5.3 | 36.32 | 27.4 |
| Moderate | 5.3–11.6 | 11.48 | 8.6 |
| High | 11.6–22.5 | 3.31 | 2.5 |
| Very high | 22.5–38.2 | 1.35 | 1 |

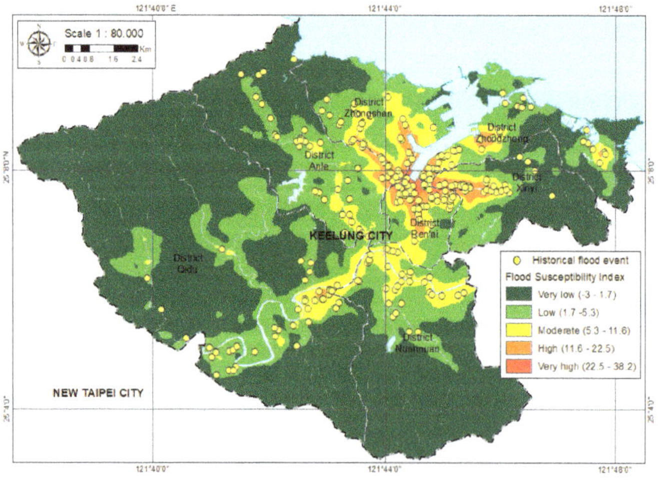


**Figure 8.** Flood sustainability map of Keelung City.

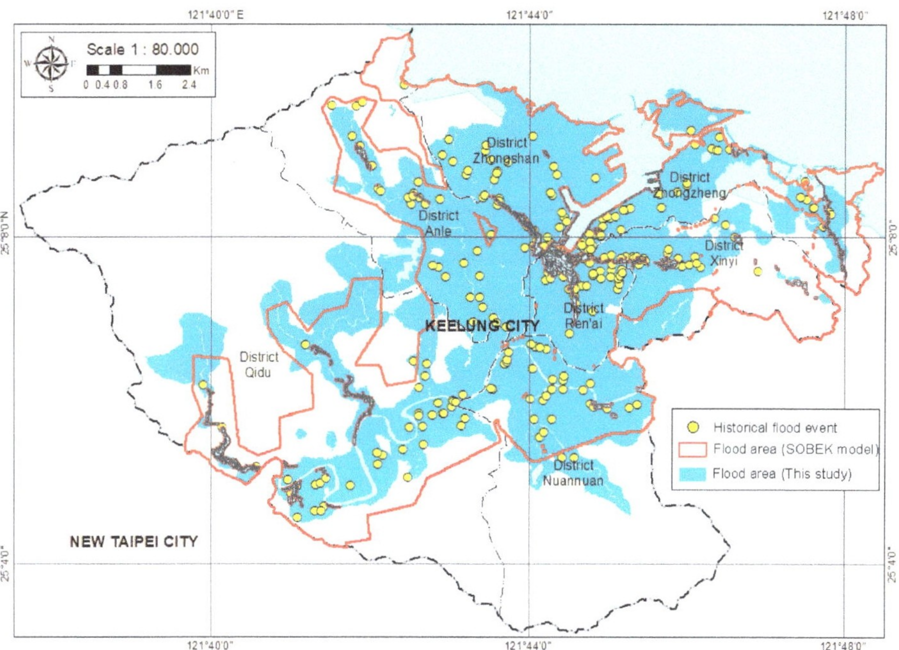


**Figure 9.** Comparison of the flood area of Keelung City.

The proposed method, based on ANN and GIS, may improve the ability to establish further precise flood models, and present results in a spatial environment. The outcomes of the research could be used to help local authorities to develop appropriate new infrastructure to protect lives and property in Keelung City.

## 4. Discussion

In this research, high elevations occurred in the western part of Qidu District and the southeastern part of Nuannuan District. Meanwhile, the lowland area covered the seashore, especially in the north of Keelung City and the central part, near the Keelung river. The flats area was located in the center of Keelung City, near the flood plain, and a very steep slope is located in the Nuannuan and Qidu District. Based on the GANN results, the most common floods were predicted to occur on the flats, and the northwest (NW), south (S), and southwest (SW) facing slopes. The floods were particularly rare on the north (N) and southeast (SE) facing slopes. Generally, the slopes of the study area are most commonly oriented to the north (N) and southeast (SE) quadrants. To produce a map of the vegetation index using the NDVI method, data from the LANDSAT 8 and high-density vegetation located in the southwest of Qidu District and eastern part of Xinyi District were used. Meanwhile, the non-vegetation area was the built-up area in the downtown region of Keelung City, especially the port area in the Ren'ai District, Xinyi District, Zhongzheng District, Zhongshan District, and southeast of Qidu District.

Finally, the high and very high prone areas were located in the northern part of the city, as well as in the central part of Keelung City alongside the river. Ren'ai District has the highest susceptibility, followed by Xinyi, Zhongzheng, and Zhongshan District. However, this very high susceptibility area is a core district of the city, which is highly populated, and the economic center of the city. The areas with a classification of very low and low potential coverage were, cumulatively, 87.9% or 116.62 km$^2$. These are located at high elevation, especially in the western and southeastern parts of Keelung City, which includes the western part of Qidu and Anle District, as well as the southeast of Nuannuan District.

The novel aspect of this work was to develop a GANN model for flood susceptibility assessment of Keelung, Taiwan. The main contribution of this work is that the proposed method, based on ANN and GIS, may improve the ability to establish further precise flood models, and present results in a spatial environment. The advantages of the GIS spatial analysis capability were integrated into the artificial neural network model. Accordingly, the proposed methodology may represent spatial continuity and the influence of parameters on flood-generating mechanisms. In addition, historical flood data from 2015 to 2019, including 307 flood events, were adopted for the comparison of the flood susceptibility on the regional spatial scale using the GIS. The finding observed in this work may provide a fundamental contribution to environmental protection engineering for flood in areas with higher occurrence and vulnerability to extreme precipitation.

## 5. Conclusions

In this research, the GANN model was developed using 10 flood causative factors. The thematic layers, including elevation, slope angle, slope aspect, flow direction, flow accumulation, TWI, NDVI, and drainage density, were generated using GIS. Rainfall data were also used as input, while historical flood events from 2015 to 2019 were the output of the GANN model.

The proposed GANN model produced satisfactory results, with a coefficient of correlation of 0.81. The susceptibility was categorized into five classes: Very low, low, moderate, high, and very high, with coverage areas of 60.5%, 27.4%, 8.6%, 2.5%, and 1%, respectively. Just 3.5% of the study area was included in the high to very high flood susceptibility zones; however, this area is a core district of the city, with a dense population, and the economic center of the city.

Furthermore, mapping flood susceptibility is crucial to mitigating flood disasters, since flood susceptibility can identify the most vulnerable areas and predict the potential locations of susceptibility, which can provide authorities responsible for emergency response and evacuation procedures with more information for planning and responses in very high susceptibility areas.

**Author Contributions:** Conceptualization and method, C.-Y.K.; verification, N.K. and C.-Y.L.; writing—original manuscript, N.K. and C.-Y.K.; data curation, C.-Y.L.; revising the manuscript, C.-Y.K. and C.-Y.L.; All authors have read and agreed to the published version of the manuscript.

**Funding:** This study was supported by the Ministry of Science and Technology (MOST), Taiwan, The Republic of China (MOST 109-2625-M-019-007).

**Data Availability Statement:** The datasets generated during the current study are available from the corresponding authors on reasonable request.

**Acknowledgments:** The authors sincerely thank the MOST for providing financial support.

**Conflicts of Interest:** The authors declare no conflict of interest.

## References

1. Merkens, J.-L.; Reimann, L.; Hinkel, J.; Vafeidis, A.T. Gridded population projections for the coastal zone under the shared socioeconomic pathways. *Glob. Planet. Chang.* **2016**, *145*, 57–66. [CrossRef]
2. Zanetti, V.; de Sousa Junior, W.; De Freitas, D. A climate change vulnerability index and case study in a Brazilian Coastal City. *Sustainability* **2016**, *8*, 811. [CrossRef]
3. Chang, H.-S.; Chen, T.-L. Spatial heterogeneity of local flood vulnerability indicators within flood-prone areas in Taiwan. *Environ. Earth Sci.* **2016**, *75*, 1484. [CrossRef]
4. Hsieh, L.-S.; Hsu, M.-H.; Li, M.-H. An assessment of structural measures for flood-prone lowlands with high population density along the Keelung River in Taiwan. *Nat. Hazards* **2006**, *37*, 133–152. [CrossRef]
5. Hirabayashi, Y.; Mahendran, R.; Koirala, S.; Konoshima, L.; Yamazaki, D.; Watanabe, S.; Kim, H.; Kanae, S. Global flood risk under climate change. *Nat. Clim. Chang.* **2013**, *3*, 816–821. [CrossRef]
6. Lee, K.T.; Ho, Y.-H.; Chyan, Y.-J. Bridge blockage and overbank flow simulations using HEC–RAS in the Keelung River during the 2001 Nari Typhoon. *J. Hydraul. Eng. ASCE* **2006**, *132*, 319–323. [CrossRef]

7. Wu, S.-J.; Yang, J.-C.; Tung, Y.-K. Risk analysis for flood-control structure under consideration of uncertainties in design flood. *Nat. Hazards* **2010**, *58*, 117–140. [CrossRef]
8. Abou Rjeily, Y.; Abbas, O.; Sadek, M.; Shahrour, I.; Hage Chehade, F. Flood forecasting within urban drainage systems using NARX neural network. *Water Sci. Technol.* **2017**, *76*, 2401–2412. [CrossRef]
9. Hsu, T.-W.; Shih, D.-S.; Li, C.-Y.; Lan, Y.-J.; Lin, Y.-C. A study on coastal flooding and risk assessment under climate change in the Mid-Western Coast of Taiwan. *Water* **2017**, *9*, 390. [CrossRef]
10. Jung, Y.; Shin, Y.; Jang, C.H.; Kum, D.; Kim, Y.S.; Lim, K.J.; Kim, H.B.; Park, T.S.; Lee, S.O. Estimation of flood risk index considering the regional flood characteristics: A case of South Korea. *Paddy Water Environ.* **2014**, *12*, 41–49. [CrossRef]
11. Qiu, J.; Wu, Q.; Ding, G.; Xu, Y.; Feng, S. A survey of machine learning for big data processing. *EURASIP J. Adv. Signal Process.* **2016**, *2016*, 67. [CrossRef]
12. Thomas, A. Memristor-based neural networks. *J. Phys. D. Appl. Phys.* **2013**, *46*, 093001. [CrossRef]
13. Fotovatikhah, F.; Herrera, M.; Shamshirband, S.; Chau, K.-w.; Faizollahzadeh Ardabili, S.; Piran, M.J. Survey of computational intelligence as basis to big flood management: Challenges, research directions and future work. *Eng. Appl. Comp. Fluid Mech.* **2018**, *12*, 411–437. [CrossRef]
14. Zhao, G.; Pang, B.; Xu, Z.; Peng, D.; Xu, L. Assessment of urban flood susceptibility using semi-supervised machine learning model. *Sci. Total Environ.* **2019**, *659*, 940–949. [CrossRef]
15. Chen, J.; Li, Q.; Wang, H.; Deng, M. A machine earning ensemble epproach based on random forest and radial basis function neural network for risk evaluation of regional flood disaster: A case study of the Yangtze River Delta, China. *Int. J. Environ. Res. Public Health* **2019**, *17*, 49. [CrossRef] [PubMed]
16. Wang, Z.; Lai, C.; Chen, X.; Yang, B.; Zhao, S.; Bai, X. Flood hazard risk assessment model based on random forest. *J. Hydrol.* **2015**, *527*, 1130–1141. [CrossRef]
17. Zhang, J.; Chen, Y. Risk assessment of flood disaster induced by typhoon rainstorms in Guangdong Province, China. *Sustainability* **2019**, *11*, 2738. [CrossRef]
18. Moel, H.D.; Alphen, J.V.; Aerts, J.C.J.H. Flood maps in Europe—Methods, availability and use. *Nat. Hazards Earth Syst. Sci.* **2009**, *9*, 289–301. [CrossRef]
19. Vos, F.; Rodriguez, J.; Below, R.; Guha-Sapir, D. *Annual Disaster Statistical Review 2009: The Numbers and Trends*; Center of Research on the Epidemiology of Disasters (CRED): Brussels, Belgium, 2010.
20. Zhang, Q.; Zhang, J.; Jiang, L.; Liu, X.; Tong, Z. Flood disaster risk assessment of rural housings—A case study of Kouqian Town in China. *Int. J. Environ. Res. Public Health* **2014**, *11*, 3787–3802. [CrossRef]
21. Kia, M.B.; Pirasteh, S.; Pradhan, B.; Mahmud, A.R.; Sulaiman, W.N.A.; Moradi, A. An artificial neural network model for flood simulation using GIS: Johor River Basin, Malaysia. *Environ. Earth Sci.* **2011**, *67*, 251–264. [CrossRef]
22. Eakin, H.; Lerner, A.M.; Murtinho, F. Adaptive capacity in evolving peri-urban spaces: Responses to flood risk in the Upper Lerma River Valley, Mexico. *Glob. Planet. Chang.* **2010**, *20*, 14–22. [CrossRef]
23. Dou, J.; Yamagishi, H.; Pourghasemi, H.R.; Yunus, A.P.; Song, X.; Xu, Y.; Zhu, Z. An integrated artificial neural network model for the landslide susceptibility assessment of Osado Island, Japan. *Nat. Hazards* **2015**, *78*, 1749–1776. [CrossRef]
24. Rahmati, O.; Falah, F.; Naghibi, S.A.; Biggs, T.; Soltani, M.; Deo, R.C.; Cerda, A.; Mohammadi, F.; Tien Bui, D. Land subsidence modelling using tree-based machine learning algorithms. *Sci. Total Environ.* **2019**, *672*, 239–252. [CrossRef]
25. Bajabaa, S.; Masoud, M.; Al-Amri, N. Flash flood hazard mapping based on quantitative hydrology, geomorphology and GIS techniques (case study of Wadi Al Lith, Saudi Arabia). *Arab. J. Geosci.* **2013**, *7*, 2469–2481. [CrossRef]
26. Sanyal, J.; Lu, X.X. Application of remote sensing in flood management with special reference to monsoon Asia: A review. *Nat. Hazards* **2004**, *33*, 283–301. [CrossRef]
27. Nandi, A.; Mandal, A.; Wilson, M.; Smith, D. Flood hazard mapping in Jamaica using principal component analysis and logistic regression. *Environ. Earth Sci.* **2016**, *75*, 465. [CrossRef]
28. Abdellatif, M.; Atherton, W.; Alkhaddar, R.; Osman, Y. Flood risk assessment for urban water system in a changing climate using artificial neural network. *Nat. Hazards* **2015**, *79*, 1059–1077. [CrossRef]
29. Aziz, K.; Rahman, A.; Fang, G.; Shrestha, S. Application of artificial neural networks in regional flood frequency analysis: A case study for Australia. *Stoch. Environ. Res. Risk Assess.* **2013**, *28*, 541–554. [CrossRef]
30. Das, S. Geospatial mapping of flood susceptibility and hydro-geomorphic response to the floods in Ulhas Basin, India. *Remote Sens. Appl. Soc. Environ.* **2019**, *14*, 60–74. [CrossRef]
31. Mair, A.; Fares, A. Comparison of rainfall interpolation methods in a mountainous region of a tropical island. *J. Hydrol. Eng.* **2011**, *16*, 371–383. [CrossRef]
32. Jiang, Z.; Huete, A.R.; Chen, J.; Chen, Y.; Li, J.; Yan, G.; Zhang, X. Analysis of NDVI and scaled difference vegetation index retrievals of vegetation fraction. *Remote Sens. Environ.* **2006**, *101*, 366–378. [CrossRef]
33. Fayne, J.V.; Bolten, J.D.; Doyle, C.S.; Fuhrmann, S.; Rice, M.T.; Houser, P.R.; Lakshmi, V. Flood mapping in the lower Mekong River Basin using daily MODIS observations. *Int. J. Remote Sens.* **2017**, *38*, 1737–1757. [CrossRef]
34. Zong, Y.; Tooley, M.J. A historical record of coastal floods in Britain frequencies and associated storm tracks. *Nat. Hazards* **2003**, *29*, 13–36. [CrossRef]
35. Islam, A.S. Improving flood forecasting in Bangladesh using an artificial neural network. *J. Hydroinform.* **2010**, *12*, 351–364. [CrossRef]

36. Buscema, M. Back propagation neural networks. *Subst. Use Misuse* **1998**, *33*, 233–270. [CrossRef]
37. Wang, L.; Zeng, Y.; Chen, T. Back propagation neural network with adaptive differential evolution algorithm for time series forecasting. *Expert Syst. Appl.* **2015**, *42*, 855–863. [CrossRef]
38. Karlik, B.; Olgac, A.V. Performance analysis of various activation functions in generalized MLP architectures of neural networks. *Int. J. Artif. Intell. Expert Syst.* **2011**, *1*, 111–122.
39. Sibi, P.; Jones, S.A.; Siddarth, P. Analysis of different activation functions using back propagation neural networks. *J. Theor. Appl. Inf. Technol.* **2013**, *47*, 1264–1268.
40. Latt, Z.Z.; Wittenberg, H. Improving flood forecasting in a developing country: A comparative study of stepwise multiple linear regression and artificial neural network. *Water Resour. Manag.* **2014**, *28*, 2109–2128. [CrossRef]
41. Ciaburro, G. *Matlab for Machine Learning*; Packt: Birmingham, UK, 2017.
42. Dang, V.-H.; Dieu, T.B.; Tran, X.-L.; Hoang, N.-D. Enhancing the accuracy of rainfall-induced landslide prediction along mountain roads with a GIS-based random forest classifier. *Bull. Eng. Geol. Environ.* **2018**, *78*, 2835–2849. [CrossRef]
43. Campolo, M.; Soldati, A.; Andreussi, P. Artificial neural network approach to flood forecasting in the River Arno. *Hydrol. Sci. J.* **2003**, *48*, 381–398. [CrossRef]
44. Taylor, R. Interpretation of the correlation coefficient: A basic review. *J. Diagn. Med. Sonogr.* **1990**, *6*, 35–39. [CrossRef]
45. Ratner, B. The correlation coefficient: Its values range between+1/−1, or do they? *J. Target. Meas. Anal. Mark.* **2009**, *17*, 139–142. [CrossRef]
46. Zhao, G.; Pang, B.; Xu, Z.; Yue, J.; Tu, T. Mapping flood susceptibility in mountainous areas on a national scale in China. *Sci. Total Environ.* **2018**, *615*, 1133–1142. [CrossRef] [PubMed]
47. Papaioannou, G.; Vasiliades, L.; Loukas, A. Multi-criteria analysis framework for potential flood prone areas mapping. *Water Resour. Manag.* **2014**, *29*, 399–418. [CrossRef]
48. Liu, Y.; Feng, G.; Xue, Y.; Zhang, H.; Wang, R. Small-scale natural disaster risk scenario analysis: A case study from the town of Shuitou, Pingyang County, Wenzhou, China. *Nat. Hazards* **2014**, *75*, 2167–2183. [CrossRef]
49. Chang, C.-H. *Integrated Platform for Application of High-Performance 2D Inundation Simulation*; Water Resources Planning Institute: Taichung, Taiwan, 2016.

International Journal of
*Environmental Research and Public Health*

*Review*

# A Review on Environmental and Social Impacts of Thermal Gradient and Tidal Currents Energy Conversion and Application to the Case of Chiapas, Mexico

**Graciela Rivera [1], Angélica Felix [1,2,*] and Edgar Mendoza [1]**

1 Engineering Institute, National Autonomous University of Mexico, 04510 Mexico City, Mexico; griverac@iingen.unam.mx (G.R.); emendozab@iingen.unam.mx (E.M.)
2 National Council on Science and Technology, 03940 Mexico City, Mexico
* Correspondence: afelixd@iingen.unam.mx; Tel.: +52-55-5623-3600

Received: 25 September 2020; Accepted: 22 October 2020; Published: 24 October 2020

**Abstract:** Despite the proved potential to harness ocean energy off the Mexican coast, one of the main aspects that have restrained the development of this industry is the lack of information regarding the environmental and social impacts of the devices and plants. Under this premise, a review of literature that could help identifying the potential repercussions of energy plants on those fields was performed. The available studies carried out around the world show a clear tendency to use indicators to assess impacts specifically related to the source of energy to be converted. The information gathered was used to address the foreseeable impacts on a hypothetical case regarding the deployment of an Ocean Thermal Energy Conversion (OTEC) plant off the Chiapas coast in Mexico. From the review it was found that for OTEC plants, the most important aspect to be considered is the discharge plume volume and its physicochemical composition, which can lead to the proliferation of harmful algal blooms. Regarding the case study, it is interesting to note that although the environmental impacts need to be mitigated and monitored, they can be somehow alleviated considering the potential social benefits of the energy industry.

**Keywords:** ocean renewable energy; OTEC; tidal current energy; environmental and social impacts

## 1. Introduction

Governments worldwide are encouraging the development of projects for electricity production from renewable and clean sources to mitigate climate change, manage the possible reduction of fossil fuels, and ensure energy security [1]. The actual challenge is to reduce the gaps of knowledge in order to provide better and more precise information to the technology producers than the presently available. The main sources of ocean energy (OE) are tidal currents, tidal range, wave energy, OTEC (ocean thermal energy conversion), and salinity gradient [2]. In particular, the energy from tides and waves is arguably considered infinite [3] and they could constitute sufficient energy sources to supply the global demand. Therefore, governments and industry have expressed strong interest in them and by extension on all ocean energy sources. In 2008, Mexico approved its "Ley para el Aprovechamiento de Energías Renovables y el Financiamiento de la Transición Energética" (Renewable energy exploitation and energy transition funding law); as of its publication, the government decided to increase the allocation of public and private resources for research, development, and innovation in renewable energy (RE). This law was revoked in 2015, but it set the basis for planning and financing instruments for RE conversion technologies and aimed for the long term to cover the country's urgent energy needs.

RE-based generation is forecast to grow 6.8% annually with prospects of reaching 37.7% participation in the total Mexican generation by 2030 [4,5].

Today, the development of OE in Mexico is regulated and supported by the 2015 "Ley de Transición Energética" (LTE-Energy Transition Law). The scopes of the LTE are established in the "Ley General de Cambio Climático" (General law on climate change), where it is stated that, by 2020, the Ministry of Energy and the Energy Regulatory Commission should have an incentive system for electricity generation with RE promotion. In February 2020, the "Estrategia de Transición para Promover el Uso de Tecnologías y Combustibles más Limpios" (Strategy for a transition to promote the use of technologies and cleaner fuels) was published. There, the importance of creating economic incentives to encourage RE development is expressed again, but the mechanisms to do so are not specified [6]. The goals of maximum participation of fossil fuels expressed in the LTE are 65% in 2024, 60% in 2035, and 50% in 2050 [7,8]. Unfortunately, up to 2020 Mexico is still behind the commitments established.

Other Mexican laws involved in RE regulation are the "Ley Orgánica de la Administración Pública Federal" (Public federal administration law), the "Ley del Servicio Público de Energía Eléctrica" (Electricity public service law), and the "Ley de la Comisión Federal de Electricidad" (National electricity company law). This confirms the national commitment to incorporate RE technologies in the production of electricity, through social and environmental responsibility. The "Plan Nacional de Desarrollo" (National development plan), in the past 12 years, has considered sustainable development, including the promotion of the use of RE sources and technologies to face the challenges regarding diversification, energy security, and strengthen the development of science and technology [9].

In summary, Mexico has the natural power availability for energy generation from ocean RE and a robust regulatory framework to promote their development. However, an element that has delayed the installation of OE extraction devices is the uncertainty regarding the potential impacts of the energy plants on the environment and the cost of monitoring programs [10]. The studies available in the literature are still scarce and most of them evaluate only the impacts of prototypes with short operating periods, which has caused the construction permissions to be postponed or denied due to lack of information. This evidences that more efforts are needed to create legal and technological environmental frameworks. The present study shows an extended review of the literature related to possible environmental impacts of the implementation of plants to take advantage of the OE in Mexico. In particular, several coastal Mexican communities are not connected to the national electricity grid, which causes electricity to be of poor quality, extremely pollutant, unstable in availability, or even nonexistent. Two regions of the country where this problem is very clear are the Baja California peninsula (central part) and the Southeast Pacific (Guerrero, Oaxaca, and Chiapas). These three states present the greater social and economic development lag in the country [11]. The Ministry of Energy mentions that the state of Chiapas has more than 55,000 people without a connection to the electricity grid [12]. However, in this area of the country, wave energy is not considered sufficient for direct harvesting due to the low energetic waves found in the tropical region. For this reason, special emphasis has been given to tidal currents (TC) and thermal gradient (OTEC) sources around the southeast Pacific Ocean region of Mexico [13,14]. Moreover, these communities have the human right to adequate housing and improvement of their social welfare.

## 2. Methods

The ocean energy technologies are still at an early stage of development; therefore, uncertainties of the potential environmental impacts are one of the main obstacles for its deployment. Hence, through bibliographic analysis key factors of the marine environment were identified. Then, the methods that would allow quantifying or classification of the changes in the physical, chemical, or biological components on the water column were selected.

A wide overview of the potential impacts of OE technologies is provided. This means that our analysis seeks to link environmental factors to socioeconomic components, resulting in the environmental transformation by the establishment of energy infrastructure in the coastal zone. It should be noted that these areas are vulnerable as a result of the increase of anthropic pressure, which demands among other services, the supply of electricity for productive activities and living.

Finally, this information was used to carry out an analysis of the possible energy extraction on the coast Chiapas, to exemplify the elements that must be considered and how the unique characteristics of the sites can generate changes in the assessment of impacts, as well as focusing on areas with a high level in social marginalization.

## 3. Impact Assessment

### 3.1. Environmental Aspects

The review is based on environmental studies of OE devices, then the type of energy extraction was narrowed to OTEC (ocean thermal energy conversion) plants and tidal energy devices due to the physical and environmental conditions on the case study. The selection of the deployment site is crucial for project investors and it should be decided considering as many environmental factors as possible. In this section, the selected factors considered relevant are those that could trigger environmental perturbations in the area selected for the OE extraction site. Although any energy conversion plant deployment has some degree of impact, the identification of these elements is part of the strategic decision for project developers. This also helps in the process of public acceptance, political support, and reliability for future investments. The following selection aims to highlight crucial factors and methodologies to decrease uncertainty regarding OTEC and tidal energy projects [15–32]. This preselection will help identifying the potential impacts of the hypothetical case on the Chiapas coast.

The literature reviewed shows that the presence of dykes for the use of tidal energy and the establishment of tidal arrays, have the potential to induce significant changes in hydrodynamics. Factors such as reduction in the average flow, deviation of the tidal flow, increase in bottom drag, and variations in the height and phase of the tides have been reported [33,34]. Hence, when the flow velocity is reduced, the sedimentation of suspended particles increases [35].

Variation in the hydrodynamics has direct and indirect impacts on local populations of seabirds, fish, and mammals [21]. To prevent the decrease in kinetic energy, it is important to limit the amount of energy that is allowed to be extracted. According to existing reports, it is recommended not to exceed 10-20% of the net flow [35], whilst Betz's limit of 60% should also be considered. Likewise, Kadiri et al. [33] documented that the acceleration of the flow in the vicinity of a turbine could lead to a resuspension of the sediments near the devices, which should be monitored to prevent modifications in the benthic communities.

The main impacts associated with OTEC technologies are deep water discharges, that can modify the trophic network. These can change the quantity and size of the species that are distributed in the areas surrounding the facilities of the plant [21]. Cilenti et al. [36] reported factors such as temperature, salinity, dissolved oxygen, chlorophyll, nutrients, and organic matter must be continuously monitored, to collect information in three periods: before establishment, during the construction, and the period of operation, to identify the potential modification of biota throughout the life cycle of the project.

In addition to abiotic factors, a local fauna inventory allows identifying changes in terrestrial and aquatic biota. Increases in the noise level produced during the construction and establishment of infrastructure can trigger the absence of key species, colonization, population decline, etc. An example of this, infrastructure suspended in the water column or on the seabed can attract organisms by changing the population structure, the colonization process occurs in the three phases of the life cycle of the project [25,37].

The increase in the level of noise produced by the construction or installation of the devices has repercussions on terrestrial and marine fauna. The monitoring of these changes is carried out by integrating several methods. Polagye et al. [35] proposed that the most feasible way to achieve monitoring at an acceptable cost is to incorporate various devices and technologies (e.g., video, sighting log, passive acoustic instruments). Davis [18] mentioned the use of sound velocity profiles collected by a CTD (conductivity, temperature, and depth sensor), where the degree of stratification of the water column can be observed and with physical–environmental modeling. In turn, Küsel et al. [38] estimated the propagation and the noise levels received in the environment). In other studies, hydrophones were placed over a long period, at a frequency of up to 50 kHz. This method was used on the central Oregon coast; the data obtained contributed to the elaboration of proposals to mitigate the possible impacts, such as population decline, alteration of migratory routes, acoustic disturbance, collision, among others [22].

Additionally, the presence of invasive species causes displacement or disappearance of local fauna, modifying the habitat permanently [37,39–41]. Therefore, it is crucial to obtain information about the species assemblages with a nonintrusive method. In several studies, remotely operated underwater vehicles (ROVs) are used to collect information on species composition, habitat associations, and population density [42–44], making it possible to identify colonization patterns in rigid structures, and use previous fauna records for comparison.

To define the monitoring method or devices it is necessary to know the foundation depth of the infrastructure [44]. For example, on the Big Russel Channel in the United Kingdom, the footage was used to compare species assemblages, thereby estimating abundance and dissimilarities between species at different sampling points [45]. Copping et al. [31] used the ecological risk model at the population level, incorporating the behavior and form of displacement to estimate fish mortality related to the establishment of turbines [25]. These models are complemented with in situ information on the distribution and behavior of the species.

Collision risk of marine mammals is a factor that slows the establishment of OE devices, this represents a crucial point in project decision-making. Cetaceans, being flagship species, are a group that increases the value of the system they live in, that is, they are species used as a symbol to attract public or government contributions to conservation programs [46]. Cetacean distribution patterns are estimated via echolocation with passive acoustic devices such as C-POD (Cetacean Acoustic Hydrophone Network). Thompson et al. [44] used this instrument within a range of 20-160 kHz for identifying dolphin distribution patterns. Despite technological support, both echolocation and group size estimates are complemented by visual identification [47].

The installation of OE devices arrays can be barriers limiting the movement of the species, leading in modifications of routes of migration. These potential variations are estimated using distribution, abundance, movement patterns, and satellite tracking data with 1-year periods under normal conditions [48]. MacKenzie [45] recommends that aerial, boat, and hydrophone monitoring should be performed for a minimum of 2 years after device placement. Finally, changes in the seabed due to drilling, excavation, installation of anchors, and power cables are also considered. Monitoring related to these activities should be carried out before and after construction to determine variations in the composition of the biota and should continue with a minimum period of 3 years [48]. It is crucial to collect information from the studies for the development of monitoring protocols [46], which contribute to the decision-making process.

### 3.2. Socioeconomic Aspects

According to Kerr et al. [47], to date, research on OE has focused on the evaluation of resources, device design, and environmental impact, thus concluding that social science research on these energies has had low priority. Changes in the landscape and the distancing of public opinion in infrastructure planning processes are factors that influence the acceptance of RE, this is linked to aspects of perception, economic impacts, or life quality [26,49].

The most commonly used components to assess socioeconomic impacts are cost–benefit analysis and environmental impact assessment. However, Uihlein and Magagna [2] mentioned that it is possible to quantify economic impacts at the regional level with empirical macroeconomic models, which allow quantifying the impacts on individual sectors, GDP (gross domestic product), the public budget, and household income. Kaldellis et al. [50] concluded that the social impacts in the pre-construction and post-construction of ocean wind energy and the coastal zone are as follows: the noise of the turbines, interruption of the landscape from the installation of the infrastructure, disturbance of the agricultural activities, and land-use change, as well as restriction of navigation routes that affect socioeconomic operations. With this, they demonstrated the importance of attaching the visual impact assessment and the marine landscape to the environmental impact assessment.

In an economic study carried out in England and Wales, Gibbson [51] revealed the reduction of the average house price between 5% and 6% in the presence of a visible wind farm at 2 km, falling to less than 2% between 2 and 4 km, in distances between 8 and 14 km, the reduction is close to zero. This trend could be reflected in the installation of energy extraction technologies in the coastal zones, whereas the modifications of the environment and the decrease in the beauty of the landscape, which is why it should be considered in socioeconomic studies.

An important social factor within technological developments is public participation during the planning and the decision-making processes since they can evoke opposition as a result of a lack of transparency through the project of OE [1]. Bonar et al. [32] suggest that the absence of public engagement and participation does not automatically lead to protests, however, the exclusion of the population in the decision-making process is inadequate since the inhabitants will be affected or benefited by these technologies.

## 4. Impacts of Specific Oceanic Energy Technologies

As noted, before, the impacts depend on the type of energy to be extracted, the shape, size, and number of devices as well as the anchors, etc. Mendoza et al. [52] offer an overview of the identification of impacts related to the biotic and abiotic variables of the environment with characteristics of the device to be installed. This section analyzes studies corresponding to the use of OTEC and TC [14–23].

### 4.1. Thermal Gradient (OTEC)

OTEC technology uses the temperature differences between the ocean surface water and cold water at a depth of 1000 m to activate a Rankine or similar cycle, and power is extracted from a gas driven turbine [53,54]. For this reason, high potential for OTEC projects exist in tropical zones. Recently, the research has focused on providing electricity and other goods to communities or islands with restricted access to national grids. OTEC plants can be closed-cycle (external working fluid), open-cycle (seawater as working fluid), and hybrid [16]. Additionally, platforms for OTEC plants can be onshore (land-based or near-shore) or offshore. There are significant cost variations in the plants due to the construction and maintenance needed by each. The floating platform installation has lower land use but requires submarine cables to carry the electricity to land, thus the cost of maintenance is higher comparatively with land-based plants [55]. The selection of the platform is crucial to determine the feasibility of the project. The plants need a pumping system as large quantities of deep cold water are carried through pipes to cool a working fluid, while surface water is used to heat it. After flowing through the system, the resulting discharge has different temperature and density than the surface water. The main impact of OTEC to the marine environment is related to the discharge plume (this volume can reach a few hundred cubic meters per second), its flow regime, and its trajectory. This plume can modify the availability of nutrients which, in turn, can favor eutrophication of the marine environment [36,56]. The difference in temperature and nutrient content could increase primary production in the surrounding environment, resulting in decreased dissolved oxygen levels and the possible proliferation of harmful algal blooms (HAB), so it is crucial to evaluate the depth of the plume discharge and the subsequent stabilization [57]. If the discharge is close to the surface, the difference in pressure will cause the release of gases, such as dissolved carbonates, which can lead to variations in the pH of the water column [48].

The increase of nutrient-rich deep ocean water on the surface may also be related to the proliferation of HAB. Giraud et al. [56] carried out evaluations of the effects of discharges from OTEC plants in the Martinique Islands; they found a variation of more than 0.3 °C at 150 m, concluding that the discharge would modify the phytoplankton assembly in the deep maximum of chlorophyll on a local scale. Additionally, the pumping system represents a threat to certain species, particularly those with little mobility that can be trapped and dragged [33,58].

A closed-cycle OTEC plant uses ammonia as working fluid due to its high thermal conductivity; tests have also been conducted with R404A, R717, R134A, among other fluids [58,59]. One concern related with the use of working fluids is the potential leakage as a result of seeping into the ocean if the pipes were damaged [58]; Golmen and Yu [60] point out that the risks related to ammonia will be low and manageable given the vast experience of handling this substance.

Garduño et al. [17] reported potential impacts related to the pipes of the OTEC plants, in depths between 100 and 150 m, such as industrial and sanitary discharges, the release of toxic coatings to the sea, drag, and compression of species.

### 4.2. Tidal Currents

The kinetic energy present in the currents can be harnessed using horizontal axis, transverse flow, or vertical axis turbines [2,61]. One of the main concerns for the placement of these devices is the local reduction of the tidal range since they can cause alterations in the vertical mixing of the water column and with it the increase of suspended particles, as well as the penetration of the light [62,63]. Increased light penetration and accretion pollutants from industrial, agriculture, or household discharges can alter water quality, which can lead to increased primary production and eutrophication [63].

The turbines used for energy extraction increase the underwater noise, causing stress and potential tissue damage to fish, marine mammals, and birds [64]. Frid et al. [61] mentioned that the effects can directly damage the sensory tissues or indirectly change the behavior of the individuals, some incidents of whale strandings have been associated to underwater noise from military activities. This information can link changes in the population dynamics of species sensitive to underwater noise and contribute to the proposal of prevention measures in the period of operation of OE devices [62]. Hammar [49] reports behavioral changes in species exposed to the noise produced by turbines at distances of 10 m, however, the evidence of this modification in behavior outside the laboratory is not conclusive. This exposes the need to carry out in situ studies during the operation of the impact determination device, to enable the definition of proposals for prevention and mitigation.

When working with tidal barrage, it is important to take into consideration whether or not there are migratory movements of organisms. Hooper and Austen [63] studied the effects on anadromous fish, due to the caused difficulty of arriving at the spawning area in freshwater, resulting in population decline. Other species with specific habitat requirements, vulnerable to changes in aquatic systems, can cause changes in the trophic network. Furthermore, the colonization of species in areas with anthropogenic disturbances must be taken into consideration and the establishment of invasive species [37]. According to Firth et al. [65] communities found in artificial structures are less diverse compared to natural habitats. Other impacts include risk of injury or collision of marine mammals and fish with the tidal barrages and turbines. To estimate the potential impact, it is common the use of encounter risk models (ERM), models of ecological risk at the population level, models of the time of population exposure (ETPM), collision risk models (CRM) and, encounter models, among others [19,20,32].

Finally, the literature mentions the use of submarine cables for the transport of energy, these emit electromagnetic fields (EMF), that could affect the behavior of marine biota. The impacts of EMFs depend on the magnitude and persistence of the field, while their effects could temporarily alter the direction of swimming or migration [66]. Gill et al. [67] mentioned that elasmobranchs, sea turtles, decapods, marine mammals, and teleost fish could present behavioral changes in the presence of emissions from submarine cables. Statistical evidence links stranding of marine mammals to high levels of EMF [66]. Therefore, laboratory studies are required to identify potential responses in organisms, the effect and impact thresholds of EMFs [30], as well as studies targeting key species.

Table 1 groups the methods for monitoring or evaluating changes in the environment as a result of the presence of OTEC and tidal current devices, as well as the minimum time recommended that monitoring must take to be reliable.

**Table 1.** Methods for monitoring potential impacts of Ocean Thermal Energy Conversion (OTEC) and tidal current (TC) into the marine environment, as well as the minimum monitoring period.

| Factors | | Technology | | |
|---|---|---|---|---|
| **Abiotic** | Method | TC | OTEC | Period |
| **Hydrodynamic** | 1D and 2D models, buoys, acoustic doppler velocimeter | x | | 1 year |
| **EMF** | Magnetometers, transects, gradiometer, calculation of the load Biot-Savart law | x | x | 1–3 years |
| **Noise** | Hydrophones and sound velocity profiles | x | x | 1 year |
| **Discharge plume** | Discharge plume model, ROMS (regional ocean models) | | x | 1 year |
| **Pollutant concentration** | Continuous stirred tank reactor | x | x | 1 year |
| **Biotic** | | | | |
| **Abundance of marine species** | Record of sightings, filming, remotely operated underwater vehicle, LIDAR, dives | x | x | 3 years * |
| **Species interaction with infrastructure** | Geometric area model of risk rates and species interaction, predator–prey encounter model, multibeam sonar, exposure time, population model | x | x | 1 year |
| **Cetaceans** | Echolocation C and T-POD, sightings records, radar | x | x | 3 years |
| **Collision risk** | Acoustic and optical equipment complemented by sighting records | x | | 3 years ** |
| **Collision risk seabirds** | Sightings records, visual recognition, radar, tracking devices | x | | 3 years ** |
| **Collision risk fish** | Ecological risk model at population level, encounter risk model, radar | x | | 3 years ** |

* Macrofauna monitoring in certain seasons of the year, monitoring may be 1 year [68]. ** Related to the sighting of species, therefore the monitoring must be simultaneous.

### 4.3. Strengths and Weaknesses of Published Literature

To corroborate de aforementioned crucial environmental factors in OTEC and tidal energy a search in SCOPUS database was made.

The displacement of water masses in the water column as a consequence of the discharges of OTEC plant is the main driver in potential changes in the marine environment. To understand and identify the variations in the OTEC plant surrounding oceanic models are used to simulate the trajectory and interactions of the OTEC discharge in the water column [17,69]. Thus, to date, there is a significant gap regarding the long-term modifications in the biological, physicochemical properties in the water column. Decision-makers play a major role in the development of the projects, in this process it is crucial to present net power generation, information on the potential environmental impacts, and cost-effective studies on the different OTEC platforms to make thermal energy competitive with other renewable energy sources.

Figure 1 indicates the number of published studies related with OTEC, tidal and environmental impact assessment. This shows that OTEC is in an early stage of development, the main subject of the articles is simulations, investigation of alternative refrigerants, and recently the potential impact of this technology, however, is theoretical information. The lack of commercial development of OTEC plants is notable lack of commercial, or even large pilot plants to date certainly has relegated the study of large-scale implementation of this technology. Instead, available studies merely introduce comprehensive descriptions of the OTEC technology [17]. Commercial scale of the projects decreases uncertainties for the future investments and provides a general estimation of cost in the different stages of the projects (construction, operation and maintenance). Additionally, in this estimation the cost of the environmental assessment in all stages has to be added.

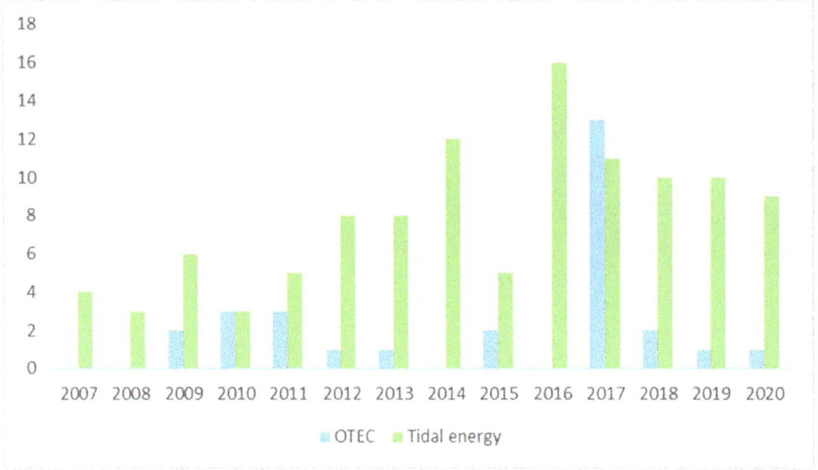


**Figure 1.** SCOPUS search of OTEC and tidal energy impacts in the environment, keywords; {OTEC} AND {environmental impact}, {tidal energy} AND {environmental impact}, {EMF} AND {marine energy}.

Tidal energy has a similar background, initially, design, estimation of energy output, optimizations of tidal turbines were the main subjects. The spatial overlapping between tidal turbines and marine mammals led to concerns with collision risk, interactions with infrastructure, and EMF potential changes in behavior are crucial for the future development of OE [70–72]. Potential negative effects have been considered in environmental assessment, at present, there is no empirical data on collision rates on operating turbines and the physical consequences, the data gathered is from numerical models that simulate coastal ocean processes, these simulations help to understand and identify potential changes in the marine food web [71,73,74]. Although modeling presents advantages to identify potential EMF effects, in-situ studies are scare making it difficult to determine alterations on the assemblage of species [53].

It is important to highlight that some studies made a quantitative evaluation (low, medium, and high), others focused on life cycle assessment and classified the potential effects in every stage of the projects [14,53,75].

In the last 3 years, the studies of OTEC projects have decreased while the number of published studies on tidal energy has remained constant. Nevertheless, in-situ information is scarce in both cases, thus delays the developments of the devices and reduces the opportunity to become cost-competitive in the market of renewable energy.

## 5. Case Study: OTEC Plant off the Coast of Chiapas, Mexico

Despite the government's push in recent years to promote renewable energies, Mexico is advancing in small steps in the generation of renewable energies and more slowly in the extraction of OE. Progress is currently being made within the Mexican Center for Innovation in Ocean Energy (CEMIE-O), where in the last 3 years the work has focused on the definition of theoretical potentials for each type of energy that can be extracted [14], as well as, in detailed studies for those areas where a particular extraction seems possible. Examples of this are the use of ocean currents for the Cozumel channel [76]; studies of the thermal gradient for the eastern Pacific [77] or the waves in the northwest of the country [78].

However, along the Mexican coast, it is possible to find small populations unconnected to the national electricity grid, this usually coincides with them being in a vulnerable condition to ocean threats. In the case of the coast of Chiapas as shown in Figure 2, 78% of the population is in poverty, this is broken down into extreme poverty with 29% and 49% with moderate poverty [79,80]. According to Borthwick [1,81] the main characteristic for an OTEC plant is to have a thermal gradient of 20 °C between the surface water and cold water from 1000 m depth. As reported by Garduño et al. [17] close to the coast of Chiapas the temperature difference is 21.45 °C; very close to the minimum usable gradient for OTEC.

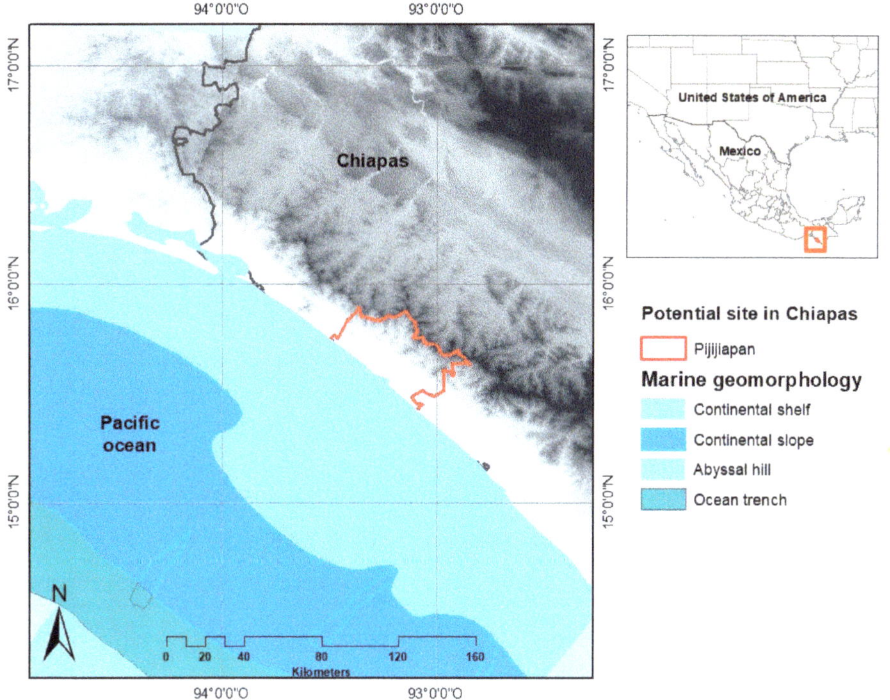


**Figure 2.** Potential site in the coast of Mexico for harnessing ocean thermal energy (OTEC).

The National Marine Renewable Energy Center of Hawaii estimated the potential on the Chiapas coast of 124.02 GW/h with a 100 MW OTEC plant and a sea surface temperature (SST) of 26.85 °C [81]. Yet, SAGARPA [82] reports an SST > 27 °C, measured from November 2017 to February 2018 obtained by satellite images of MODIS-AQUA of NOAA. The reported months are the coldest of the year, indicating the possibility of higher available power than the reported estimates of HINMREC. This corresponds with the results of autumn and winter thermal difference between 0 and 1000 m depth described by García et al. [77], in both seasons the differences were 25.17 and 23 °C, respectively, they used NOAA, NODC, ODV, and the Mexican Navy (SEMAR) database to analyze historical mean gradient in Mexico between surface and 1000 m depth.

Hernández-Fontes et al. [14] presented theoretical results of the availability of OTEC on the Mexican coast. In Chiapas, the authors estimated yearly percentages of >100, >150, and >200 MW of available power. Although electric power changes as a consequence of seasonal temperature variations, one of the best sites for the operation of the OTEC offshore plant is on the Chiapas coast [13]. However, the water pumping area is far from the coast and offshore plant studies are scarce to determine the feasibility.

The search for new energy sources is focused not only on being renewable, but also on being compatible with sustainable development. The government of Chiapas is committed to the conservation of its natural resources, so its development plan prioritizes environmental sustainability. In addition to this commitment, there is an interest in improving the quality of life of its population, and thus decrease its high level of marginalization.

*Potential Impacts of an OTEC Plant on the Chiapas Coast*

From the analysis of the different studies cited in this article, the parameters that allow identifying changes in the physicochemical and biological structure of the water column were determined. These could lead to negative environmental impacts in the marine area, as well as in the coastal area during the three phases of the OTEC plant (construction, establishment, and operation), the impacts identified are summarized in Table 2.

**Table 2.** Methods to identify changes in the water column and the terrestrial part of the Chiapas coast.

| Abiotic | | Biotic | |
|---|---|---|---|
| **Parameter** | **Method** | **Parameter** | **Method** |
| Temperature | Multiparameter, NOAA data or use of MLD (mixed layer depth), CTD | Abundance of species | Diversity and species richness |
| Salinity | Multiparameter, MLD, refractometer, CTD | NOM-059-SEMARNAT-2010 | Geographic Information System (GIS) |
| Dissolved oxygen | Multiparameter, Winkler's method, CTD | Mangrove monitoring | Centered quadrant method |
| Nutrient | $NO^{2-}$ Bendschneider method, $NO^{3-}$ Stickland and Parsons method, $NH^{4+}$ Koroleff method, orthophosphates method described by Murphy and Riley and total phosphorus Menzel and Corwin 13C/15N isotope technique | Chelonium distribution | Distribution data, quantification of nests and nesting females, and collection of morphological data |
| Chlorophyll | Spectrophotometry, satellite images | Vegetation analysis | 1. NDVI (Normalized Vegetation Index) 2. SAVI (Soil Adjusted Vegetation Index) |
| Turbidity | Secchi disk or turbidimeter | Benthic fauna | Ekman dredge, nucleator, dives sampling |
| Suspended organic matter | Titration procedure | Primary production | Light/dark bottles, 14 C and satellite images |

Despite the lack of information on the effects of discharges in the surrounding areas of OTEC plants, there is knowledge of discharges of nutrients caused by other anthropogenic activities, for example, the $NO_3^-$ ion from residual discharges affects the aquatic invertebrates due to increased concentration and exposure time [83,84]. With this information, it is possible to compare the effects of the discharge plume of the OTEC plant on the surrounding environment.

Modifications of these factors would have a direct effect on the marine community [85–88]. Additionally, the Chiapas coast is a zone of upwelling, this has a relationship with the natural presence of HAB by the contribution of nutrients [85,86]. Notwithstanding, OTEC plume discharges could increase the frequency of these blooms, however, there are no studies that confirm this relationship.

It is crucial to monitor the changes associated with the HAB on the Chiapas coast since the increase of this community directly affects the fisheries, which are one of the main forms of income for the population. In addition, it also represents an alert for public health, even at a national level, different institutions in the health, production, and research sectors implement control methods for seafood contaminated with toxins to prevent the risk of poisoning [85,87].

As an offshore OTEC plant requires electric power transmission via marine cables, this can cause disturbances in the surrounding environment, as well as an increase in the capital cost for its establishment. The short-term environmental effects associated with cables include physical disturbances of the habitat as a result of their installation, resuspension of sediments. Together with the long-term effects (operational phase); heat emission, species colonization, and emission of electromagnetic fields [74].

The prediction of impacts by offshore plants can be supported by information on the construction of oil platforms, this may help with minimizing risks and better estimating costs. Nevertheless, the presence of the plant could cause social disagreement because of the visual impact on the landscape. This has been addressed by Gibbson [51] who mentions the decrease in property prices as a result of the presence of offshore wind farms.

Chiapas has mostly rural municipalities where more than 50% of the population lives in communities with less than 2500 inhabitants [88]. The high degree of social backwardness manifests itself in the coastal area, where irregular human settlements with less than 100 habitants predominate. Most of these groups are disconnected from the national electricity grid. Here, microgrids are an alternative energy supply for isolated communities, even the island mode would be appropriate [89]. This method of energy supply could strongly increase social welfare.

As a first step, an evaluation of consumption should be performed to identify the benefited communities [88]; in this case, the towns would be made up of El Fortín, Playa Cocos, and Las Conchas. These localities are located outside the polygon of the La Encrucijada biosphere reserve, therefore the installation of a microgrid does not pose a threat to the environment. The beneficiary population is of approximately 431 habitants. These localities are incorporated into the Program for the Development of Priority Zones, which seeks to provide basic housing services in localities with high levels of social backwardness in the country, however, information on consumption and electricity supply is incipient. Assuming that a 5–10 MW microgrid had the viable capacity to supply the aforementioned population, it is important to take into account that its control is decentralized and the maximum use of energy is limited [90].

With the supply of these three populations, it is possible to propose the extension for neighboring settlements, in such a way that it seeks to increase social welfare. In addition, this would allow areas for small-scale tourism; for example, Costa Azul Chocohuital previously presented problems due to the lack of infrastructure and services, minimizing the growth of the tourism sector. The expansion of energy supply would provide perks to both the tourism sector to increase hotel occupancy and basic housing services for the population. They are potential applications of the by-products of OTEC plants in the economic activities, for example, the use of deep water in aquaculture in this matter is necessary to select species adapted to low temperatures as salmonids their maturation is between 9 years at 13 °C and 13 years at 18 °C in the growth-fattening phases [91], also Masutani and Takahashi [92] mention the cultivation of oysters, lobster, abalone, kelp, and nori in aquaculture.

In the coast of Chiapas, 44.2% of the population carries out activities in the primary sector [93], mainly agriculture, livestock, and fishing. There are 606 artisanal fishermen distributed in four cooperatives, the main product is shrimp and scale [94]. For this reason, it is not possible to say with certainty that it is possible to integrate the by-products into the economic activities of the area, because fishermen are not familiar with the aquaculture of the aforementioned species. Furthermore, as a technology in development, studies on the use of these by-products are incipient, therefore, it cannot be affirmed that Mexican aquaculture would benefit from the supply of deep water.

Consequently, it is imperative to perform an analysis of the viability of using by-products from an OTEC plant, through the incorporation of courses, workshops, and capacitation to motive the involvement of the inhabitants.

The creation of jobs for the population close to the area where the energy plant will be displayed is one of the criteria to promote its placement. However, the characteristics of this job offer needs to include training given that the education level of the population aged 15 and over is of incomplete basic education (55.76%), and 13.47% are illiterate. In the stages of construction and establishment of an OTEC plant, it is possible to recruit workers from the population, however, in the operation phase, equipment and workers with specific knowledge of plant management and maintenance are necessary.

The direct impact on society is exclusively the supply of energy, and there would be no creation of any type of employment or the use of any added material. The direct impacts, related to visual obstruction, could be observed if they occur near towns with a high influx of tourism. Nonetheless, tourism is low compared to other areas of Oaxaca and the biggest limitation would be the rejection by the intrusion of external companies due to unacceptable practices of its predecessors.

Figure 3 shows a summary of the main aspects to be monitored before, during, and after the deployment of an OTEC plant in order to identify and assess the environmental and socioeconomic impacts it would produce off the Chiapas coast.

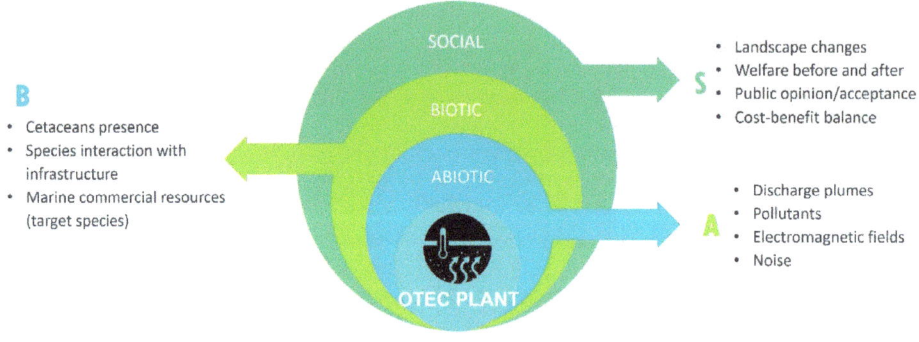


**Figure 3.** Aspects to be monitored for environmental and social impact assessment of ocean energies, OTEC Plant case.

## 6. Conclusions

This study presented a first approach to the identification of basic factors (abiotic and biotic) that ocean energy projects associated with OTEC and TC should consider before, during, and after their deployment. The selection of these elements is a priority in the recognition of the problems that may arise in the specific area to be intervened and its current state. This information may contribute to the elaboration of monitoring protocols that allow the reduction or absence of negative impacts on the environment.

From the literature reviewed, two main conditions were found: (a) that an accurate assessment of environmental and socioeconomic impacts of ocean energy plants is not possible yet, due to the lack of installed facilities and (b) this does not mean that the development of the ocean energy industry should be left to a trial and error process. In a similar way to other coastal and marine projects, it is clear that monitoring is the path to understand the environment (its dynamics and resilience) and its response to the different drivers of change that the energy plants would produce along their useful life. Obviously, the main concern for investors is that systematic, permanent monitoring is an expensive activity so it must be planned and executed carefully. In this work an identification of the aspects to be monitored for tidal currents and OTEC plants was presented.

As shown in Table 1, the bibliographic compilation allowed defining the important variables to be monitored to detect changes in the biotic and abiotic factors associated with TC and OTEC. The modifications of these factors could lead to indicators for monitoring the response of the environment and can be used to support the impact assessment. The most important information presented, together with the variables, is the recommended method for data gathering and the minimum monitoring time recommended to produce time series long enough to detect changes and responses as well as undesired effects. Following these recommendations is seen as the path to effectively mitigate and correct any negative impact.

Notably, test sites play a key role in obtaining information from deployed devices, acting as centers for testing methods and data analysis focused on OE research programs. With this data, the effects can be extrapolated to different sites of interest and thereby promote the transition towards alternative energy.

The social and economic axis, as observed in the bibliographic search, is mainly related to cost–benefit studies, most published works do not consider the economic needs of the populations and their perspective in the potential changes in the surrounding environment. This review of information highlights the importance of conducting local studies of the benefits, social and economic, that the generation of OE will have on the population, to ingrate them in the decision process, and avoid possible future problems that put the project at risk. It is worth noting that monitoring is also applicable to socioeconomic variables. In this work, gathering of information regarding social welfare, services availability, public opinion on the energy plant and a general cost–benefit balance is proposed as the data framework for assessing the impacts of ocean energy on the local communities.

Despite having the optimal thermal gradient for the harnessing of thermal energy on the Chiapas coast, the extension of the continental shelf represents a limitation, coupled with the lack of studies of offshore plants that increase the uncertainty of their viability of maintenance costs. Environmentally, changes in the population structure of the water column and distribution of key species should be considered. Notwithstanding, this type of development must have a construction and production cost study to determine the economic viability of the establishment of the power plant.

**Author Contributions:** Conceptualization, A.F., G.R. and E.M.; methodology, G.R. and A.F.; formal analysis, G.R.; investigation, G.R.; resources, A.F. And E.M.; writing—original draft preparation, G.R.; writing—review and editing, A.F. and E.M.; supervision, A.F.; funding acquisition, E.M. and A.F. All authors have read and agreed to the published version of the manuscript.

**Funding:** This research was supported by the Mexican National Council of Science and Technology (CONACYT) and the Secretary of Energy (SENER/Sustenatbilidad Energética) of Mexico through the Mexican Center of Innovation in Ocean Energies (CEMIE-Océano) under contract 249795.

**Acknowledgments:** The authors wish to thank the Mexican National Council of Science and Technology (CONACYT) and the Secretary of Energy (SENER/Sustenatbilidad Energética) of Mexico through the Mexican Center of Innovation in Ocean Energies (CEMIE-Océano).

**Conflicts of Interest:** The authors declare no conflict of interest. The funders had no role in the design of the study; in the collection, analyses, or interpretation of data; in the writing of the manuscript, or in the decision to publish the results.

## References

1. Borthwick, G. Marine renewable energy seascape. *Engineering* **2016**, *2*, 69–78. [CrossRef]
2. Uihlein, A.; Magagna, D. Wave and tidal current energy—A review of the current state of research beyond technology. *Renew. Sustain. Energy Rev.* **2016**, *58*, 1070–1081. [CrossRef]
3. Siddiqui, M.A.; Ahmed, S.M.; Munir, M.A.; Hussain, S.M.; Randhawa, J. Ocean Energy: The Future of Renewable Energy Generation. 2015. Available online: https://www.researchgate.net/publication/280937085_Ocean_Energy_The_Future_of_Renewable_Energy (accessed on 15 November 2017).
4. INEEL (Instituto Nacional de Electricidad y Energias Limpias). Available online: https://www.ineel.mx/cemie-oceano.html (accessed on 15 November 2017).
5. SENER (Secretaria de Energia). Available online: https://base.energia.gob.mx/Prospectivas18-32/PER_18_32_F.pdf (accessed on 10 July 2020).
6. DOF (Diario Oficial de la Federacion). 2020. Available online: https://www.dof.gob.mx/nota_detalle.php?codigo=5585823&fecha=07/02/2020 (accessed on 19 october 2020).
7. DOF (Diario Oficial de la Federacion). Available online: http://www.diputados.gob.mx/LeyesBiblio/pdf/LGCC_130718.pdf (accessed on 10 July 2020).

8. SENER (Secretaria de Energia). Available online: https://www.gob.mx/cms/uploads/attachment/file/62949/Prospectiva_del_Sector_El_ctrico_2013-2027.pdf (accessed on 10 July 2020).
9. DOF (Diario Oficial de la Federacion). Available online: https://www.dof.gob.mx/nota_detalle.php?codigo=5565599&fecha=12/07/2019 (accessed on 10 July 2020).
10. Greaves, D.; Pérez, C.; Magagna, D.; Conley, D.; Bailey, I.; Simas, T.; Holmes, B.; O'Hagan, A.M.; O'Callaghan, J.; Torre-Encino, Y.; et al. *SOWFIA Enabling Wave Power: Streamlining Processes for Progress*; Plymouth Universit: Plymouth, UK, 2013.
11. CONEVAL (Consejo Nacional de Evaluacion de la Politica de Desarrollo Social). Available online: https://www.coneval.org.mx/Medicion/IRS/Paginas/Índice-de-Rezago-social-2010.aspx (accessed on 12 October 2020).
12. SENER (Secretaria de Energia). Available online: https://datos.gob.mx/busca/dataset/regiones-sin-electricidad (accessed on 12 October 2020).
13. García, A.; Rodríguez, Y.; Garduño, P.; Hernández, E. General Criteria for Optimal Site Selection for the Installation of Ocean Thermal Energy Conversion (OTEC) Plants in the Mexican Pacific. In *Ocean Therm Energy Conversion—Past, Present Progress*; Kim, A.S., Kim, H.J., Eds.; IntechOpen: London, UK, 2020.
14. Hernández-Fontes, J.V.; Felix, A.; Mendoza, E.; Rodríguez, Y.; Silva, R. On the marine energy resources of Mexico. *J. Mar. Sci. Eng.* **2019**, *7*, 191. [CrossRef]
15. Scottish Natural Heritage. Assessing Collision Risk between Underwater Turbines and Marine Wildlife. 2016. Available online: https://www.nature.scot/sites/default/files/2017-09/Guidance%20Note%20-%20Assessing%20collision%20risk%20between%20underwater%20turbines%20and%20marine%20wildlife.pdf (accessed on 15 November 2017).
16. Collins, N. *Assessment of Potential Ecosystem Effects from Electromagnetic Fields (EMF) Associated with Subsea Power Cables and TISEC Devices in Minas Channel*; Fundy Ocean Resources Center for Energy: Halifax, NS, Canada, 2012.
17. Garduño, E.P.; García, A.; Rodríguez, Y.; Bárcenas, J.F.; Alatorre, M.A.; Cerezo, E.; Guadalupe, J.; Romero, V.M.; Silva, R. *Conversión de Energía Térmica Oceánica (OTEC), Estado del Arte*; CEMIE-O, Universidad Autonoma de Campeche: Mexico City, Mexico, 2017; p. 136.
18. Davis, A. *Potential Impacts of Ocean Energy Development on Marine Mammals in Oregon*; Oregon State University: Corvallis, OR, USA, 2012.
19. Kreting, L.; Elsaesser, B.; Kennedy, R.; Smyth, D.; O'Carroll, J.; Savidge, G. Do changes in current flow as a result of arrays of tidal turbines have an effect on benthic communities? *PLoS ONE* **2016**, *11*, e0161279.
20. Comfort, C.M.; Vega, L. *Environmental Assessment for Ocean Thermal Energy Conversion in Hawaii: Available Data and a Protocol for Baseline Monitoring*; IEEE: Waikoloa, HI, USA, 2011; pp. 1–8.
21. Knight, C.; McGarry, S.; Hayward, J.; Osman, P.; Behrens, S. A review of ocean energy converters, with an Australian focus. *AIMS Energy* **2014**, *2*, 295–320. [CrossRef]
22. Haverson, D.; Bacon, J.; Smith, H.C.M.; Venugopal, V.X.Q. Modelling the hydrodynamic and morphological impacts of tidal stream development in Ramsey sound. *Renew. Energy* **2018**, *126*, 876–887. [CrossRef]
23. Witt, M.J.; Sheehan, E.V.; Bearhop, S.; Broderick, A.C.; Conley, D.C.; Cotterell, S.P.; Crow, E.; Grecian, W.J.; Halsband, C.; Hodgson, D.J.; et al. Assessing wave energy effects on biodiversity the Wave Hub experience. *Philos. Trans. R. Soc.* **2012**, *370*, 502–529. [CrossRef] [PubMed]
24. Wilde, P.; Sandusky, J.; Jassby, A. *Assessment and Control of OTEC Ecological Impacts*; U.S. Department of Energy: Washington, DC, USA, 1978.
25. Dolman, S.J.; Green, M.; Simmonds, M.P. *Marine Renewable Energy and Cetaceans*; Whale and Dolphin Conservation Society: Chippenham, UK, 2006.
26. Bender, A.; Francisco, F.G.A.; Sundberg, J. A review of methods and models for environmental monitoring of marine renewable energy. In Proceedings of the European Wave and Tidal Energy Conference, Cork, Ireland, 27 August–1 September 2017.
27. Jia, Y.; Nihous, G.C.; Rajagopalan, K. An evaluation of the large-scale implementation of Ocean Thermal Energy Conversion (OTEC) Using an Ocean General Circulation model with low-complexity atmospheric feedback effects. *J. Mar. Sci. Eng.* **2018**, *6*, 12. [CrossRef]
28. Nihous, G. A preliminary investigation of the effect of Ocean Thermal Energy Conversion (OTEC) effluent discharge options on global OTEC resources. *J. Mar. Sci. Eng.* **2018**, *6*, 25. [CrossRef]

29. Wood, J.; Joy, R.; Sparling, C. *Harbor Seal—Tidal Turbine Collision Risk Models: An Assessment of Sensitivities*; SMRU Consulting: Washington, DC, USA, 2016.
30. Wilson, B.; Batty, R.S.; Daunt, F.; Carter, C. *Collision Risks between Marine Renewable Energy Devices and Mammals, Fish and Diving Birds*; Centre for Ecology & Hydrology: Oban, UK, 2007.
31. Copping, A.; Sather, N.; Hanna, L.; Whitting, J.; Zydleswki, G.; Stainesm, G.; Gill, A.; Hutchinson, I.; O'Hagan, A.; Simas, T.; et al. *Annex IV 2016 State of the Science Report: Environmental Effects of Marine Renewable Energy Development Around the World*; OES: Richland, WA, USA, 2018.
32. Bonar, P.A.J.; Bryden, I.G.; Borthwick, A.G.L. Social and ecological impacts of marine energy development. *Renew. Sustain. Energy Rev.* **2015**, *47*, 486–495. [CrossRef]
33. Kadiri, M.; Ahmadian, R.; Bockelmann-Evans, B.; Rauen, W.; Falconer, R. A review of the potential water quality of tidal renewable energy systems. *Renew. Sustain. Energy Rev.* **2012**, *16*, 329–341. [CrossRef]
34. El-Gezinry, T.M.; Bryden, I.G.; Couch, J.S. Environmental impact assessment for tidal energy schemes: An exemplar case study of the Strait of Messina. *J. Mar. Eng. Technol.* **2009**, *8*, 39–48. [CrossRef]
35. Polagye, B.; Copping, A.; Suryan, R.; Kramer, S.; Brown-Saracino, J.; Smith, C. *Instrumentation for Monitoring around Marine Renewable Energy Converters: Workshop Final Report*; U.S. Deparment of Energy: Seattle, WA, USA, 2014.
36. Cilenti, L.; Dario, R.; Dentamaro, G.; Di Lecce, V.; Guaragnella, C.; Cardellichio, A.; Mancinelli, G.; Petruzzelli, D.; Quarto, A.; Soldo, D.; et al. Seawater distributed monitoring system: A proposal for architecture and data format. In Proceedings of the IEEE International Conference on Environmental Engineering, Milan, Italy, 12–14 March 2018.
37. Laidig, T.E.; Krigsman, L.M.; Yoklavich, M.M. Reactions of fishes to two underwater survey tools, a manned submersibles and remotely operated vehicle. *Fish Bull* **2013**, *111*, 4–67. [CrossRef]
38. Küsel, E.T.; Mellinger, D.K.; Thomas, L.; Marques, T.A.; Moretti, D.; Ward, J. Cetacean population density estimation from fixed sensors using passive acoustics. *J. Acoust. Soc. Am.* **2011**, *129*, 3610–3622. [CrossRef]
39. Consoli, P.; Esposito, V.; Battaglia, P.; Altobelli, C.; Perzia, P.; Romeo, T.; Canese, S.; Andoloro, F. Fish distribution and habitat complexity on Banks of the strait of Sicily (central Mediterranean Sea) from remotely operated vehicle (ROV) explorations. *PLoS ONE* **2016**, *11*, e0167809. [CrossRef]
40. Culloch, R.; Bennet, F.; Bald, J.; Menchaca, I.; Jessop, M.; Simas, T. *Report on Potential Emerging Innovative Monitoring Approaches, Identifying Potential Reductions in Monitoring Costs and Evaluations of Existing Long-Term Datasets*; RiCORE Project: Aberdeen, Scotland, 2015.
41. Bald, J.; Curtin, R.; Díaz, E.; Fontán, A.; Franco, J.; Garmendia, J.M.; González, M.; Liriondo, A.; Liria, P.; Menchaca, I.; et al. *Guía Para la Elaboración de Estudios de Impacto Ambiental de Proyectos de Energías Renovables Marinas, Informe Técnico Realizado en el Marco del Proyecto Nacional*; Azti Tecnalia: Acciona España, España, 2013.
42. Sheehan, E.V.; Gall, S.C.; Cousens, S.L.; Atrill, M. Epibenthic assessment of renewable tidal energy site. *Sci. World J.* **2013**, *8*. [CrossRef]
43. Isasi-Catalán, E. Los conceptos de especies indicadoras, paraguas, bandera y claves: Su uso y abuso en la ecología de la conservación. *Interciencia* **2011**, *36*, 31–38.
44. Thompson, P.M.; Brookes, K.L.; Cordes, L. Integrating passive acoustic and visual data to model spatial patterns of occurrence in coastal dolphins. *J. Mar. Sci.* **2015**, *72*, 651–660. [CrossRef]
45. MacKenzie, S. Techniques for Marine Biological Baseline Data Collection at Offshore Renewable Energy Developments and How Best Applty These to Guernsey Waters. Master's Thesis, Plymouth University, Plymouth, UK, 2013.
46. Shumchenia, E.J.; Smith, S.L.; McCann, J.; Carnevale, M.; Fugate, G.; Kenney, R.D.; King, J.W.; Paton, P.; Schwartz, M.; Spaulding, M.; et al. An adaptive framework for selecting environmental monitoring protocols to support ocean renewable energy development. *Sci. World J.* **2012**, 1–23. [CrossRef]
47. Kerr, S.; Watts, L.; Colton, J.; Conway, F.; Hull, A.; Johnson, K.; Jude, S.; Kannen, A.; MacDougall, S.; McLachlan, C.; et al. Establishing an agenda for social studies research in marine renewable energy. *Energy Policy* **2014**, *67*, 694–702. [CrossRef]
48. Dreyer, A.; Polis, H.J.; Jenkins, L.D. Changing tides: Acceptability, support, and perceptions of tidal energy in the United States. *Energy Res. Soc. Sci.* **2017**, *29*, 72–83. [CrossRef]
49. Hammar, L. Will ocean energy harm marine ecosystem? In *System Perspectives on Renewable Power*; Sandén, B., Ed.; Chalmers University of Technology: Gothenburg, Sweden, 2014; p. 84.

50. Kaldellis, J.J.; Apostolou, D.; Kapsali, M.; Kondoli, E. Environmental and social footprint of offshore wind energy. *Renew. Energy* **2016**, *92*, 543–556. [CrossRef]
51. Gibbson, S. Gone with the wind: Valuing the visual impacts of wind turbines through houses prices. *J. Environ. Econ. Manag.* **2015**, *72*, 177–196.
52. Mendoza, E.; Lithgow, D.; Flores, P.; Felix, A.; Simas, T.; Silva, R. A framework to evaluate the environmental impact of ocean energy devices. *Renew. Sustain. Energy Rev.* **2019**, *112*, 440–449. [CrossRef]
53. Paredes, M.G.; Padilla-Rivera, A.; Güereca, L.P. Life cycle assessment of ocean energy technologies: A systematic review. *J. Mar. Sci. Eng.* **2019**, *7*, 322. [CrossRef]
54. Soukissian, T.H.; Denaxa, D.; Karathanasi, F.; Prospathopoulos, A.; Sarantakos, K.; Iona, A.; Georgantas, K.; Mavrakos, P. Marine Renewable Energy in the Mediterranean Sea: Status and perspectives. *Energies* **2017**, *10*, 1512. [CrossRef]
55. Cunningham, J.; Magdol, Z.; Kinner, N. *Ocean Thermal Energy Conversion: Assessing Potential Physical, Chemical and Biological Impacts and Risk*; University of New Hampshire-NOAA: Durham, UK, 2010.
56. Giraud, M.; Garcon, V.; de la Broise, D.; L'Helguen, S.; Sudre, J.; Boye, M. Potential effects of deep seawater discharge by an ocean thermal energy conversion plant on the marine microorganisms in oligotrophic waters. *Sci. Total Environ.* **2018**, *639*, 1–33.
57. Khosravi, A.; Syri, A.; Assad, M.E.H.; Malekan, M. Thermodynamic and economic analysis of a hybrid ocean thermal energy conversion/photovoltaic system with hydrogen-based energy storage system. *Energy* **2019**, *172*, 304–319. [CrossRef]
58. Devault, D.A.; Péné-Annette, D. Analysis of the environmental issues concerning the deployment of an OTEC power plant in Martinique. *Environ. Sci. Pollut. Restor.* **2017**, *24*, 25582–25601. [CrossRef]
59. Ganic, E.N.; Wu, J. On the selection of working fluids for OTEC power plants. *Energy Convers. Manag.* **1980**, *20*, 9–22. [CrossRef]
60. Golmen, L.G.; Yu, J.C.S. OTEC in the TROPOS multipurpose platform concept. In *OTEC Matters*; Dessne, P., Golmen, L., Eds.; University of Boras: Borås, Sweden, 2015; pp. 50–58.
61. Frid, C.; Andonegi, E.; Depestele, J.; Judd, A.; Rihan, D.; Rogers, S.I.; Kenchington, E. The environmental interactions of tidal and wave energy generation devices. *Environ. Impact Assess Rev.* **2012**, *32*, 133–139. [CrossRef]
62. European Comission. Available online: http://publications.europa.eu/resource/cellar/3a4f6411-6777-11e7-b2f2-01aa75ed71a1.0001.01/DOC_1 (accessed on 12 March 2017).
63. Hooper, T.; Austen, M. Tidal barrages in the UK: Ecological and social impacts, potential mitigation and tools to support barrage planning. *Renew. Sustain. Energy Rev.* **2013**, *23*, 289–298. [CrossRef]
64. Halvorsen, M.B.; Carlson, T.J.; Copping, A.E. *Effects of Tidal Turbine Noise on Fish Task 2.1.3.2: Effects on Aquatic Organisms: Acoustics/Noise—Fiscal Year 2011—Progress Report—Nvironmental Effects of Marine and Hydrokinetic Energy*; U.S. Deparment of Energy: Richland, WA, USA, 2011.
65. Firth, L.B.; Thompson, R.C.; White, F.J.; Schofield, M.; Skov, M.W.; Hoggart, S.P.G.; Jackson, J.; Knights, A.M.; Hawkins, S. The importance of water-retaining features for biodiversity on artificial intertidal coastal defence structures. *Divers. Distrib.* **2013**, *19*, 1275–1283. [CrossRef]
66. Fisher, C.; Slater, M. *Effects of Electromagnetic Fields on Marine Species: A Literature Review*; Oregon Wave Energy Trust: Oregon, OR, USA, 2010.
67. Gill, A.B.; Gloyne-Philips, I.; Kimber, J.; Sigray, P. Marine renewable energy, electromagnetic (EM) fields and EM-sensitive animals. In *Marine Renewable Energy Technology and Environmental Interactions*; Shields, A.M., Payne, A.I.L., Eds.; Springer: Dordrecht, UK, 2014; pp. 61–79.
68. LaFrance, M.; English, P.; King, J.; Khan, A. *Benthic Monitoring during Wind Turbine Installation and Operation at the Block Island Wind Farm, Rhode Island*; BOEM: Englewood, CO, USA, 2018.
69. Finney, K. Ocean thermal energy conversion. *Guelph. Eng. J.* **2008**, *1*, 17–23.
70. Onoufriou, J.; Brownlow, A.; Moss, S.; Hastie, G.; Thompson, D. Empirical determination of severe trauma in seals from collisions with tidal turbine blades. *J. Appl. Ecol.* **2019**, *56*, 1712–1724. [CrossRef]
71. Williamson, B.J.; Blondel, P.; Armstrong, E.; Bell, P.S.; Hall, C.; Waggitt, J.J.; Scott, B.E. A Self-Contained Subsea Platform for Acoustic Monitoring of the Environment around Marine Renewable Energy Devices-Field Deployments at Wave and Tidal Energy Sites in Orkney, Scotland. *J. Ocean Eng.* **2016**, *41*, 67–81.

72. Baker, A.L.; Craighead, R.M.; Jarvis, E.J.; Stenton, H.C.; Angeloudis, A.; Mackie, L.; Avdis, A.; Piggot, M.; Hill, J. Modelling the impact of tidal range energy on species communities. *Ocean Coast Manag.* **2020**, *193*, 105221. [CrossRef]
73. Scherelis, C.; Penesis, I.; Hemer, M.A.; Cossu, R.; Wright, J.T.; Guihen, D. Investigating biophysical linkages at tidal energy candidate sites; A case study for combining environmental assessment and resource characterisation. *Renew. Energy* **2020**, *159*, 399–413. [CrossRef]
74. Taormina, B.; Bald, J.; Want, A.; Thouzeau, G.; Lejart, M.; Desroy, N.; Carlier, A. A review of potential impacts of submarine power cables on the marine environment: Knowledge gaps, recommendations and future directions. *Renew. Sustain. Energy Rev.* **2018**, *96*, 380–391. [CrossRef]
75. Zhang, X.; Zhang, L.; Yuan, Y.; Zhai, Q. Life cycle assessment on wave and tidal energy systems: A review of current methodological practice. *Int. J. Environ. Res. Public Health* **2020**, *17*, 1604. [CrossRef]
76. Alcérreca-Huerta, J.C.; Encarnacion, J.I.; Ordoñez-Sánchez, S.; Callejas-Jiménez, M.; Barroso, G.G.D.; Allmark, M. Energy yield assessment from ocean currents in the insular shelf of Cozumel Island. *J. Mar. Sci. Eng.* **2019**, *7*, 147. [CrossRef]
77. García, A.; Cueto, Y.; Silva, R.; Mendoza, E.; Vega, L.A. Determination of the potential thermal gradient for the Mexican Pacific Ocean. *J. Mar. Sci. Eng.* **2018**, *6*, 20.
78. Ocampo-Torres, F.J. Wave Power Resources assessment in Northeast Mexico. In Proceedings of the Pan American Marine Energy Conference, San Jose, Costa Rica, 24–29 January 2020.
79. SEMARNAT (Secretaria de Medio Ambiente y Recursos Naturales). *Política Nacional de Mares y Costas de México. Gestión Integral de las Regiones más Dinámicas del Territorio Nacional*; Comisión Intersecretarial para el Manejo Sustentable de Mares y Costas (CIMARES): Mexico City, Mexico, 2011; p. 65.
80. CONAFOR (Comision Nacional Forestal). *Programa de Inversión de la Región Istmo-Costa en el Estado de Chiapas*; CONAFOR: Chiapas, Mexico, 2016.
81. HINMREC (Hawaii National Renewable Energy Center). Available online: http://hinmrec.hnei.hawaii.edu/hinmrecftp/AnnualTempDiff.html (accessed on 11 November 2017).
82. SAGARPA (Secretaria de Agricultura y Desarrollo Rural). Available online: https://www.gob.mx/cms/uploads/attachment/file/325216/Temperatura_superficial_marina_del_Pac_fico_Mexicano10nov17_02_feb_18.pdf (accessed on 8 June 2017).
83. García-Mendoza, E.; Quijano-Scheggia, S.; Olivos-Ortiz, A.; Núñez-Vázquez, E.J. *Florecimientos Algales Nocivos en México*; CICESE: Ensenada, México, 2016; p. 438.
84. Camargo, J.A.; Alonso, A. Contaminación por nitrógeno inorgánico en los ecosistemas acuáticos: Problemas medioambientales, criterios de calidad del agua, e implicaciones del cambio climático. *Ecosistemas* **2007**, *16*, 98–110.
85. Gárate-Lizárraga, I.; Pérez-Cruz, B.; Díaz-Ortíz, J.A.; López-Silva, S.; González-Armas, R. Distribución del dinoflagelado Pyrodinium bahamense en la costa pacífica de México. *Rev. Lat. Ambient Cienc.* **2015**, *6*, 2666–2669.
86. Ronsón, J. Análisis retrospectivos y posibles causas de las mareas rojas tóxicas en el litoral del sureste mexicano (Guerrero, Oaxaca, Chiapas). *Cienc Mar.* **1999**, *3*, 49–55.
87. Band-Schmidt, C.J.; Bustillos-Guzmán, J.J.; López-Cortés, D.J.; Núñez-Vázquez, E.; Hernández-Sandoval, F. El estado actual del estudio de florecimientos algales nocivos en México. *Hidrobiológica* **2011**, *21*, 381–413.
88. Cota, R.; Velázquez, N.; González, E.; Aguilar, A. Microrred aislada para una comunidad pesquera de Baja California, México: Caso de estudio. In Proceedings of the IV Congreso Iberoamericano Sobre Microrredes con Generación Distribuida de Renovables, Concepción, Chile, 25–28 October 2016; p. 8.
89. Marrero, S. Estudio de Eficiencia Energética y Estabilidad de una Micro-red en La Restiga, isla de El Hierro. Master's Thesis, Universidad de las Palmas de Gran Canaria, Las Palmas de Gran Canaria, Spain, 2015.
90. Hossain, E.; Kabalci, E.; Bayindir, R.; Perez, R. Microgrid testbeds around the word: State of art. *Energy Conserv. Manag.* **2014**, *86*, 132–153. [CrossRef]
91. Flores, H.; Vergara, A. Efecto de reducir la frecuencia de la alimentación en la supervivencia, crecimiento, conversión y conducta alimenticia en juveniles de salmón del Atlántico Salmo salar (Linnaeus, 1758): Experiencia a nivel productivo. *Lat. Am. J. Aquat. Res.* **2012**, *40*, 536–544. [CrossRef]
92. Masutani, S.M.; Takahashi, P.K. Ocean Thermal Energy Conversion (OTEC). In *Encyclopedia of Electrical and Electronics Engineering*, 2nd ed.; Steele, J., Turekian, K.K., Thorpe, S.A., Eds.; Elsevier ltd: Oxford, UK, 2001; pp. 167–173.

93. ONU. *Índices Básicos de las Ciudades Prosperas, Medición, Nivel Básico*; Pijijiapan, ONU HABITAT: Chiapas, Mexico, 2018; p. 130.
94. SAGARPA (Secretaria de Agricultura y Desarrollo Rural). *Manifestación de Impacto Ambiental, Modalidad Particular para el Proyecto*; Obras de Dragado y Escolleras de Barra de Santiago Iolomita Municipio de Pijijiapan: Chiapas, Mexico, 2010; p. 8.

International Journal of
*Environmental Research and Public Health*

*Article*

# A Software for Calculating the Economic Aspects of Floating Offshore Renewable Energies

**Laura Castro-Santos [1,*] and Almudena Filgueira-Vizoso [2]**

1 Departamento de Enxeñaría Naval e Industrial, Escola Politécnica Superior, Universidade da Coruña, Esteiro, 15471 Ferrol, Spain
2 Departamento de Química, Escola Politécnica Superior, Universidade da Coruña, Esteiro, 15471 Ferrol, Spain; almudena.filgueira.vizoso@udc.es
* Correspondence: laura.castro.santos@udc.es

Received: 25 November 2019; Accepted: 24 December 2019; Published: 27 December 2019

**Abstract:** The aim of this work is to develop a software to calculate the economic parameters so as to determine the feasibility of a floating offshore renewable farm in a selected location. The software can calculate the economic parameters of several types of offshore renewable energies, as follows: one renewable energy (floating offshore wind—WindFloat, tension leg platform (TLP), and spar; floating wave energy—Pelamis and AquaBuoy), hybrid offshore wind and wave systems (Wave Dragon and W2Power), and combined offshore wind and waves with different systems (independent arrays, peripherally distributed arrays, uniformly distributed arrays, and non-uniformly distributed arrays). The user can select several inputs, such as the location, configuration of the farm, type of floating offshore platform, type of power of the farm, life-cycle of the farm, electric tariff, capital cost, corporate tax, steel cost, percentage of financing, or interest and capacity of the shipyard. The case study is focused on the Galicia region (NW of Spain). The results indicate the economic feasibility of a farm of floating offshore renewable energy in a particular location in terms of its costs, levelized cost of energy (LCOE), internal rate of return (IRR), net present value (NPV), and discounted pay-back period. The tool allows for establishing conclusions about the dependence of the offshore wind resource parameters, the main distances (farm–shore, farm–shipyard, and farm–port), the parameters of the waves, and the bathymetry of the area selected.

**Keywords:** feasibility study; offshore wind; levelized cost of energy (LCOE); wave energy; software

## 1. Introduction

Floating offshore renewable energies are those that are installed in deep waters (more than 50 m). They are composed by several main components, namely: energy generators, floating offshore platforms, mooring, anchoring, and electric systems. The main difference between these floating platforms and the fixed platforms (up to 50 m of depth; monopiles [1], tripiles, jackets [2], etc.) [3] is that these last ones have platforms fixed to the seabed using several devices; therefore, they do not have mooring and anchoring systems.

In this context, the most developed technology is offshore wind systems, although wave energy will have great development in the future too. Floating offshore wind energy has been developed over the last years, mainly in Europe and Japan [4,5]. Hywind Scotland was the first commercial floating offshore wind farm installed in the world. It was installed in 2017 in Scottish waters [6] by Statoil. It is based on a spar floating platform called Hywind, which was previously probed in Norway in 2009 [7]. It has five spar platforms that were built in the Navantia-Fene shipyard in Fene (A Coruña, Spain). This shipyard is the leader in floating offshore wind building—it built one platform for the WindFloat Atlantic project (Windplus) in Portugal in 2019 (the first floating offshore wind farm in the Iberia

Peninsula [8]), and it is starting to build five platforms for the Kinkardine project (Aberdeen, U.K.), consisting of 9.5 MW platforms [9].

Floating offshore renewable energies will have a great future. However, nowadays, they are still in development, mainly wave energy systems and hybrid technology, and need to increase their unitary power in order to be more competitive with conventional offshore wind. Therefore, it is important to determine the best areas where a floating offshore renewable energy farm can be installed, and calculate their economic feasibility in order to make decisions about the final installation of these technologies. There are some studies about the types of restrictions of the areas where offshore wind energy is installed [10], restrictions for wave energy [11], mapping the electric system of offshore ports [12], and a comparison between onshore and offshore wind [7].

There is a lot of information about offshore platforms, namely: semisubmersible [13–15], tensioned leg platform (TLP) [16,17], and spar [18]. Arapogianni [19] analysed the main floating systems and differences between the grid connected systems (Hywind of Statoil and WindFloat of Principle Power) and the concepts under development (Advanced Floating Turbine of Nautica WindPower; Aero-generator X of Wind Power Ltd. Arup, Azimut of Consortium of Spanish Wind Energy Industry lead by Gamesa; Blue H TLP of Blue H; DeepCWind floating wind of the consortium of the University of Maine, AEWC, Seawall, Maine Maritime Academy, Technip, NREL and MARIN; Deepwind, which is an EU project; DIWET Semisub of Pole Mer; EOLIA of Acciona Energy; IDEOL of IDEOL; GICON TLP of GICON; Hexicon platform of Hexicon; HiPRwind, an EU project; Karmoy of Sway; Ocean Breeze of Xanthus Energy; W2Power of Pelagic Power; Pelastar of Glosten Associates; Poseidon Floating Power of the Floating Power; Sea Twirl of Sea Twirl; Trifloater Semisub of Gusto; Vertiwind of Technip and Nenuphar; WindSea Floater of Force technology NLI; Winflo of Nass and Wind and DCNS; ZEFIR Test Station of Catalonia Institute for Energy Research; and Haliade of Alstom). Perhaps the most representative is the research developed by Jonkman and Matha [19,20], where a spar, a semisubmersible, and a TLP (tensioned leg platform) were calculated, because it compares the three main types of floating offshore wind platforms. In this context, Sclavounos et al. [21,22] also considered the taught leg buoy concept. Collu et al. compared fixed and floating structures for a 5-MW wind turbine [23].

However, all of these studies did not take into account the economic aspects of such technologies, although Wind Europe has established the importance of cost reductions [24] in order to create competitive technologies comparable with onshore renewable energies. In addition, presently, there is not any software that allows the user to develop the economic calculation of the offshore renewable energy of several choices (wind, waves, and wind and waves), which is the objective of the present work.

The aim of this paper is to create a software to calculate the most important parameters of the economic feasibility of a floating offshore renewable farm in a selected location.

## 2. Software Characteristics

### 2.1. Description of the Software

The software can calculate the economic parameters of several types of offshore renewable energies, as follows: one renewable energy (floating offshore wind—WindFloat, TLP, and spar; floating wave energy—Pelamis [25] and AquaBuoy [26]), hybrid offshore wind and wave systems (Wave Dragon [27] and W2Power [28]), and combined offshore wind and waves with different systems (independent arrays, peripherally distributed arrays, uniformly distributed arrays, and non-uniformly distributed arrays). The user can select several inputs, namely: location, configuration of the farm, type of floating offshore platform, type of calculation of the wave's energy, power of the farm, life-cycle of the farm, electric tariff, capital cost, corporate tax, steel cost, percentage of financing, interest, and capacity of the shipyard. The case study is focused on the Galicia region, located in the North-West of Spain. The economic results are as follows: the cost of each phase of the life-cycle of the project, the total cost of the life-cycle of the farm, the internal rate of return (IRR), the net present value (NPV), the discounted payback period (DPBP), and the levelized cost of energy (LCOE). The results indicate the economic feasibility of a farm of floating offshore renewable energy in a particular location.

The objective of the created software (W2EC by LCS "Wind and Wave Energy farm economic Calculator" by Laura Castro Santos) is to calculate several economic parameters of a floating offshore renewable energy farm in a location. The formulation of costs has been previously developed [28]. This software has been created for the locations of Galicia, the Galicia and Cantabric region, and Portugal, because these are the input maps data that we have available. However, if the data of other locations is obtained, new areas of analysis can be included. Therefore, the software can be used for any location that the user wants. The software has been registered.

The user can select the inputs wanted in order to calculate the economic maps of the location selected. The programming language is MATLAB®(MathWorks, Natick, MA, USA), and the software is compatible with Microsoft Windows®(Microsoft, Redmond, WA, USA).

Firstly, the software calculates the main costs of the life-cycle of the offshore renewable energy farm (definition of the concept, design and development, manufacturing, installation [13], exploitation, and dismantling), and then the total life-cycle cost of the farm is calculated [29,30]. The novelty of this work is to develop a software using an easy interface and considering several types of offshore renewable energies. Therefore, the calculation of the costs will be conditioned by the type of energy selected, among other factors.

In addition, the map inputs (scale parameter of the offshore wind resource, shape parameter of the offshore wind resource, distance from farm to shore, distance from farm to shipyard, distance from farm to port, height of waves, period of waves, and bathymetry) generate a map of the energy produced in the specific location selected.

### 2.2. Inputs of the Software

In this context, the inputs of the software are (see Tables 1 and 2 and Figure 1):

**Table 1.** Inputs of the software I. TLP- tensioned leg platform.

| Input | Types |
|---|---|
| Location | Galicia<br>Galicia and Cantabric region<br>Portugal |
| Configuration of the farm | One renewable energy<br>Two renewable energies-independent arrays<br>Two renewable energies-peripherally distributed array<br>Two renewable energies-uniformly distributed array<br>Two renewable energies-non-uniformly distributed array |
| Type of floating platform | WindFloat<br>Pelamis<br>AquaBuoy<br>Wave Dragon<br>W2Power<br>Poseidon<br>TLP<br>Spar |
| Type of calculation of the wave's energy | T and H<br>Matrix |

Figure 2 shows the input variables (location, configuration of the farm, floating platform, calculation of energy waves, total power of the farm (Ptotalfarm), number of years of the life-cycle (Nfarm), electric tariff, corporate tax, % financing, % interest, capital cost, cost of steel (Csteel), and number platforms per year) that are used to calculate the total cost of the farm (Ctotal) and the input maps related to the selected location (shape and scale of the wind parameter, bathymetry, and period and height of waves), whose value is different depending on the point (k) of the defined grid.

**Table 2.** Inputs of the software II.

| Inputs |
| --- |
| Power of the farm (MW) |
| Life-cycle of the farm (years) |
| Electric tariff (€/MWh) |
| Capital cost (%) |
| Corporate tax (%) |
| Steel cost (€/ton) |
| Percentage of Financing (%) |
| Interest (%) |
| Capacity of the shipyard (platform/year) |

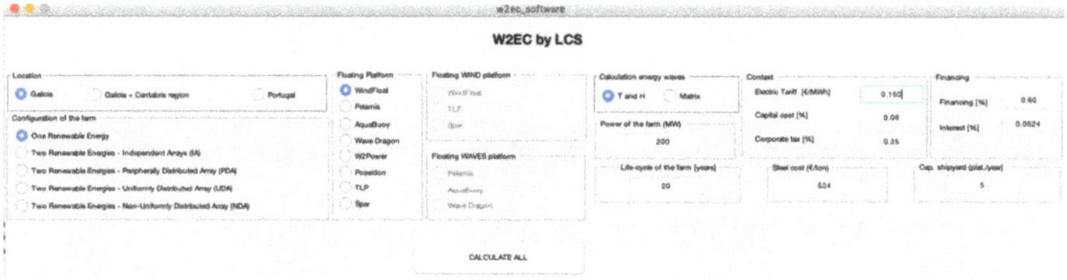

**Figure 1.** Software interface.

The software calculates the economic parameters of the project, considering the energy produced and the total life-cycle cost such as input variables. The economic results obtained are as follows:

Internal rate of return of the financed project (IRR FP; %).
Net present value of the financed project (NPV FP; M€).
Discounted payback period of the financed project (DPBP FP; years).
Levelized cost of energy (LCOE; €/MWh).

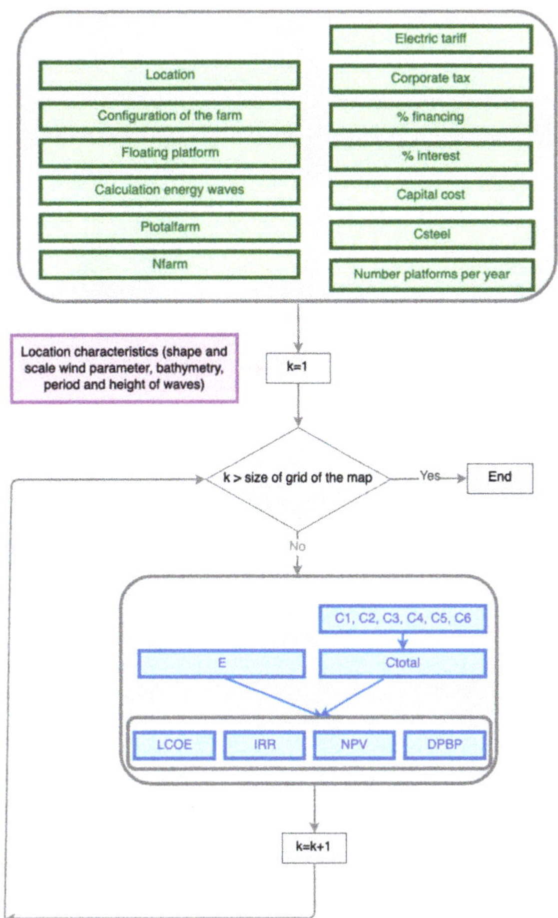


**Figure 2.** Flow diagram of the software. Input variables being the following: the total power of the farm (Ptotalfarm), number of years of the life-cycle (Nfarm), cost of steel (Csteel), the number of points that the input maps have (k), energy produced (E), total cost of the farm (Ctotal), defining cost (C1), designing and developing cost (C2), manufacturing cost (C3), installing cost (C4), exploiting cost (C5), and dismantling cost (C6).

### 2.3. Criteria and Protocol

Regarding the location, the user selects this input and he has three options (Galicia, Galicia + Cantabric region and Portugal). These choices condition the map files (in .mat) that the software selects for acting related to scale parameter of the offshore wind resource (Figure 3a), shape parameter of the offshore wind resource (Figure 3b), distance from farm to shore (Figure 3c), distance from farm to shipyard (Figure 3d), distance from farm to port (Figure 3e), height of waves (Figure 3f), period of waves (Figure 3g) and bathymetry (Figure 3h).

Regarding the type of floating platform, the criteria is that they are divided in wind (WindFloat [31], spar [21] and TLP (Tensioned Leg Platform) [21]), waves (Pelamis [25], AquaBuoy [32] and Wave Dragon [33]) or hybrid (W2Power [28] and Poseidon [34]) (see Figure 4). Depending on the type of platform selected, the protocol to calculate the cost of manufacturing the platform, its generator, its mooring and its anchoring, will change. In addition, it also affects to the installation, maintenance and decommissioning cost.

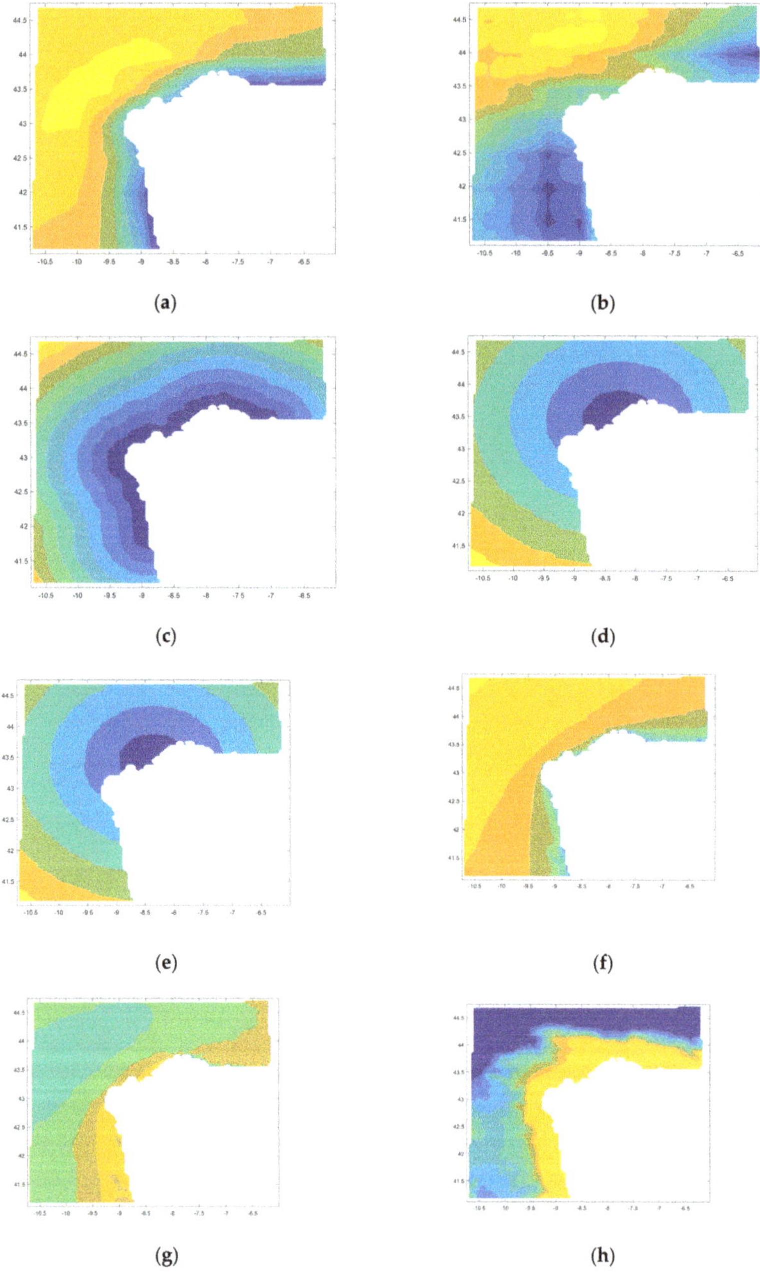


**Figure 3.** Map inputs: scale parameter of the offshore wind resource (**a**) in m/s, shape parameter of the offshore wind resource (**b**), distance from farm to shore in m (**c**), distance from farm to shipyard in m (**d**), distance from farm to port in m (**e**), height of waves in m (**f**), period of waves in s (**g**), and bathymetry in m (**h**).

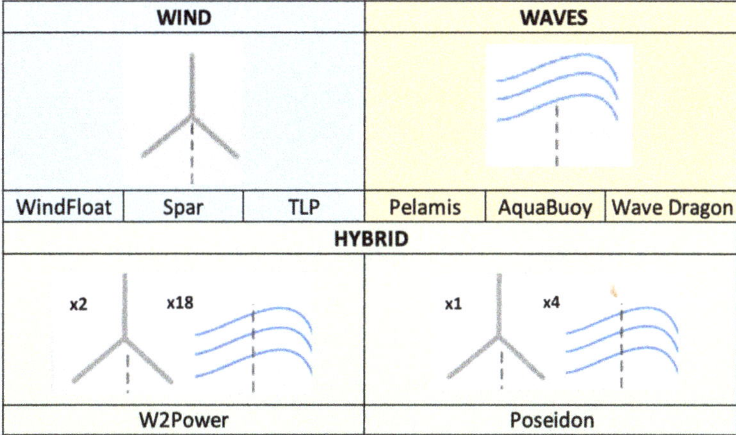


**Figure 4.** Type of floating platform inputs: wind (WindFlaot, spar, TLP), waves (Pelamis, AquaBuoy, Wave Dragon) and Hybrid (W2Power, Poseidon).

Regarding the configuration of the farm figure shows the four types of inputs: one renewable energy (Figure 5a), two renewable energies and independent arrays (Figure 5b), two renewable energies and peripherally distributed array. Figure 5c two renewable energies and uniformly distributed array (Figure 5d) and two renewable energies and non-uniformly distributed array (Figure 5e). They will condition the length, type and size of the electric cable between platforms, which has a great influence on costs.

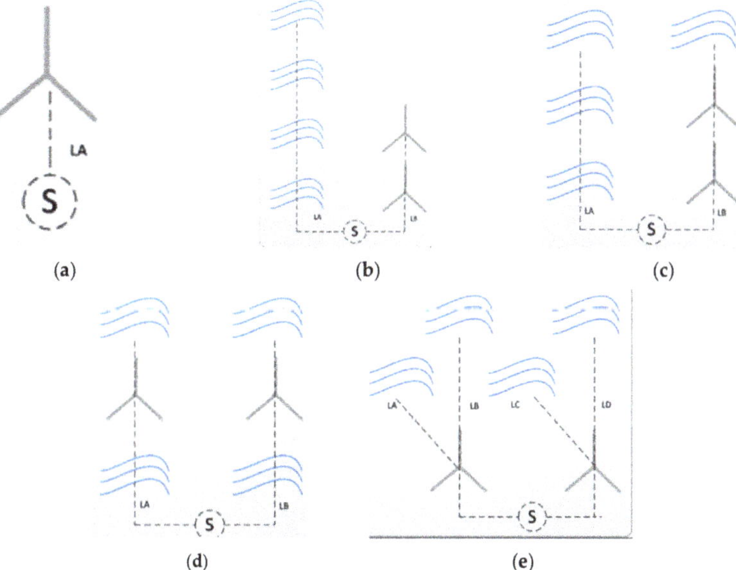


**Figure 5.** Configuration of the farm inputs [35]: one renewable energy (**a**), two renewable energies and independent arrays (**b**), two renewable energies and peripherally distributed arrays (**c**), two renewable energies and uniformly distributed arrays (**d**), and two renewable energies and non-uniformly distributed arrays (**e**).

Considering the type of calculation of the wave's energy, there are two main types, namely: T and H (period and height of waves, respectively), and matrix. The first consideration takes into account a general equation for the calculation of the energy generated by a wave energy converter considering the period of waves ($T_{wa}$), the height of waves ($H_{wa}$), the number of hours per year (NHAT), the density of water ($\rho$), the gravity ($g$), the main dimension ($D_{wa}$), and the efficiency of the wave generator ($\eta_{effiency}$). This equation is used when there is not enough information of the area selected to create the matrix (see Equation (1) [35]). Therefore, this calculation is an approximation to the real value of the energy produced by the wave farm.

$$E_{1wa} = NHAT \cdot \frac{2}{64} \cdot \frac{\rho \cdot g^2}{\pi} \cdot T_{wa} \cdot H^2_{wa} \cdot D_{wa} \cdot \eta_{effiency} \quad (1)$$

The second type considers the power matrix of the technology. Table 3 shows an example of the power matrix of the Pelamis wave energy converter) and the matrix of the location, which is more difficult to obtain.

**Table 3.** Pelamis power matrix (in kW) [36].

| Te(s)Hs(m) | Power Matrix (in kW) | | | | | | | | | | | | | | | | |
|---|---|---|---|---|---|---|---|---|---|---|---|---|---|---|---|---|---|
| | **5** | **5.5** | **6** | **6.5** | **7** | **7.5** | **8** | **8.5** | **9** | **9.5** | **10** | **10.5** | **11** | **11.5** | **12** | **12.5** | **13** |
| 0.5 | 0 | 0 | 0 | 0 | 0 | 0 | 0 | 0 | 0 | 0 | 0 | 0 | 0 | 0 | 0 | 0 | 0 |
| 1 | 0 | 22 | 29 | 34 | 37 | 38 | 38 | 37 | 35 | 32 | 29 | 26 | 23 | 21 | 0 | 0 | 0 |
| 1.5 | 32 | 50 | 65 | 76 | 83 | 86 | 86 | 83 | 78 | 72 | 65 | 59 | 53 | 47 | 42 | 37 | 33 |
| 2 | 57 | 88 | 115 | 136 | 148 | 153 | 152 | 147 | 138 | 127 | 116 | 104 | 93 | 83 | 74 | 66 | 59 |
| 2.5 | 89 | 138 | 180 | 212 | 231 | 238 | 238 | 230 | 216 | 199 | 181 | 163 | 146 | 130 | 116 | 103 | 92 |
| 3 | 129 | 198 | 260 | 305 | 332 | 240 | 332 | 315 | 292 | 266 | 240 | 219 | 210 | 188 | 167 | 149 | 132 |
| 3.5 | 0 | 270 | 345 | 415 | 438 | 440 | 424 | 404 | 377 | 362 | 326 | 292 | 260 | 230 | 215 | 202 | 180 |
| 4 | 0 | 0 | 462 | 502 | 540 | 546 | 530 | 499 | 475 | 429 | 384 | 366 | 339 | 301 | 267 | 237 | 213 |
| 4.5 | 0 | 0 | 544 | 635 | 642 | 648 | 628 | 590 | 562 | 528 | 473 | 432 | 382 | 356 | 338 | 300 | 266 |
| 5 | 0 | 0 | 0 | 739 | 726 | 726 | 707 | 687 | 670 | 607 | 557 | 521 | 472 | 417 | 369 | 348 | 328 |
| 5.5 | 0 | 0 | 0 | 750 | 750 | 750 | 750 | 750 | 737 | 667 | 658 | 586 | 530 | 496 | 446 | 395 | 355 |
| 6 | 0 | 0 | 0 | 0 | 750 | 750 | 750 | 750 | 750 | 750 | 711 | 633 | 619 | 558 | 512 | 470 | 415 |
| 6.5 | 0 | 0 | 0 | 0 | 750 | 750 | 750 | 750 | 750 | 750 | 750 | 743 | 658 | 621 | 579 | 512 | 481 |
| 7 | 0 | 0 | 0 | 0 | 0 | 750 | 750 | 750 | 750 | 750 | 750 | 750 | 750 | 676 | 613 | 584 | 525 |
| 7.5 | 0 | 0 | 0 | 0 | 0 | 0 | 750 | 750 | 750 | 750 | 750 | 750 | 750 | 750 | 686 | 622 | 593 |
| 8 | 0 | 0 | 0 | 0 | 0 | 0 | 0 | 750 | 750 | 750 | 750 | 750 | 750 | 750 | 750 | 690 | 625 |

On the other side, the power farm and the life-cycle of the farm are introduced by the user, depending on his needs. The electric tariff, capital cost, corporate tax, and steel cost are values that depend on the market being analyzed. The percentage of financing and the interest are determined by the financing company. Finally, the capacity of the shipyard depends on the shipyard where the platforms were built, mainly its size, and works in development in the moment of building the platforms.

Finally, considering the outputs, the equations used are shown in Table 4, as follows [36–38]: C1, in €, is the defining cost; C2 in € is the designing and developing cost; C3, in €, is the manufacturing cost; C4, in €, is the installing cost; C5, in €, is the exploiting cost; C6, in €, is the dismantling cost, $C_t$, in €, is the cost of the correspondent year, $E_t$, in MWh/year, is the energy produced, $r$, in %, is the capital cost; $CF_t$ is the cash flow; $t$, in years, is the life-cycle of the project; and $G_0$, in €, is the initial investment. The project will be economically feasible if the net present value is positive, the internal rate of return is higher than the capital cost, and the levelized cost of energy has low values.

**Table 4.** Output criteria.

| Output | Equation |
|---|---|
| Total cost | $C_{total} = C1 + C2 + C3 + C4 + C5 + C6$ |
| Levelized cost of energy | $LCOE = \frac{\sum_{t=0}^{N_{farm}} \frac{C_t}{(1+r)^t}}{\sum_{t=0}^{N_{farm}} \frac{E_t}{(1+r)^t}}$ |
| Net present value | $NPV = -G_0 + \sum_{t=1}^{n} \frac{CF_t}{(1+r)^t}$ |
| Internal rate of return | $-G_0 + \sum_{t=1}^{n} \frac{CF_t}{(1+IRR)^t} = 0$ |

## 3. Case of Study and Results

### 3.1. Case of Study

The case of study of the present paper is the Galician region, located in the North-West of Spain (in red in Figure 6), which has very good conditions in terms of offshore wind and offshore wave resources.

**Figure 6.** Galicia region (NW of Spain), in red.

The four cases studied in this paper differ, depending on their type of floating offshore renewable energy platforms (Table 5), as follows: wind platform (WindFloat) [16,39] in Case 1, wind platform (WindFloat) and wave platform (AquaBuoy) in Case 2, wave platform (AquaBuoy) [26] in Case 3, and a hybrid platform that mixes wind and wave energy in the same platform (W2Power [28]) in Case 4.

**Table 5.** Characteristics of the case studies.

| Input | Case of Study 1 | Case of Study 2 | Case of Study 3 | Case of Study 4 |
|---|---|---|---|---|
| Location | Galicia | Galicia | Galicia | Galicia |
| Configuration of the farm | One Renewable Energy | Two Renewable Energies–Independent Arrays (IA) | One Renewable Energy | One Renewable Energy |
| Floating platform | WindFloat | WindFloat and AquaBuoy | AquaBuoy | W2Power |
| Calculation energy waves | - | T and H | T and H | T and H |

For the case study of this paper, the inputs of the four cases of study are shown in Table 6.

**Table 6.** Inputs.

| Input | Value | Units |
|---|---|---|
| Power of the farm | 200 | MW |
| Life-cycle of the farm | 20 | years |
| Electric tariff | 150 | €/MWh |
| Capital cost | 8% | - |
| Corporate tax | 25% | - |
| Steel cost | 524 | €/ton |
| Percentage of financing | 60% | - |
| Interest | 5.24% | - |
| Capacity of the shipyard | 5 | Platforms/year |

### 3.2. Results

The software created calculates the energy produced by a floating offshore renewable energy farm. In the particular cases of study of this paper, the energy produced by a floating offshore renewable energy farm composed by offshore wind turbines (Figure 7a), offshore wind and waves (different platforms) (Figure 7b), offshore wave energy (Figure 7c) and offshore hybrid platform (wind and waves in the same platform) (Figure 7d), has different values depending on the type of technology selected. The values go from 79,500,000 MWh/year to 701,000,000 M€/year for case 1; from 58,890,000 MWh/year to 497,340,000 M€/year for case 2; from 7,087,500 MWh/year to 121,110,000 M€/year for case 3; and from 37,843,000 MWh/year to 312,220,000 M€/year for case 4.

On the other hand, the software gives results regarding the costs (C1, C2, C3, C4, C5, C6, and Ctotal) and the economic feasibility of the floating offshore renewable energy farm (LCOE, IRR, NPV, and DPBP).

Figures 8–11 are the maps of results for case 1, case 2, case 3 and case 4 respectively. It is important to notice that results of all the maps depend on the shape and scale parameters of the offshore wind resource, the distance from farm to shore, distance from farm to shipyard, distance from farm to port, the height and the period of waves and the bathymetry of the location of study (in this case the Galician region). This fact is shown in Figures 8–11. Of course, the maps of results depend on the inputs, therefore this software is valid for all the locations that user wants.

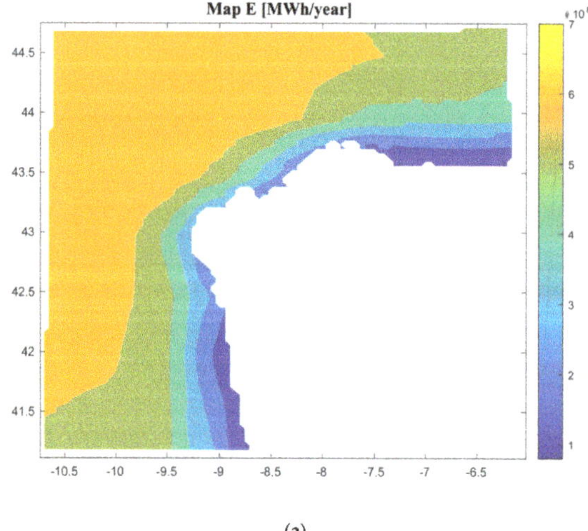


(a)

**Figure 7.** *Cont.*

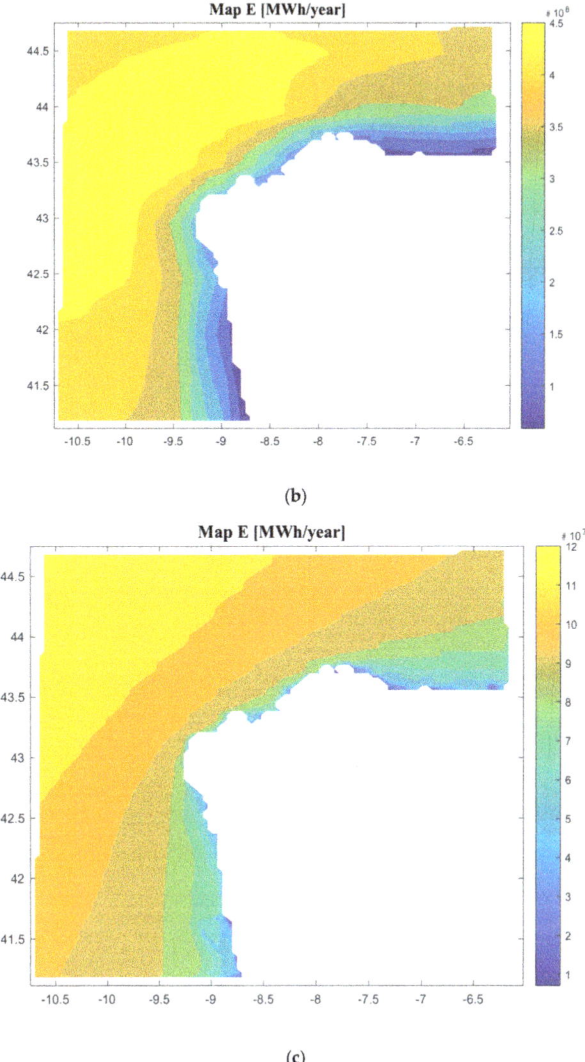


(c)

**Figure 7.** *Cont.*

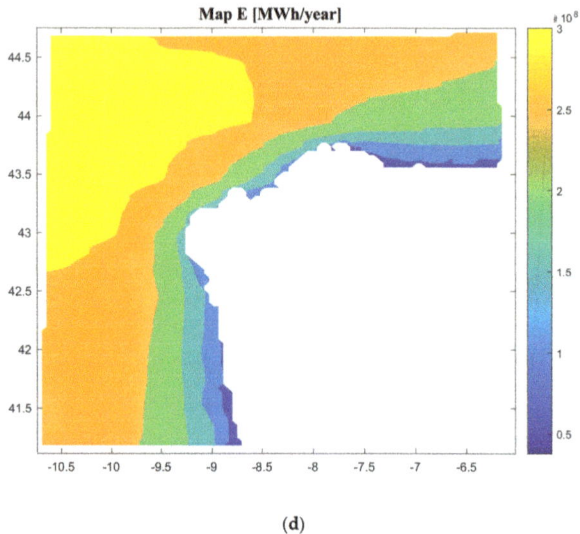


(d)

**Figure 7.** Results of the energy produced for Case 1 (**a**); Results of the energy produced for Case 2 (**b**); Results of the energy produced for Case 3 (**c**); Results of the energy produced for Case 4 (**d**).

Regarding case 1, results depends on the location and they go (see Figure 8): from 6.94 M€ to 126.52 M€ for C1; 0.46 M€ for C2; from 403.02 M€ to 847.23 M€ for C3; from 14.85 M€ to 272.11 M€ for C4; from 241.68 M€ to 409.15 M€ for C5; from 1.44 M€ to 37.23 M€ for C6; from 674.21 M€ to 1650 M€ for Ctotal; from 100.31 €/MWh to 882.93 €/MWh for LCOE; from 11 years to 22 years for DPBP; from −142.40% to 14.75% for IRR; and from −535.50 € to 330.85 € for NPV.

Regarding Case 2, the results depend on the location, and they go (see Figure 9) from 4.85 M€ to 503.08 M€ for C1, 0.34 M€ for C2, from 367.45 M€ to 676.09 M€ for C3, from 9.92 M€ to 180.32 M€ for C4, from 314.59 M€ to 482.57 M€ for C5, from 20.72 M€ to 37.46 M€ for C6, from 720.97 M€ to 1835.70 M€ for Ctotal, from 138.22 €/MWh to 1186.30 €/MWh for LCOE, from 18 years to 22 years for DPBP, from −176.90% to 7.83% for IRR, and from −840.31 € to 59.06 € for NPV.

Regarding case 2, results depends on the location and they go (see Figure 9): from 4.85 M€ to 503.08 M€ for C1; 0.34 M€ for C2; from 367.45 M€ to 676.09 M€ for C3; from 9.92 M€ to 180.32 M€ for C4; from 314.59 M€ to 482.57 M€ for C5; from 20.72 M€ to 37.46 M€ for C6; from 720.97 M€ to 1835.70 M€ for Ctotal; from 138.22 €/MWh to 1,186.30 €/MWh for LCOE; from 18 years to 22 years for DPBP; from -176.90% to 7.83% for IRR; and from −840.31 € to 59.06 € for NPV.

Regarding Case 4, the results depend on the location, and they go (see Figure 11) from 4.14 M€ to 167.97 M€ for C1, 0.17 M€ for C2, from 222.77 M€ to 350.75 M€ for C3, from 7.37 M€ to 186.64 M€ for C4, from 355.56 M€ to 521.30 M€ for C5, from 11.60 M€ to 29.81 M€ for C6, from 603.83 M€ to 1210.40 M€ for Ctotal, from 167.85 €/MWh to 1340.10 €/MWh for LCOE, more than 22 years for DPBP, from −182.39% to −3.57% for IRR, and from −465.38 € to −39.78 € for NPV.

Therefore, considering all of these results, the user can select what is the best technology depending on the location. In this sense, in terms of LCOE, the best technology for the case study is the Case 1 (offshore wind—WindFloat), because it has the smallest LCOE (100.31 €/MWh), and the worst technology is Case 3 (wave energy-AquaBuoy), with a minimum value of 756.39 €/MWh. On the other hand, in terms of IRR and NPV, the best value is also for Case 1 with 14.75% and 330.85 €, respectively, and the worst is Case 3, with values of −173.52% and −581.42 €, respectively.

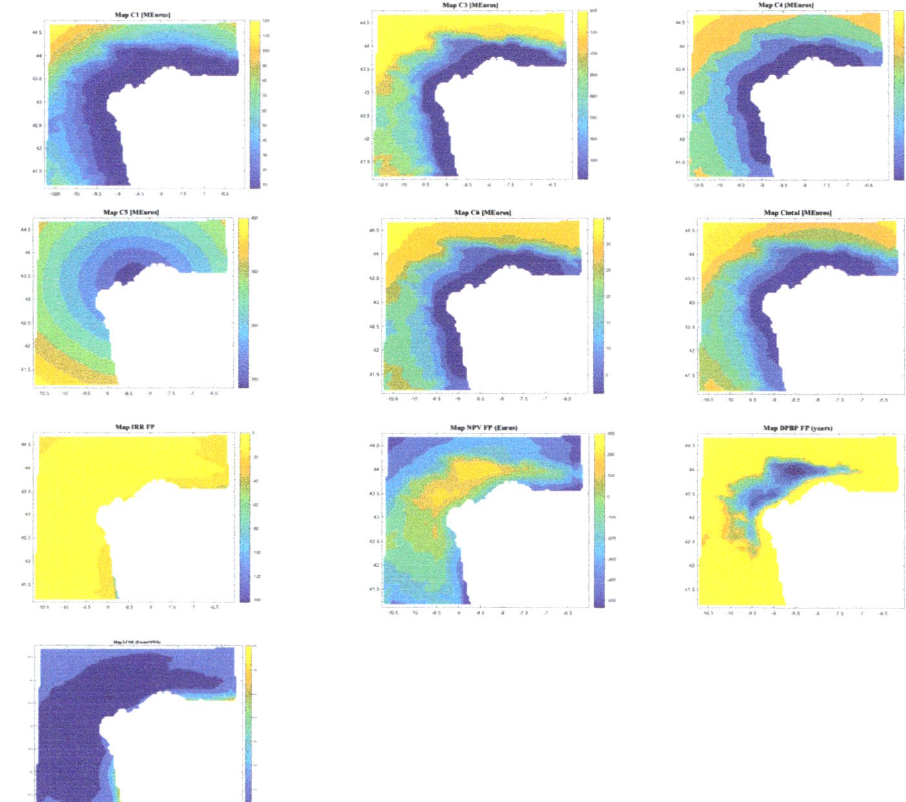


**Figure 8.** Results (C1, C1, C3, C4, C5, C6, Ctotal, LCOE, DPBP, IRR, NPV) for Case 1.

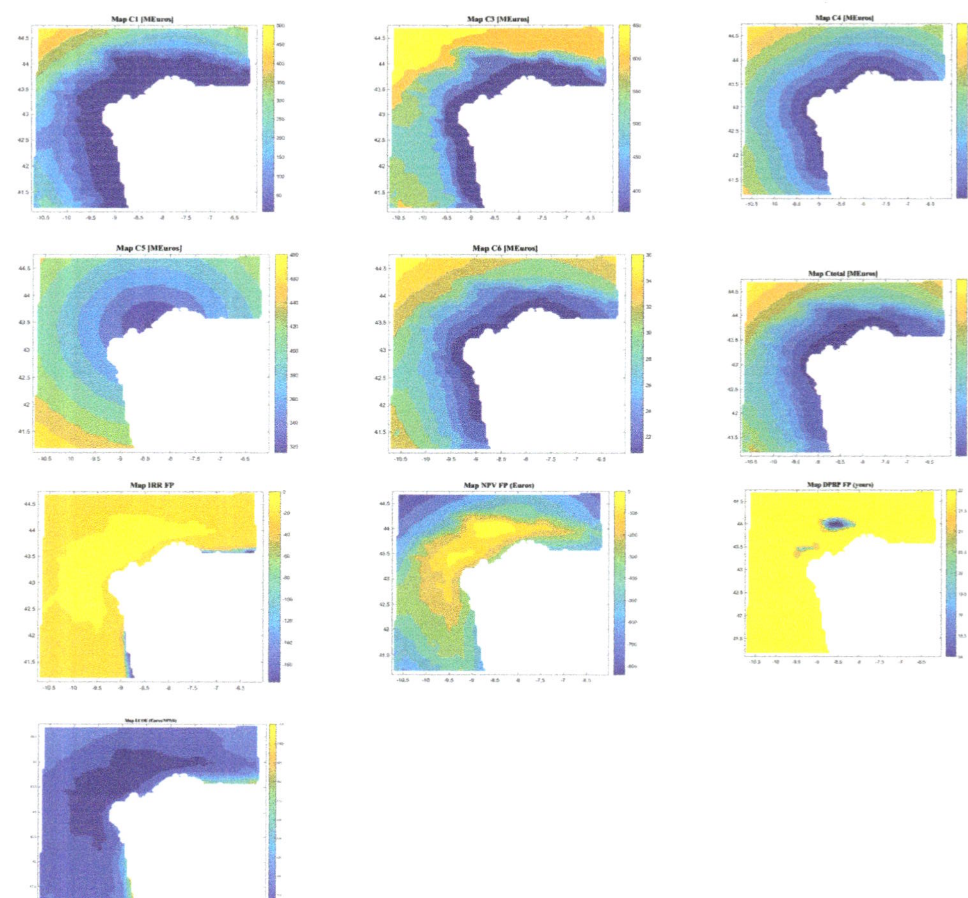


**Figure 9.** Results (C1, C1, C3, C4, C5, C6, Ctotal, LCOE, DPBP, IRR, NPV) for Case 2.

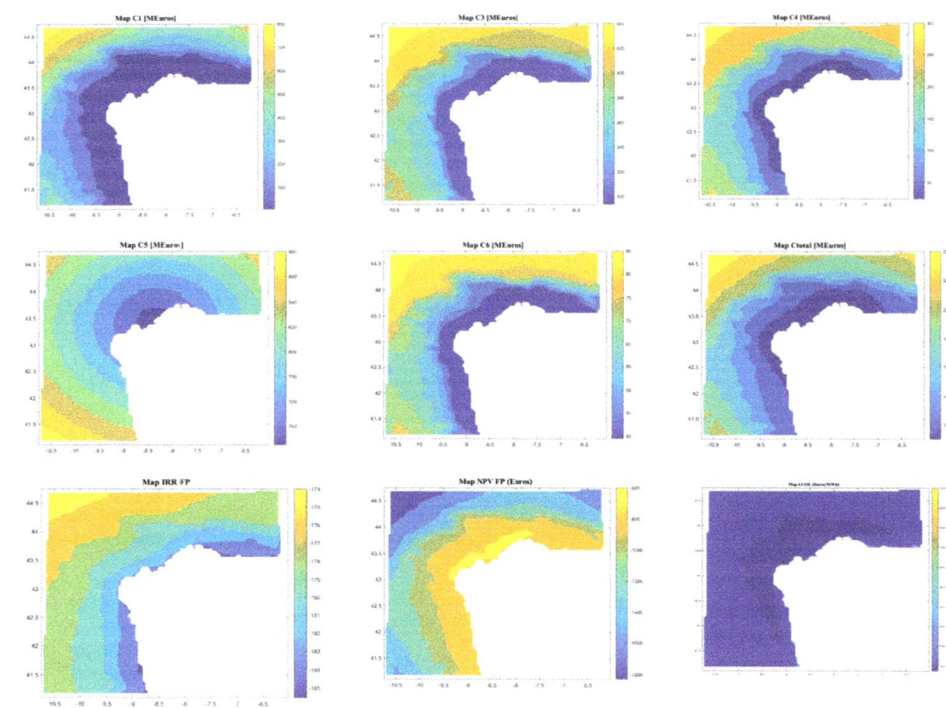


**Figure 10.** Results (C1, C1, C3, C4, C5, C6, Ctotal, LCOE, DPBP, IRR, NPV) for Case 3.

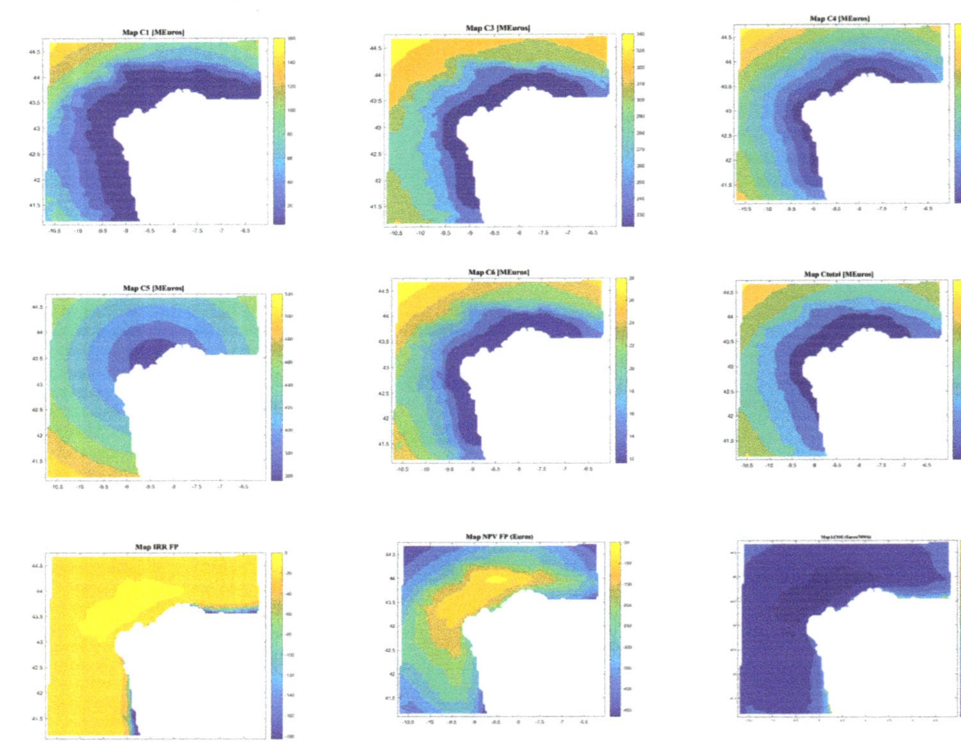


**Figure 11.** Results (C1, C1, C3, C4, C5, C6, Ctotal, LCOE, DPBP, IRR, NPV) for Case 4.

## 4. Conclusions

The objective of this work was to present a software to calculate the most important parameters of the feasibility of a floating offshore renewable farm in a selected location. This tool can be valuable for enterprises or public entities that want to know the best places where an offshore renewable energy farm can be located, because this can compile the software for the economic calculation of the most important types of offshore renewable energies existing in the world.

The software can calculate the economic parameters of several types of offshore renewable energies, namely: one renewable energy (floating offshore wind—WindFloat, TLP, and spar; floating wave energy—Pelamis and AquaBuoy), hybrid offshore wind and wave systems (Wave Dragon and W2Power), and combined offshore wind and waves with different systems (independent arrays, peripherally distributed arrays, uniformly distributed arrays, and non-uniformly distributed arrays).

The user can select several inputs, namely: location, configuration of the farm, type of floating offshore platform, type of calculation of the wave's energy, power of the farm, life-cycle of the farm, electric tariff, capital cost, corporate tax, steel cost, percentage of financing, interest, and capacity of the shipyard.

The economic results are as follows: the cost of each phase of the life-cycle of the project, the total cost of the life-cycle of the farm, the internal rate of return (IRR), the net present value (NPV), the discounted payback period (DPBP), and the levelized cost of energy (LCOE).

The case study considered here analysed the Galician region (North-West of Spain), using four alternatives depending on the technology, namely: one renewable energy (offshore wind, WindFloat; wave energy, AquaBuoy; and a hybrid system with offshore wind and waves, W2Power) and two renewable energies installed separately in the same farm (offshore wind, Windfloat and wave energy, AquaBuoy). The software gives the results for the case that the user selects. In this particular case study, the technology that is more economically feasible is the offshore wind (WindFloat) with the best value of LCOE (100.31 €/MWh), IRR (14.75%, and NPV (330.85 €), and the technology with the worst economic values is the wave energy (AquaBuoy), with results of 756.39 €/MWh (LCOE), −173.52% of IRR, and −581.42 € of NPV.

The results depend on the shape and scale parameters of the offshore wind resources, the distance from farm to shore, distance from farm to shipyard, distance from farm to port, the height and the period of waves, and the bathymetry of the location of study. The results indicate the economic feasibility of a farm of floating offshore renewable energy in a particular location.

**Author Contributions:** Introduction: L.C.-S., A.F.-V.; software characteristics: L.C.-S.; case of study and results: L.C.-S.; conclusions: L.C.-S., A.F.-V. All authors have read and agreed to the published version of the manuscript.

**Funding:** This research received no external funding.

**Conflicts of Interest:** The authors declare no conflicts of interest.

## References

1. Kallehave, D.; Byrne, B.W.; LeBlanc Thilsted, C.; Mikkelsen, K.K. Optimization of monopiles for offshore wind turbines. *Philos. Trans. R. Soc. A Math. Phys. Eng. Sci.* **2015**, *373*, 20140100. [CrossRef] [PubMed]
2. Henderson, A.R.; Witcher, D. Floating offshore wind energy - a review of the current status and an assessment of the prospects. *Wind Eng.* **2010**, *34*, 1–16. [CrossRef]
3. Wind Europe. *Offshore Wind in Europe*; Wind Europe: Brussels, Belgium, 2018.
4. Henderson, A.R.; Leutz, R.; Fujii, T. Potential for floating offshore wind energy in japanese waters. In Proceedings of the 12th International Offshore and Polar Engineering Conference, Kitakyushu, Japan, 26–31 May 2002; International Society of Offshore and Polar Engineers: Mountain View, CA, USA, 2002; Volume 3, pp. 505–512.
5. Kosugi, A.; Ogata, R.; Kagemoto, H.; Akutsu, Y.; Kinoshita, T. A feasibility study on a floating wind farm off Japan coast. In Proceedings of the 12th 2002 International Offshore and Polar Engineering Conference, Kitakyushu, Japan, 26–31 May 2002.

6. Ulazia, A.; Gonzalez-Roji, S.J.; Ibarra-Berastegi, G.; Carreno-Madinabeitia, S.; Saenz, J.; Nafarrate, A. Seasonal Air Density Variations over the East of Scotland and the Consequences for Offshore Wind Energy. In Proceedings of the 7th International IEEE Conference on Renewable Energy Research and Applications (ICRERA 2018), Paris, France, 14–17 October 2018; Volume 5, pp. 261–265.
7. Myhr, A.; Bjerkseter, C.; Ågotnes, A.; Nygaard, T.A. Levelised cost of energy for offshore floating wind turbines in a lifecycle perspective. *Renew. Energy* **2014**, *66*, 714–728. [CrossRef]
8. Repsol Repsol. Available online: http://www.repsol.es (accessed on 26 December 2019).
9. Navantia Navantia. Available online: http://www.navantia.es (accessed on 26 December 2019).
10. Schillings, C.; Wanderer, T.; Cameron, L.; van der Wal, J.T.; Jacquemin, J.; Veum, K. A decision support system for assessing offshore wind energy potential in the North Sea. *Energy Policy* **2012**, *49*, 541–551. [CrossRef]
11. Castro-Santos, L.; Prado Garcia, G.; Estanqueiro, A.; Justino, P.A.P.S. The Levelized Cost of Energy (LCOE) of wave energy using GIS based analysis: The case study of Portugal. *Int. J. Electr. Power Energy Syst.* **2015**, *65*, 21–25. [CrossRef]
12. Itiki, R.; Di Santo, S.G.; Costa, E.C.M.; Monaro, R.M. Methodology for mapping operational zones of VSC-HVDC transmission system on offshore ports. *Int. J. Electr. Power Energy Syst.* **2017**, *93*, 266–275. [CrossRef]
13. Castro-Santos, L.; Filgueira-Vizoso, A.; Lamas-Galdo, I.; Carral-Couce, L. Methodology to calculate the installation costs of offshore wind farms located in deep waters. *J. Clean. Prod.* **2018**, *170*, 1124–1135. [CrossRef]
14. Aubault, A.; Cermelli, C.; Roddier, D. Windfloat: a floating foundation for offshore wind turbines. Part III: Structural analysis. In Proceedings of the 28th ASME International Conference on Ocean, offshore and Arctic Engineering (OMAE 2009), Honolulu, HI, USA, 31 May–5 June 2009; pp. 1–8.
15. Maciel, J. *The WindFloat Project*; EDP: Lisbon, Portugal, 2010.
16. Matha, D. *Model Development and Loads Analysis of an Offshore Wind Turbine on a Tension Leg Platform with a Comparison to Other Floating Turbine Concepts*; National Renewable Energy Laboratory (NREL): Golden, CO, USA, 2010.
17. Oguz, E.; Clelland, D.; Day, A.H.; Incecik, A.; López, J.A.; Sánchez, G.; Almeria, G.G. Experimental and numerical analysis of a TLP floating offshore wind turbine. *Ocean Eng.* **2018**, *147*, 591–605. [CrossRef]
18. Driscoll, F.; Jonkman, J.; Robertson, A.; Sirnivas, S.; Skaare, B.; Nielsen, F.G. Validation of a FAST Model of the Statoil-hywind Demo Floating Wind Turbine. *Energy Procedia* **2016**, *94*, 3–19. [CrossRef]
19. Athanasia, A. Deep offshore and new foundation concepts. In *DeepWind2013*; DeepWind: Trondheim, Norway, 2013; p. 17.
20. Jonkman, J.; Matha, D. Dynamics of offshore floating wind turbines — analysis of three concepts. *Wind Energy* **2011**, *14*, 557–569. [CrossRef]
21. Jonkman, J.; Matha, D. *A Quantitative Comparison of the Responses of Three Floating Platforms*; National Renewable Energy Laboratory (NREL): Stockholm, Sweden, 2010.
22. Sclavounos, P.D.; Lee, S.; DiPietro, J. Floating offshore wind turbines: Tension leg platform and taught leg buoy concepts supporting 3–5 mw wind turbines. In Proceedings of the European Wind Energy Conference (EWEC), Warsaw, Poland, 20–23 April 2010; pp. 1–7.
23. Collu, M.; Kolios, A.A.J.; Chabardehi, A.; Brennan, F. *Marine Renewable and Offshore Wind Energy*; RINA—Royal Institution of Naval Architects: London, UK, 2010; pp. 63–74.
24. Wind Europe. Driving Cost Reductions in Offshore Wind. Available online: https://windeurope.org/about-wind/reports/driving-cost-reductions-offshore-wind/ (accessed on 27 November 2017).
25. Pelamis Wave Power Pelamis. Available online: http://www.pelamiswave.com (accessed on 26 December 2019).
26. Weinstein, A.; Fredrikson, G.; Jane, M.; Group, P.; Denmark, K.N.R. *AquaBuOY—The Offshore Wave Energy Converter Numerical Modeling and Optimization*; IEEE: Piscataway, NJ, USA, 2003; pp. 1854–1859.
27. Castro-Santos, L.; Silva, D.; Bento, A.R.; Salvação, N.; Soares, C.G. Economic feasibility ofwave energy farms in Portugal. *Energies* **2018**, *11*, 1–16. [CrossRef]
28. Pelagic Power AS W2Power. Available online: http://www.pelagicpower.no/ (accessed on 17 December 2016).
29. Castro-Santos, L.; Martins, E.; Soares, C.G. Methodology to calculate the costs of a floating offshore renewable energy farm. *Energies* **2016**, *9*, 324. [CrossRef]
30. Castro-Santos, L.; Filgueira-Vizoso, A.; Carral-Couce, L.; Formoso, J.Á.F. Economic feasibility of floating offshore wind farms. *Energy* **2016**, *112*, 868–882. [CrossRef]

31. Castro-Santos, L.; Diaz-Casas, V. Life-cycle cost analysis of floating offshore wind farms. *Renew. Energy* **2014**, *66*, 41–48.
32. Dunnett, D.; Wallace, J.S. Electricity generation from wave power in Canada. *Renew. Energy* **2009**, *34*, 179–195. [CrossRef]
33. Page, W.D.W. Wave Dragon Technology. Available online: http://www.wavedragon.net (accessed on 7 September 2016).
34. Floating Power Plant AS Poseidon Floating Power. Available online: http://www.floatingpowerplant.com/ (accessed on 17 February 2016).
35. Castro-Santos, L.; Martins, E.; Guedes Soares, C. Economic comparison of technological alternatives to harness offshore wind and wave energies. *Energy* **2017**, *140*, 1121–1130. [CrossRef]
36. Silva, D.; Bento, A.R.; Martinho, P.; Guedes Soares, C. High Resolution local wave energy modelling in the Iberian Peninsula. *Energy* **2015**, *91*, 1099–1112. [CrossRef]
37. Silva, D.; Rusu, E.; Guedes Soares, C. Evaluation of Various Technologies for Wave Energy Conversion in the Portuguese Nearshore. *Energies* **2013**, *6*, 1344–1364. [CrossRef]
38. Castro-Santos, L.; Martins, E.; Guedes Soares, C. Cost assessment methodology for hybrid floating offshore renewable energy platforms. *Renew. Energy* **2016**, *97*, 866–880. [CrossRef]
39. Roddier, D.; Cermelli, C. Windfloat: A floating foundation for offshore wind turbines. Part I: Design basis and qualification process. In Proceedings of the 28th ASME International Conference on Ocean, Offshore and Arctic Engineering (OMAE 2009), Honolulu, HI, USA, 31 May–5 June 2009; pp. 1–9.

International Journal of
*Environmental Research and Public Health*

*Article*

# Carbon Footprint of a Port Infrastructure from a Life Cycle Approach

**Rodrigo Saravia de los Reyes, Gonzalo Fernández-Sánchez, María Dolores Esteban * and Raúl Rubén Rodríguez**

Department of Civil Engineering, Universidad Europea de Madrid, 28670 Villaviciosa de Odón, Madrid, Spain; rsaraviadelosreyes@gmail.com (R.S.d.l.R.); gonzalo.fernandez@ext.universidadeuropea.es (G.F.-S.); raulruben.rodriguez@universidadeuropea.es (R.R.R.)

* Correspondence: mariadolores.esteban@universidadeuropea.es

Received: 7 September 2020; Accepted: 4 October 2020; Published: 12 October 2020

**Abstract:** One of the most important consequences caused by the constant development of human activity is the uncontrolled generation of greenhouse gases (GHG). The main gases ($CO_2$, $CH_4$, and $N_2O$) are illustrated by the carbon footprint. To determine the impact of port infrastructures, a Life Cycle Assessment approach is applied that considers construction and maintenance. A case study of a port infrastructure in Spain is analyzed. Main results reflect the continuous emission of GHG throughout the useful life of the infrastructure (25 years). Both machinery (85%) and materials (15%) are key elements influencing the obtained results (117,000 Tm $CO_2$e).

**Keywords:** greenhouse gases emissions; port infrastructure; carbon footprint; life cycle assessment

## 1. Introduction

In the world, there is a high demand for achieving a balance between daily activities and natural resources [1]. The concept of sustainability aims to improve environmental conditions by promoting the balance between human development, the surrounding environment, and the resources. Although it is essential to recognize that the sustainability trend has improved, there is still a lot of work ahead, and the development of new methodologies [2,3] that encompass the development of innovative techniques is necessary, as well as tools that analyze and evaluate the present and future state of natural resources.

Human activities are measured in economic and social costs. However, there is a measurement [4] that is often overlooked, perhaps as it is not directly detected, which is the environmental cost. Human development generates an environmental cost that is represented by greenhouse gases (GHG). Recent trends position GHG as one of the main aspects to be mitigated by the extensive and continuous emissions generated [5].

GHG are made up of different gases, with carbon dioxide ($CO_2$), methane ($CH_4$) and nitrous oxide ($N_2O$) being the most important ones. The factors that influence radiative forcing are the following: Radiation power, and average time the gas molecule remains in the atmosphere [6]. These two factors make up the global warming potential (GWP) of each gas, whose unit is the kilogram of equivalent $CO_2$. In other words, each gas analyzed is transformed to kg of $CO_2$ units using an equation sanctioned by practice.

The concentration of atmospheric $CO_2$ exceeded 417 parts per million (PPM) in 2019, a fact that sets a historical record. The observatory that carried out that research was Mauna Loa in Hawaii (National Oceanic and Atmospheric Administration—NOAA) [7]. From the statistics provided by NOAA, it may be concluded that the $CO_2$ emission trend is exponential and that these results are exceeded each year (Table 1).

**Table 1.** $CO_2$ emissions trend over the years.

| Year | $CO_2$ (Parts per Million—PPM) | Growth (PPM) |
|---|---|---|
| 1960 | 315 | 0.71 |
| 1970 | 325 | 1.13 |
| 1980 | 340 | 1.70 |
| 1990 | 350 | 1.16 |
| 2000 | 370 | 1.23 |
| 2010 | 390 | 2.43 |
| 2019 | 410 | 2.60 |

The cause of the uncontrolled growth rests unquestionably on human activity. The burning of fossil fuels for energy production, transport, and industry has increased in recent years [8,9]. In the unexpected situation such as the pandemic caused by COVID-19, the impact of anthropogenic activity was observed, through a weekly reduction in greenhouse gases of 30% during the period of confinement [7].

Until present, the procedure that is most effective in determining the levels of emissions generated is measuring the carbon footprint (CF) [10]. This concept quantifies the carbon dioxide ($CO_2$) emissions generated, produced directly and/or indirectly by an activity, or by the set of activities during the life cycle of a product [11]. One of the most important advantages of the CF calculation is that it is possible to adapt it to any project [12]. Therefore, all sectors should be obliged to perform CF calculations in the near future.

The above should even be made more so in the construction sector, which has a significant impact on the environment due to large earth movements, the treatment of compound materials, and land modification [5]. Currently, the construction sector is responsible for 20% of the emissions generated worldwide, being the first in materials consumption [13].

Life Cycle Analysis (LCA) is a methodology that allows the calculation, evaluation, and interpretation of the generated emissions during the life time of an infrastructure, thereby showing the GHG produced during all the project phases [14,15].

In the civil engineering sector, seaports are the infrastructures where the generation of GHG is highest. On the other hand, ports are considered dynamic centers within the international market where numerous services and benefits are integrated [16,17]. It should be noted that, although the literature in relation to these infrastructures is not very abundant, it has been shown that the greatest impact is produced by large machinery and ships than by the construction of infrastructure. Some references showing the previous statements include: A study on the annual climate change and primary use of Swedish transport infrastructure, including roads, railways, airports, and fairway channels [18]; evaluation of the latest trends in the cargo handling equipment industry in ports [19]; the life-cycle emissions from port electrification in the case study of the Port of Los Angeles [20].

In the absence of research, and therefore of precise conclusions that accurately describe the reality of these infrastructures, a study was carried out that focused on the Life Cycle Analysis of a port, which considered both the phases of construction and operation. This manuscript summarizes this research. The selected project was carried out in Spain, and the infrastructure considered in the research consists of the construction and maintenance of a quay of approximately 32,000 $m^2$, being characterized by the use of caissons.

On the other hand, it is important to consider that most of the cases in which this methodology has been applied are outside the scope of the civil engineering sector [5,21–23], and it is, therefore, a challenge to apply it to port infrastructures. However, there are very specific cases that illustrate the large emissions that are generated in this sector.

Previous research typically focused on the individual performance of materials [24,25] and the behavior of different materials in the overall infrastructure [26–30]. GHG research in maritime engineering is scarce, which is the key aspect to understand the importance of this research. However, it has been essential to consider the approaches developed in this sector [31–33] to carry out this study.

## 2. Objectives

This study mainly seeks to know the emission ranges produced during the entire operational life of the selected infrastructure. This involves analyzing all phases (construction and maintenance), as well as all the elements (materials and machinery). This approach permits the characterization of the problem of greenhouse gases in port infrastructures, an aspect that has never been studied up until now. The main objectives of this research are:

- Characterization of the GHG problem in maritime infrastructures.
- Calculation of the emissions produced in a maritime project that includes both construction and maintenance.
- Identification of aspects with the highest GHG emissions.

## 3. Scope

The scope established in this research corresponds to the Life Cycle Analysis of a port construction. It is important to mention that both the construction and maintenance have been considered. Although each construction is unique, various studies related to roads [13,27–29,34], waste management [22], or the specific elaboration of materials [24,26] have been taken into account to build a powerful methodology that begins to illustrate the problem of GHG emissions in ports. The whole of the infrastructure life cycle has been considered, which includes materials extraction, production, transportation, port construction, and maintenance (operation and conservation), but not the port dismantling. The transport and energy upstream chain was also considered.

The elements included in this LCA can initially be classified into three groups: Machinery, materials, and natural systems. As there are still no references that reflect a clear and concise calculation for the seabed, this element has not been considered in this research. Considering natural systems, Posidonia Oceanica is an algae considered as protected species in the Mediterranean Sea (Figure 1) [35]; laws establish a prohibition regarding the disruption of that species, which are so important in the absorption of $CO_2$ [36–38].

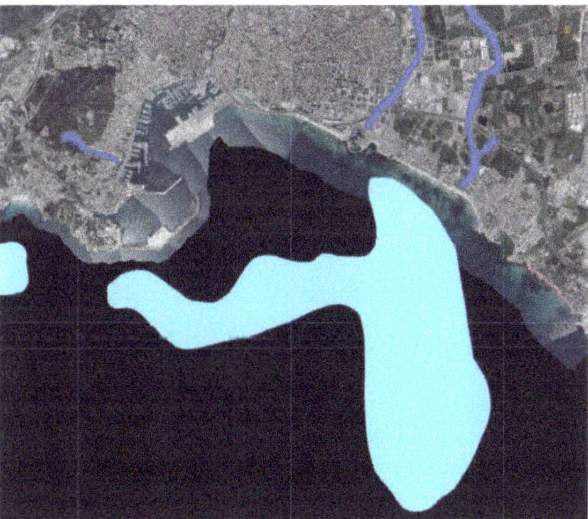

**Figure 1.** Posidonia meadow (in blue color) close to the construction area.

In relation to the machinery, it was decided to analyze only the use of the machines during construction and maintenance. That means that the manufacturing of the machinery itself is out of the scope of this research, an aspect in which emissions are also generated.

Table 2 summarizes the most important information and LCA concepts included in this research for studying the GHG emissions in construction and maintenance of a port infrastructure [15].

**Table 2.** Summary of Life Cycle Analysis (LCA) methodology variables in a port infrastructure.

| Concept | Assignment in Selected Project |
|---|---|
| Location | Spain, Europe |
| Life Cycle of the Port | 50 years |
| Product system | Construction and maintenance |
| System limits | Work units |
| Impact category | Climate change assessment |
| Main GHG | $CO_2$, $CH_4$, $N_2O$ |
| Functional unit | Quay |
| GHG Major Parties | Dredging, dock, superstructure, esplanade, pavements |
| GHG minority items | Drainage, facilities, waste management, and health and safety |
| Main GHG agents | Machinery, materials, and natural systems |

## 4. Methodology

The methodology that has been used for the management information data is based on CO2NSTRUCT [14], which has an in-depth database containing more than 300 data elements (materials, off-road machinery, transport machinery, electricity mixes, energy sources, and types of waste), and which includes European and Spanish data.

The materials use a three-step approach following the CO2NSTRUCT approach: Pre-production and production (related to construction stage), and installed material (related to maintenance stage).

- The pre-production emission is related to energy consumption, process emissions, and transportation to final factory (e.g., the machinery used for the extraction of raw material, transport machinery to the processing facilities) [29].
- Production is based on the emissions generated for the final processing of materials. In this case, the emissions make reference to the final product established in the work units necessary for the construction of the work [29].
- Installed materials make reference to artificial surfaces created after the construction that can be carbonated during the life cycle such as concrete that can absorb up to 3800 kg $CO_2/m^3$ in a period of 100 years according to Galan et al. [39].

On-site activities focus on the construction of the port infrastructure, as well as the maintenance activities. The machinery is classified into two groups:

- Transport machinery: Machinery used to transport the material to the site [40,41]. EMEP/Copert methodology tier 3 was used.
- Construction machinery or off-road machinery: Specific machinery to carry out the activity established in the work units, involved in the construction activities [40,41]. It used EMEP methodology tier 3.

As the construction scope is well-defined in the as-built project, in this research, a maintenance plan for a port was presented considering all the characteristics of the specific project and location conditions. Although the maintenance plan is based on estimations from good lessons and experts, the calculation is done in a similar way as in the construction stage.

Waste management (related to energy consumption in its treatment and transportation from site) and environmental systems were also initially considered [28]. Energy consumption, from the so-called pre-combustion (source production and distribution in electricity or fuel production and distribution), and its combustion with national emission factors allow us to obtain the data for energy emissions related to materials, machinery, and transportation. For more information about this methodology, the in-depth detailed methodology from [14] was followed. Emission factors of the electrical mix, materials, and fuel combustion of the off-road machinery were updated with the last published data following national and European reports (EMEP/EEA air pollutant emission inventory) from the system and database referenced.

## 5. Study Case

### 5.1. Construction Stage

The case study selected was located in Spain. The quay construction began in late 2017, with the project lasting a duration of 37 months, and the total cost of the project was around 35 million Euros. More than 150 off-road machinery were identified on the as-built construction project. The main activities of the port expansion considered for this case study were:

- Dock extension: 215 + 130 + 24 m, with submerged concrete blocks for passenger traffic. Eleven floating drawers of around 34 m length with granular filling cemented on breakwater bench, with a superstructure including bollards and fenders.
- Enlargement of facing to 100 m, also with a dock with submerged concrete blocks.
- Harbor dike extension: 115 + 410 m, with concrete drawers in breakwater benches.
- Esplanade: Reinforced concrete plates with tongue and groove modular elements. A bituminous mix finish.

The most important materials in the construction stage are concrete and granular materials, with some important activities such as the esplanade and pavement of the dike and dock. Most of sand and gravel came from dredging. This recycling strategy will reduce the GHG footprint by reusing this kind of material used to fulfill the drawers. Off-road machinery age is considered on an average of around 4 years. Natural systems were initially considered but most of the parts were directly above the sea, so just a small impact on land-use and change in land-use is included.

### 5.2. Maintenance Stage

A maintenance and conservation plan for the port was defined. Using existing guides and manuals [42–44], and our own experience, a maintenance plan was designed tailored to project conditions (Figure 2). The type and number of inspections were detailed in the maintenance plan. The actions to be carried out are listed in Table 3, including the following information in each operation: Area where the task is going to be carried out, main material, surface, thickness and volume affected, frequency of carrying out the task and percentage affected, number of operations during the life time of the infrastructure, ratio of operations in the life time, and ratio of material volume divided by the frequency.

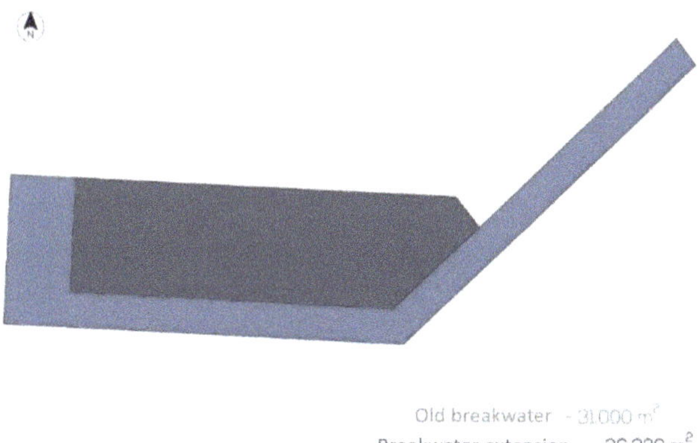


**Figure 2.** Distribution of the maintenance area to carry out the relevant work.

**Table 3.** Inspections and action during the maintenance phase.

| Maintenance Plan: Operations | | | | | | | | | | | |
|---|---|---|---|---|---|---|---|---|---|---|---|
| N | Operation | Area | Surface ($m^2$) | Main Material | Thickness (m) | Volume ($m^3$) | Frequency (years) | Estimated Improvement | Operation during Useful Life | Operations/ Useful Life | Material Volume/ Frequency |
| 1 | Pavement improvement | New Dock | 36,232 | Bituminous mixture | 0.06 | 2174 | 6 | 100% | 4 | 0.16 | 8696 |
| 2 | Pavement improvement | Old dock | 31,000 | Concrete | 0.2 | 6200 | 12 | 100% | 2 | 0.08 | 12,400 |
| 3 | Caissons improvement | New Dock | - | Concrete | - | 31,658 | 10 | 3% | 3 | 0.12 | 2849 |
| 4 | Caissons improvement | Old dock | - | Concrete | - | 150,000 | 10 | 3% | 3 | 0.12 | 13500 |
| 5 | Breakwater replacement | New Dock | - | Rockfill | - | 5100 | 4 | 6% | 6 | 0.24 | 1836 |

## 6. Results

### 6.1. General Results

The results obtained in this research are aligned with the expected trend included in [29] (see Figure 3). In fact, Figure 3 shows that main emissions occur during the construction phase (close to the 100% of the total emissions), with the emissions from the operation and maintenance phases making up only 1% of the total emissions. Furthermore, the proportion of $CO_2$ emissions is the greatest, leaving the other GHG with almost zero proportions in total. The result of Tm of $CO_2e$ for the entire project is 117,000 Tm, that is, only taking into account the construction and maintenance phases.

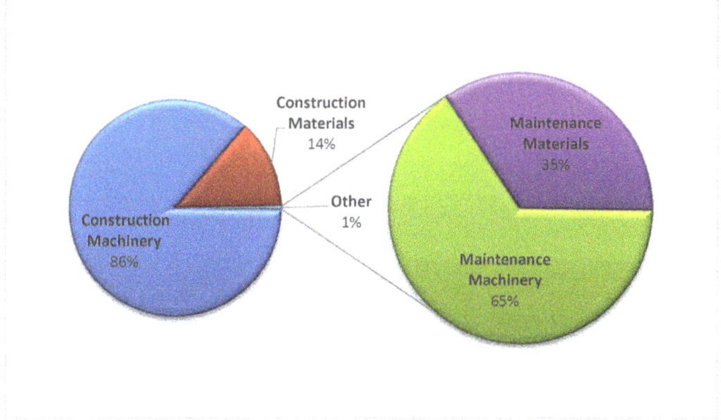


**Figure 3.** Distribution of the phases analyzed in this research: Results in percentage (%).

Machinery, and more specifically, the construction or off-road machinery, is the most influential element. On the contrary, the machinery used for the transport is the minority component. In addition, 14% of emissions correspond to materials during the construction phase. Within this component, pre-production is the phase with the greatest impact with almost 90% of total $CO_2$ material emissions.

The activity in the construction stage, which is the highest contributor to the GHG, is the construction of the Esplanade. The influence of the off-road machinery (Table 4) is such that it makes this activity stand out.

**Table 4.** Emissions of off-road machinery in construction stage.

| | Off-Road Machinery | | |
|---|---|---|---|
| Main Activity | Kg $CO_2$ | Kg $CH_4$ | Kg $N_2O$ |
| 1. Demolition | 216,000 | 4.00 | 20 |
| 2. Dock | 15,800,000 | 300 | 1000 |
| 3. Superstructure | 2,550,000 | 60 | 300 |
| 4. Esplanade | 72,500,000 | 2700 | 800 |
| 5. Flooring | 5,520,000 | 60 | 300 |
| 6. Drainage | 448,000 | 10 | 50 |
| 7. Installations | 128,000 | 4 | 10 |
| 8. Waste management | 759,000 | 20 | 100 |
| Total results | 97,900,000 | 3160 | 9780 |
| Kg $CO_2e$ results | 97,900,000 | 265,000 | 2,580,000 |
| Kg $CO_2e$ total | | 98,423,000 | |

In the construction stage, it is clearly observed that the off-road machinery is the element that generates the most emissions. Transportation is three orders of magnitude lower. Table 5 shows the breakdown of the results obtained at the construction stage.

**Table 5.** Total $CO_2$e results obtained in the construction phase.

| Total Results Tm $CO_2$e: Construction | | | |
|---|---|---|---|
| **Total Machinery** | **101,000** | **Tm $CO_2$e** | **87%** |
| Off Road Machinery | 101,000 | Tm $CO_2$e | 99.85% |
| Transport Machinery | 156 | Tm $CO_2$e | 0.15% |
| **Total Materials** | **15,600** | **Tm $CO_2$e** | **13%** |
| Preproduction | 13,600 | Tm $CO_2$e | 87.18% |
| Production | 2000 | Tm $CO_2$e | 12.82% |
| Total Construction Stage | 116,000 | Tm $CO_2$e | 100% |

As maintenance work is important and almost nullified by the importance of construction, this section is established to interpret the values obtained during this phase. The general trend, although with lower values, is the same as in the construction phase where the off-road machinery generates almost all emissions. It is worth mentioning that the distribution between the elements is a little more balanced. Table 6 breaks down the results obtained for the maintenance stage of off-road machinery, transportation, and materials.

**Table 6.** Total $CO_2$e results obtained in the maintenance stage.

| Total Results Tm $CO_2$e: Maintenance | | | |
|---|---|---|---|
| **Total Machinery** | **339** | **Tm $CO_2$e** | **65%** |
| Off Road Machinery | 339 | Tm $CO_2$e | 99.92% |
| Transport Machinery | 0.26 | Tm $CO_2$e | 0.08% |
| **Total Materials** | **181** | **Tm $CO_2$e** | **35%** |
| Preproduction | 11.9 | Tm $CO_2$e | 6.58% |
| Production | 169 | Tm $CO_2$e | 93.42% |
| Total Maintenance Stage | 520 | Tm $CO_2$e | 100% |

### *6.2. Scenarios*

From the case study and obtained results, some sensitivity analysis was made according to the detail of data used in the study. Following some scenarios created by [45] from road projects, four scenarios were studied in depth to illustrate the emission ranges of a maritime construction and to observe the sensitivity of the proposal and results. The four scenarios used were:

- Scenario 1 is based on variables depending on off-road machinery as it is the most important element. The age of the off-road machinery has been changed from 4 years from the base case to 20 years.
- Scenario 2, following the previous case, is that the technology of the off-road machinery has been modified, and so has the transport machinery, to the most actual one from that which was in the construction project to check the sensitivity to this variable.
- Scenario 3, going in depth in the machinery element, is that biodiesel fuel is established instead of diesel. Although the performance can be lower, the emissions in the case of using biodiesel B20 fuel values are 20% less than in the case of using diesel fuel.
- Scenario 4 is focused on materials. This is because machinery is considered "well-calculated" as in most actual European guidelines, tier 3 is used. But what about materials? We have used national

data, but it is not the actual exact information from the construction of the port infrastructure. To measure the sensitivity to this element, a known database has been used (ICE, the Inventory of Carbon Energy [34]) for measuring the sensitivity of global results changing the raw data of elements used.

The scenarios served to obtain a broader vision of the importance of the factors that participate in this process. As can be seen in Table 7 and in Figure 4, large changes are only detected in scenario 3 and in scenario 4.

**Table 7.** Summary of the results obtained in each scenario, with the % of variation in scenarios in relation to the base case.

| Case | Construction Stage | | | |
|---|---|---|---|---|
| | Machinery | | Materials | |
| | Off Road | Transport | Preproduction | Production |
| Base case (Tm of $CO_2$e) | 101,000 | 156 | 13,600 | 2000 |
| Scenario 1 | 0.00001% | 0.00001% | - | - |
| Scenario 2 | 0.00001% | 0.00001% | - | - |
| Scenario 3 | −7% | −8% | - | - |
| Scenario 4 | - | - | +216% | |

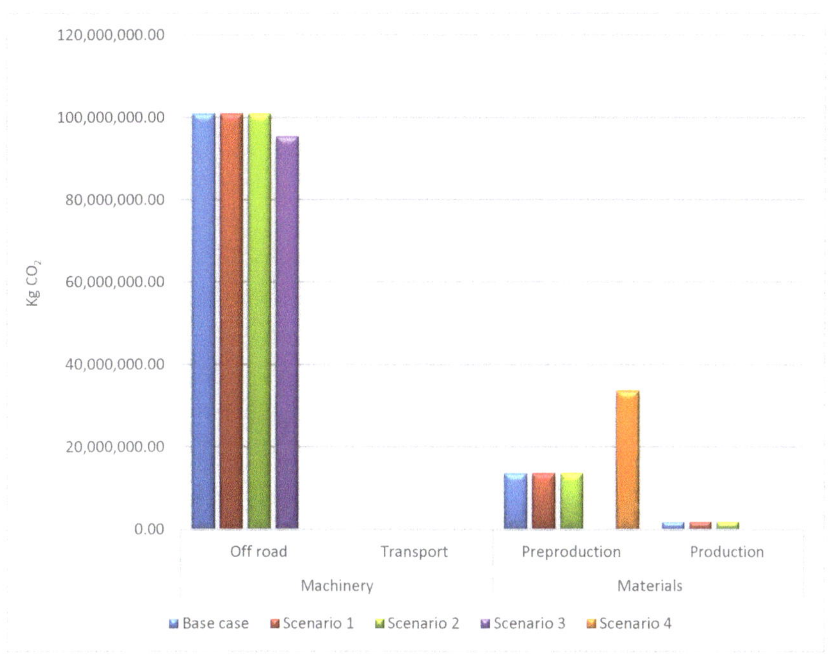


**Figure 4.** Results of $CO_2$ kg emissions obtained in each scenario: Construction phase.

Scenario 1 and 2 are considered near to the base case results. New technologies are more focused on reducing $NO_x$ and PPM than on $CO_2$, and age is not a main factor when the initial average is 4 years. Scenario 3 shows important changes that can open new work scenarios focused on renewable energies. The use of biodiesel as fuel reduces emissions by approximately 7% in the construction phase and 8% in the maintenance phase. The results can be viewed in Figure 5.

**Figure 5.** Results of kg $CO_2$e emissions obtained in scenario 3 compared to the base result.

In the case of scenario 4, the information database changes completely. The Inventory of Carbon Energy [34] database is used, which uses a more general methodology as it treats pre-production and production as the same stage (Figure 6).

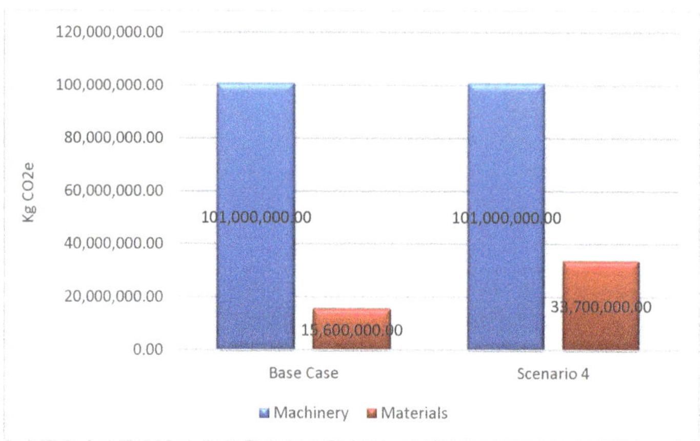


**Figure 6.** Results of $CO_2$e kg emissions obtained in scenario 4 compared to the base result.

It must be remembered that materials only account for about 14% of all project emissions. In this scenario, 33,700 Tm $CO_2$e is obtained (Figure 6), which means that the materials increase 216% (Table 7). Therefore, the total distribution in this scenario is more balanced, as can be seen in Figure 7, which shows that the distribution of emissions in scenario 4 is around 75% due to machinery, and 25% due to materials.

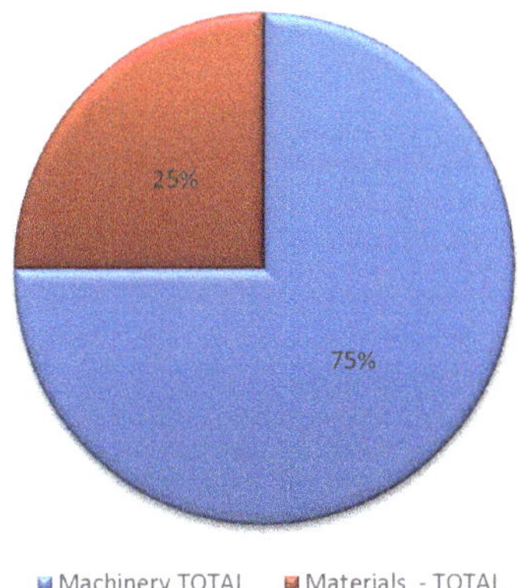


**Figure 7.** Total greenhouse gases (GHG) distribution in scenario 4.

## 7. Discussion

This research begins to reflect the large volumes of GHG emissions produced in the construction sector. Although there are references that analyze this phenomenon in other areas of the construction sector, this document is the pioneer in evaluating and interpreting the GHG emissions in maritime construction. Although various references were consulted [13,21,23–25,30,46,47], it should be noted that the entire process has been deeply influenced by the research carried out in the analysis of greenhouse gas emissions throughout the life cycle of roads [29].

It is also important to note that the study is carried out on a real construction project, a maritime infrastructure that expands the service of the port. Therefore, all the work units analyzed have been part of the construction of this structure.

On the other hand, although this aspect is based on estimation, a maintenance plan was prepared according to the project conditions, thus giving the opportunity to obtain a global perspective of the GHG emission levels generated throughout the operational life of the infrastructure, 25 years. Adding the maintenance phase to the research allows the achievement of a more complete and differential study as there are few studies that take into account this phase in the calculation, despite it being in the engineering sector.

Element analysis is also one of the strengths of this study. In machinery, more than 200 items were analyzed in the work, and therefore, a detailed search of the specifications of each machine in the catalogs formalized by the major brands was carried out. It can be established that approximately 90% of the data processed in this element have been accurately referenced, estimating only the performance of the remaining 10%. Different models were analyzed, fitted to the characteristics established in the work units, thus giving the possibility of obtaining the average performance of each machine. Regarding materials, the database of the most influential reference in this research was used. It should also be noted that a considerable percentage of these have been updated, thus adapting them to the project conditions.

The importance of this research is especially shown in the interpretation of the results [39]. The nonexistence of studies that analyze the GHG emissions generated in maritime constructions make this research clarify its results in comparison to others of a diverse nature.

According to the overall results, no clear differences were observed with other projects [29,48] as the distribution is approximately the same (80% machinery—20% materials). However, this ratio in the maintenance phase is less. Clearly, having fully created a maintenance plan through estimates, it has concentrated the distribution.

The extensive bibliographical search that was carried out made possible the creation of an extensive database, which proved essential for the calculation stage. The distribution of the results, both in the construction stage and in the maintenance stage, continue the previous trend established in other research, where the highest proportion of emissions is due to the off-road machinery, approximately 85%. These results are similar to those of other studies of different infrastructure projects [14].

Finally, the variety of scenarios reproduced shows the power and the capacity for improvement of current construction projects. As previously indicated, as machinery is the aspect with the greatest weight in this study, the most influential technical characteristics in the result were detected, with fuel being the key factor. By improving the efficiency of biodiesel, more positive results can be obtained in the immediate future.

In scenario 1 and scenario 2, it can be seen how the most advanced technologies focus on reducing local emissions rather than global ones, as has been shown in similar studies [48].

However, it is interesting to highlight that the methodology according to the fuel developed in scenario 3 is considerably distanced from the base result, so, although fuel is an important aspect to consider, it is better and more accurate to develop the methodology that encompasses the technical characteristics. In scenario 4, in which another database was used, the results agree with the base result; however, its value is approximately double. Despite the fact that the base case methodology is more specific and detailed, the ICE database has a greater data record, which allows more material batches to be calculated, thus obtaining a more bulky result.

Finally, it is interesting to check the results obtained in past studies. It is worth highlighting that the results obtained in this study (base case) are very similar to those published in related literature.

## 8. Conclusions

A serious problem such as the high carbon footprint generated in construction projects must be visualized in order to determine and evaluate the consequences of the environmental impact produced. Emission levels have been exceeded each year, and only drastic reductions in this trend have been observed in a situation as strange as the COVID-19 pandemic, which has completely stopped the world. The implementation of procedures that contribute at least to the visualization of this phenomenon must be immediate. The LCA not only fulfills this function, but due to its adaptability, it also indicates the critical elements that make up the construction project and the maintenance activities.

Probably, the circumstance that caused the low presence of LCA procedures in the world of maritime construction is the peculiarity of these infrastructures. It should be noted that this research has not considered the impact on the natural systems. However, it is necessary to take into account the presence in this location of the Posidonia meadows, an endemic species with high percentages of $CO_2$ absorption and that, due to this characteristic, among others, it is totally illegal to modify its area.

An inventory was developed where all the elements that make up this construction project were analyzed (maintenance works included). This strategy allowed the development of a pioneering work in the calculation of the carbon footprint in maritime infrastructure through LCA. Therefore, the most critical activities and elements of the system were identified, in addition to presenting new alternatives using the scenarios established in this research. The most characteristic results are shown below (Table 8):

**Table 8.** Main results of this study.

| | Construction | Maintenance | Cost (€) | Surface ($m^2$) | kg $CO_2e$ | kg $CO_2e/m^2$ |
|---|---|---|---|---|---|---|
| | Tm $CO_2e$ | | | | | |
| Base Case | 117,000 | 520 | 34,060,648.00 € | 130,510 | 3.43 | 899 |

The main objectives planned were totally reached. We now have an initial order of magnitude for this kind of infrastructure, but the research and number of case studies should be continued in the future for different port typologies and areas to obtain reliable ratios of GHG emissions. The initial ratio of GHG emissions in construction and maintenance is calculated, where construction stage is the most important and off-road machinery is the main element from an LCA approach, as similar studies have concluded in infrastructure projects (ports were greatly left out of these studies).

**Author Contributions:** Conceptualization, R.S.d.l.R. and G.F.-S.; formal analysis, R.S.d.l.R.; investigation, R.S.d.l.R.; methodology, R.S.d.l.R., G.F.-S., M.D.E. and R.R.R.; supervision, G.F.-S., M.D.E. and R.R.R.; validation, G.F.-S., M.D.E. and R.R.R.; writing—original draft, R.S.d.l.R.; writing—review and editing, G.F.-S., M.D.E. and R.R.R. All authors have read and agreed to the published version of the manuscript.

**Funding:** This research received no external funding.

**Conflicts of Interest:** The authors declare no conflict of interest.

## References

1. Alavedra, P.; Dominguez, J.; Gonzalo, E.; Serra, J. Sustainable construction: The state of the art. (In Spanish: La construcción sostenible: El estado de la cuestión). *Inf. Constr.* **1997**, *49*, 41–47. [CrossRef]
2. Christini, G.; Fetsko, M.; Hendrickson, C. Environmental management systems and ISO 14001 certification for construction firms. *J. Constr. Eng. Manag.* **2004**, *130*, 330–336. [CrossRef]
3. Novara, P. Introduction to Development Tools (In Spanish: Introducción a las Herramientas de Desarrollo) Report, 1–12. 2010. Available online: http://zinjai.sourceforge.net/Anexo1.pdf (accessed on 5 June 2020).
4. Espíndola, C.; Valderrama, J.O. Carbon footprint. Part 1: Concepts, estimation methods and methodological complexities (In Spanish: Huella del carbono. Parte 1: Conceptos, métodos de estimación y complejidades metodológicas). *Inf. Tecnol.* **2012**, *23*, 163–176. [CrossRef]
5. Curran, M.A. *Environmental Life-Cycle Assessment*; McGraw-Hill: New York, NY, USA, 1996.
6. Intergovernmental Panel on Climate Change (IPCC). Energy. IPCC Guidance on good practice and managing uncertainty (In Spanish: Energía. orientación del IPCC sobre las buenas prácticas y la gestión de la incertidumbre). In *Los Inventarios Nacionales de Gases de Efecto Invernadero*; IPCC: Geneva, Switzerland, 2007; pp. 1–103.
7. Earth System Research Laboratories. Trends in Atmospheric Carbon Dioxide. 2020. Available online: https://www.esrl.noaa.gov/gmd/ccgg/trends/mlo.html (accessed on 20 May 2020).
8. Aichele, R.; Felbermayr, G. Kyoto and the carbon footprint of nations. *J. Environ. Econ. Manag.* **2012**, *63*, 336–354. [CrossRef]
9. International Energy Agency. Global Energy Review 2020. Iea 2020, 52. Available online: https://www.iea.org/reports/global-energy-review-2020 (accessed on 10 May 2020).
10. Matthews, H.S.; Hendrickson, C.T.; Weber, C.L. The importance of carbon footprint estimation boundaries. *Environ. Sci. Technol.* **2008**, *42*, 5839–5842. [CrossRef] [PubMed]
11. Wiedmann, T.; Minx, J. A definition of carbon footprint. *Science* **2007**, *1*, 1–11. [CrossRef]
12. Pandey, D.; Agrawal, S.B.; Pandey, J.S. Carbon footprint: Current methods of estimation. *Environ. Monit. Assess.* **2010**, *178*, 135–160. [CrossRef]
13. Yu, M.; Wiedmann, T.; Crawford, R.; Tait, C. The carbon footprint of Australia's construction sector. *Procedia Eng.* **2017**, *180*, 211–220. [CrossRef]
14. Barandica, J.M.; Fernández-Sánchez, G.; Berzosa, Á.; Delgado, J.A.; Acosta, F.J. Applying life cycle thinking to reduce greenhouse gas emissions from road projects. *J. Clean. Prod.* **2013**, *57*, 79–91. [CrossRef]
15. Hauschild, M.Z. Introduction to LCA methodology. In *Life Cycle Assessment: Theory and Practice*; Springer: Cham, Switzerland, 2017.
16. González Laxe, F. Ports and maritime transport: Axes of a new global articulation (In Spanish: Puertos y transporte marítimo: Ejes de una nueva articulación global). *Rev. Econ. Mund.* **2005**, *12*, 123–148.
17. Rúa, C. *Ports i Maritime Transport (In Spanish: Los Puertos en el Transporte Marítimo)*; Universitat Politecnica de Catalunya: Barcelona, Spain, 2006; pp. 3–12.
18. Liljenström, C.; Toller, S.; Åkerman, J.; Björklund, A. Annual climate impact and primary energy use of Swedish transport infrastructure. *Eur. J. Transp. Infrastruct. Res.* **2019**, *19*. [CrossRef]
19. Vujičić, A.; Zrnic, N.; Jerman, B. Ports sustainability: A life cycle assessment of zero emission cargo handling equipment. *J. Mech. Eng.* **2013**, *59*, 547–555. [CrossRef]
20. Kim, J.; Rahimi, M.; Newell, J.; Newell, J.P. Life-Cycle emissions from port electrification: A case study of cargo handling tractors at the port of Los Angeles. *Int. J. Sustain. Transp.* **2011**, *6*, 321–337. [CrossRef]

21. Recchia, L.; Cappelli, A.; Cini, E.; Pegna, F.G.; Boncinelli, P. Environmental sustainability of pasta production chains: An integrated approach for comparing local and global chains. *Resources* **2019**, *8*, 56. [CrossRef]
22. Arnell, M.; Rahmberg, M.; Oliveira, F.; Jeppsson, U. Multi-objective performance assessment of wastewater treatment plants combining plant-wide process models and life cycle assessment. *J. Water Clim. Chang.* **2017**, *8*, 715–729. [CrossRef]
23. Olsen, S.I.; Miseljic, M. Life cycle assessment in nanotechnology—Issues in impact assessment and case studies. In Proceedings of the Symposium on Nanosafety 2011—LEITAT Technological Center, Barcelona, Spain, 4–5 May 2011.
24. Grimaud, G.; Perry, N.; Laratte, B. Life cycle assessment of aluminium recycling process: Case of Shredder cables. *Procedia CIRP* **2016**, *48*, 212–218. [CrossRef]
25. Jónsson, Á.; Tillman, A.-M.; Svensson, T. Life cycle assessment of flooring materials: Case study. *Dr. Vid Chalmers Tek. Hogsk.* **2006**, *32*, 245–255. [CrossRef]
26. Xing, S.; Zhang, X.; Jun, G. Inventory analysis of LCA on steel- and concrete-construction office buildings. *Energy Build.* **2008**, *40*, 1188–1193. [CrossRef]
27. Biswas, W. Carbon footprint and embodied energy assessment of a civil works program in a residential estate of Western Australia. *Int. J. Life Cycle Assess.* **2014**, *19*, 732–744. [CrossRef]
28. Barandica, J.M.; Delgado, J.A.; Berzosa, Á.; Fernández-Sánchez, G.; Serrano, J.M.; Zorrilla, J.M. Estimation of $CO_2$ emissions in the life cycle of roads through the disruption and restoration of environmental systems. *Ecol. Eng.* **2014**, *71*, 154–164. [CrossRef]
29. Berzosa González, Á. *Analysis of Greenhouse Gas Emissions throughout the Life Cycle of Roads. (In Spanish: Análisis de las Emisiones de Gases de Efecto Invernadero a lo Largo del Ciclo de Vida de las Carreteras)*; Universidad Complutense de Madrid: Madrid, Spain, 2013; pp. 7–8.
30. Ahn, C.; Xie, H.; Lee, S.H.; Abourizk, S.; Peña-Mora, F. Carbon footprints analysis for tunnel construction processes in the preplanning phase using collaborative simulation. In Proceedings of the Construction Research Congress 2010: Innovation for Reshaping Construction Practice, Banff, AB, Canada, 8–10 May 2010; pp. 1538–1546. [CrossRef]
31. Villalba, G.; Gemechu, E.D. Estimating GHG emissions of marine ports—The case of Barcelona. *Energy Policy* **2011**, *39*, 1363–1368. [CrossRef]
32. Broekens, R.; Escarameia, M.; Cantelmo, C.; Woolhouse, G. Quantifying the carbon footprint of coastal construction—A new tool HRCAT. In Proceedings of the 7th International Coastal Management Conference, Arcachon, France, 24–28 June 2012; pp. 253–262. [CrossRef]
33. Patrizi, N.; Pulselli, R.M.; Neri, E.; Niccolucci, V.; Vicinanza, D.; Contestabile, P.; Bastianoni, S. Lifecycle environmental impact assessment of an overtopping wave energy converter embedded in breakwater systems. *Front. Energy Res.* **2019**, *7*, 1–10. [CrossRef]
34. Hammond, G.P.; Jones, C.I. Embodied energy and carbon in construction materials. *Proc. Inst. Civ. Eng. Energy* **2008**, *161*, 87–98. [CrossRef]
35. IPCC. *Chapter 3. Mobile Combustion.* Available online: https://www.ipcc-nggip.iges.or.jp/public/2006gl/spanish/pdf/2_Volume2/V2_3_Ch3_Mobile_Combustion.pdf (accessed on 15 May 2020).
36. Bauzà, J. The Decree that will Establish the before and after for Posidonia (In Spanish: El decreto que fijará un antes y un después para la Posidonia). Available online: https://www.diariodemallorca.es/mallorca/2018/07/01/decreto-fijara-despues-posidonia/1327221.html (accessed on 18 May 2020).
37. Consejo de Gobierno. *Illes Balears. Section I. General Provisions GOVERNING COUNCIL Decree 25/2018 of July 27, on the Conservation of Posidonia Oceanica in the Balearic Islands (In Spanish: Sección I. Disposiciones Generales Consejo de Gobierno Decreto 25/2018 de 27 de Julio, Sobre la Conservación de la Posidonia Oceanica en las Illes Balears)*; Consejo de Gobierno: Madrid, Spain, 2018; pp. 25994–26042.
38. Infraestructura de Datos Espaciales de las Islas Baleares. Habitats in the Balearic Islands (In Spanish: Hábitats en las Islas Baleares). Available online: https://portalideib.caib.es/portal/home/webmap/viewer.html?useExisting=1&layers=cc47fef5e3b846e48d4e4b6015b40fcc (accessed on 12 May 2020).
39. Galan, I.; Andrade, C.; Prieto, M.; Mora, P.; Lopez, J.C.; Sanjuan, M.A. Study of the $CO_2$ sink effect of cement-based materials (In Spanish: Estudio del efecto sumidero de $CO_2$ de los materiales de base cemento). *Cem. Hormig.* **2010**, *939*, 70–83.

40. EEA. EMEP/EEA Air Pollutant Emission Inventory Guidebook 2019: Technical Guidance to Prepare National Emission Inventories (3.B Manure Management). Available online: https://doi.org/10.2800/293657 (accessed on 1 June 2020).
41. Winther, M.; Dore, C.; Lambrecht, U.; Norris, J.; Samaras, Z.; Zierock, K.H. Non road mobile machinery 2016. In *EMEP/EEA Air Pollutant Emission Inventory Guidebook*; EEA: Copenhagen, Denmark, 2019; pp. 1–81.
42. American Society of Civil Engineers (ASCE). *Standard Practice Manual for Underwater Investigations*; ASCE: Reston, VA, USA, 2001; p. 400.
43. Asociación Técnica de Puertos y Costas. *Management of Conservation in the Port Environment (In Spanish: Gestión de la Conservación en el Entorno Portuario)*; Organismo Público Puertos del Estado: Madrid, Spain, 2012.
44. Asociación Técnica de Puertos y Costas. *Guide to Repair Concrete Structures in the Marine Environment (In Spanish: Guía de Reparación de Estructuras de Hormigón en Ambiente Marino)*; Technical Report, 2018; Asociación Técnica de Puertos y Costas: Madrid, Spain, 2018.
45. Fernández-Sánchez, G.; Berzosa, Á.; Barandica, J.M.; Cornejo, E.; Serrano, J.M. Opportunities for GHG emissions reduction in road projects: A comparative evaluation of emissions scenarios using CO2NSTRUCT. *J. Clean. Prod.* **2015**, *104*, 156–167. [CrossRef]
46. Cantisani, G.; Di Mascio, P.; Moretti, L. Comparative life cycle assessment of lighting systems and road pavements in an italian twin-tube road tunnel. *Sustainability* **2018**, *10*, 4165. [CrossRef]
47. Cavalett, O.; Slettmo, S.N.; Cherubini, F. Energy and environmental aspects of using eucalyptus from Brazil for energy and transportation services in Europe. *Sustainability* **2018**, *10*, 4068. [CrossRef]
48. Berzosa, Á.; Barandica, J.M.; Fernández-Sánchez, G. A new proposal for greenhouse gas emissions responsibility allocation: Best available technologies approach. *Integr. Environ. Assess. Manag.* **2013**, *10*, 95–101. [CrossRef]

International Journal of
*Environmental Research and Public Health*

MDPI

*Article*

# Community-Based Portable Reefs to Promote Mangrove Vegetation Growth: Bridging between Ecological and Engineering Principles

**Sindhu Sreeranga, Hiroshi Takagi * and Rikuo Shirai**

Department of Transdisciplinary Science and Engineering, School of Environment and Society, Tokyo Institute of Technology, Meguro, Tokyo 152-8550, Japan; sindhu.s.aa@m.titech.ac.jp (S.S.); shirai.r.aa@m.titech.ac.jp (R.S.)
* Correspondence: takagi@ide.titech.ac.jp

**Abstract:** Despite all efforts and massive investments, the restoration of mangroves has not always been successful. One critical reason for this failure is the vulnerability of young mangroves, which cannot grow because of hydrodynamic disturbances in the shallow coastal water. For a comprehensive study bridging ecological and engineering principles, a portable community-based reef is proposed to shield mangroves from waves during the early stages of their growth. A series of field observations were conducted on Amami Oshima Island (Japan), to observe the growth of young mangroves and their survival rate under moderate wave conditions. The evolution of young mangroves was also observed in the laboratory under a controlled indoor environment. At the research site, it was confirmed that, after six months of germination, young mangroves could withstand normal high waves. Laboratory-grown plants were lower in height and had fewer leaves compared with the native mangroves on Amami. Based on these results, an economical reef system was designed. For this purpose, the Ahrens formula for the design of a low-crested reef breakwater was revisited. The results showed that a 50-cm-high reef constructed with 15-kg stones can protect mangroves that are a few months old and effectively promote early mangrove growth.

**Keywords:** young mangroves; mangrove restoration; portable reef design; field observation; Amami Oshima

check for updates

**Citation:** Sreeranga, S.; Takagi, H.; Shirai, R. Community-Based Portable Reefs to Promote Mangrove Vegetation Growth: Bridging between Ecological and Engineering Principles. *Int. J. Environ. Res. Public Health* **2021**, *18*, 590. https://doi.org/10.3390/ijerph18020590

Received: 26 November 2020
Accepted: 8 January 2021
Published: 12 January 2021



## 1. Introduction

Mangrove forests are the most productive ecosystems on the planet among various marine ecosystems [1]. The leaves and roots that filter the salt from seawater enable mangroves to survive in the high tide, while they absorb oxygen for photosynthesis during low tide [2]. Mangroves play a crucial role in protecting coastal regions by reducing the damage caused by tsunamis, storm surges, and tropical cyclones and in saving human settlements. Several studies on the 2004 Indian Ocean tsunami reported that the loss of human life was significantly lower in the presence of mangrove forests, although it was also dependent on the distance and elevation of human settlements from the coastline [3–5]. An interview-based survey revealed that local people in the Philippines believed that mangroves protected their lives from the historical event of Typhoon Haiyan in 2013 [6].

The dynamic interaction between the mangrove system and ocean waves is not fully understood, it is believed that mangroves reduce wave energy and promote sedimentation [7–9]. Mangroves work as a barrier, leading to changes in flow direction resulting in vegetative surface friction, inducing wave energy dissipation and damping [10]. The advantages of mangrove forests are referred to as "ecological resilience" for their ability to absorb hydrodynamic disturbance [11]. Mangroves are also considered to be a green infrastructure that contributes to disaster prevention through flood regulation, erosion control, sediment trapping, nutrient recycling, wildlife habitat, and nurseries [1,12–16].

The mangrove area has been reduced by 45% in the past 23 years, shrinking its geographical coverage from 137,760 km$^2$ [17] to 81,484 km$^2$ [18] worldwide. The declining rate is most significant in developing countries in Asia [19,20]. If such degradation continues, mangroves may disappear in the next 100 years [21]. This degradation is triggered by multiple factors, such as urban development, industrialization, agricultural land expansion, timber and charcoal production, and shrimp farming [16,20,22,23]. In Indonesia, Thailand, and Malaysia, the loss of mangrove forests is triggered by the conversion of land to aquaculture, agriculture, and salt production [23,24]. Tin mining and wood harvesting are also major causes of mangrove degradation in Thailand [24]. Half of the cleared mangroves in Southeast Asia and in South and Central America were due to fish and shrimp aquaculture [25,26]. Conservation actions have been implemented globally to compensate for the loss from deforestation, aquaculture, urban development, industrialization, and shrimp farming. Large-scale mangrove plantations were created in many countries of South and Southeast Asia by nongovernmental organizations and nonprofit organizations such as Wetland International (WI) and the International Union for the Conservation of Nature (IUCN) [22,27]. The Mangroves for the Future initiative was set up through the collaboration of multiple international agencies, such as the United Nations Development Programme, United Nations Food and Agriculture Organisation, and IUCN, to promote the sustainable conservation of coastal ecosystems [27]. The collective efforts of these programs have emphasized the sustainability of coastal ecosystems. The cost of mangrove restoration projects has varied from 1 to 10 million USD, as observed in projects that have taken place in Pakistan, Indonesia, Vietnam, the Philippines, and Senegal [27].

In addition to these global initiatives, scientific communities are trying to develop new ideas for protecting coastal areas from coastal hazards by incorporating the mangrove ecosystem for ecological disaster risk reduction (Eco-DRR). Eco-DRR is an effort composed of the restoration, conservation, and sustainable management of ecosystems to reduce the risk of disaster [11]. The idea of rehabilitating mangroves on a hybrid raised platform proposed recently [12] is expected to lead to new strategies for disaster risk reduction. Ideally, mangrove replantation and conservation should be implemented as a community-based approach (CBA) to improve the preparedness of the local community in response to coastal disasters [28]. However, it has been reported that the success rate of large-scale restoration is not necessarily high [27,29], creating a difficult situation for further dissemination of the mangrove rehabilitation program globally [30,31]. A success rate of only 10–20% was achieved in a community-based restoration program in the Philippines because of inappropriate species and site selection [30,32], while 40% of mangrove seedlings vanished and a 60% success rate was achieved in a similar attempt in Sri Lanka [33]. The principal reasons for these failures are thought to be physical factors (e.g., unusually high waves and less sediment supply [30]) and biological factors, such as the death of seedlings resulting from the dense growth of algae, sapling damage by insects, eating away of young seedlings by aramid crabs, and increase in predation rates by crabs on mangrove plants [31]. Biochemical factors, such as deficiencies of carbon, nitrogen, phosphorus, and other organic matter in sediments, can lead to failure to maintain healthy mangrove seedlings [34]. It is difficult to measure the success and failure of mangrove rehabilitation efforts because of inadequate documentation, particularly when a project fails [32].

Hydrological factors, such as tides, wind-generated waves, and currents, significantly influence the growth phase of young mangrove plants. Wave actions are higher in the wetland rehabilitation sites, causing flooding and damaging young mangrove seedlings [31]. Waves uproot the seedlings, mainly where propagules did not root firmly on loosely deposited sediments [32,35–37]. In Colombia, 93% of seeds died during the initial four months during a prolonged period of flooding. However, the high mortality ratio was not necessarily caused by the inundation, but the uprooting of seedlings in soft sediments as a result of wave actions was also responsible [35].

Simple countermeasures have been implemented to protect mangrove conservation areas by local communities, e.g., constructing barriers made of rocks, logs, and sand bars to attenuate wave actions and trap sediments [31]. Portable and inexpensive materials are also preferable because they can be used for construction by local communities. Wooden piles were also tested in Thailand and Vietnam; however, the effectiveness of these methods has not been sufficiently proved [38]. As a negative effect, they may even cause erosion and sediment destabilization, affecting the natural mangrove settlements [31,39]. Ecological engineering perspectives and applications are considered important steps in restoration [40]. However, the National Research Council of United States, stated that there are barriers to implementing coastal engineering principles in mangrove restoration projects because they are usually costly [40,41]. WI's Mangrove Action Project identified the failed planting techniques and emphasized the necessity of a new approach based on lessons learned from the failed projects [42].

As previously reported, mangrove reforestation did not adequately incorporate engineering principles. The aim of present study was to bridge the ecological and engineering approaches. For example, a stone dike was used for mangrove reforestation (Figure 1). However, in the end, the expansion of the mangrove forest was stopped owing to the presence of the dike. Hence, the size of the stone should be carefully designed so that the dike can be demolished at a later stage. In this study, a CBA called a "portable reef" was developed to protect mangrove plants from hydraulic disturbances. The reef was designed to achieve low-cost coastal protection by placing portable rubble or blocks in front of mangrove plantation areas. It may take several months for mangroves to grow sufficiently to withstand high waves. Therefore, a barrier must sustain its function, at least during the first several months after plantation [12]. Accordingly, a portable reef for only the very early stage of the plantation would have a simple structure. Once mangrove plants grow sufficiently, the portable reef can be dismantled and relocated to other locations for another community activity use. To confirm the feasibility of this concept, the present study has two parts: (i) field and laboratory observations were conducted to understand the basic ecology and growth rate of young mangroves, and (ii) the minimal design requirement for an efficient portable reef was identified.

**Figure 1.** Successful mangrove plantation in Chonburi, Thailand—however, the rubble dike stopped the expansion of the forest [photo taken by one of the authors].

## 2. Materials and Methods

Mangrove growth was observed in the field and the laboratory for approximately six months. The findings were used in designing a portable reef with an emphasis on reducing stone weight, which is essential for community-based construction.

### 2.1. Field Survey

Field surveys were conducted in May 2019 and December 2019, and seaward expansion of young mangrove shrubs was found on sediment deposition in the tidal inlets of Amami Oshima Island (hereinafter "Amami"), Japan. The month of May is the postpollination period, in which the species *Kandelia obovata* in the primeval mangrove forest produces enormous seeds, whereas December is cold, and mangrove growth is not vibrant [43]. A large colony of natural mangroves (*K. obovata*) was identified in Sumiyo Bay of Amami near the inlets of the Yakugachi River and Sumiyo River (Figure 2). As observed in the field survey, mangrove seedlings were transported from the mainstream of the primeval forest and settled on the shallow mudflat with the fast-developing root system. While many mangrove forests are facing degradation in Japan, the primeval mangrove forest of Amami has been expanding for several decades [44]. This mangrove forest is located in areas where the flow of rivers is mild, and the coast is shallow and calm. When the tide recedes, the tidal flat becomes a place where organic matter from the river and the ocean is deposited, providing a habitat for a variety of animals. In addition to the favorable environmental conditions, the expansion of the area may also partially be attributed to the breakwater constructed at the bay mouth, protecting the inner bay area from offshore high waves (Figure 2).

During the survey conducted in May, 40 mangrove propagules were collected to measure the size (Figure 3). Half were used for the plantation test in the Amami primeval mangrove forest, and the rest were transported to a laboratory in Tokyo. A topography survey was conducted to measure the ground level within the mangrove zone. The salinity level and water temperature were recorded. An aerial survey was conducted using a drone (Phantom 4 Pro; DJI Technology Co. Ltd., Shenzhen, China) to observe mangrove shrub density in near-shore, midshore, and offshore regions. Drone images were validated with field observations to confirm the densities of the mangrove plants. The elevation of the mudflat was measured using laser range finders (TruPulse 360; Laser Technology Inc., Centennial, CO, USA) (Figure 4). The predominant mangrove species observed in the study site was *K. obovata*, which is a dwarf-type tree often found in India, Singapore, Cambodia, Malaysia, the Philippines, Indonesia, Myanmar, Bangladesh, Thailand, and Vietnam [45]. The genetic and phenotypic segregation suggests that the species *K. candelin* originated from some parts of China and Japan, and it is now classified as a new species, *K. obovata* [46]. This species is often found in the intertidal region of an estuary, which is frequently inundated by tides, like other Rhizophora mangrove species [40]. *K. obovata* in the Amami region produces seeds, particularly between the months of May and August, which are suspended by tidal flow to colonize themselves in new locations.

During the December 2019 survey, the number of surviving mangrove plants, plant height, number of leaves, root length, stem thickness, and stem color were investigated. The plant age was estimated based on the growth rate between the two surveys. These parameters were then compared with those of laboratory-grown plants under a controlled environment. An in situ manual wave-generating test using a paper board was also conducted to test the failure limitation of young mangrove plants against waves. Additionally, the salinity level, water temperature, turbidity, current velocity (FP111; YSI Inc., Yellow Springs, OH, USA), and water depth at high tides were measured during this second campaign. Water levels were recorded using pressure gauges for approximately 2 h (DEFI2-D10; JFE Co. Ltd., Chiba, Japan). The measured parameters were further considered in the conceptual design of the portable reef.

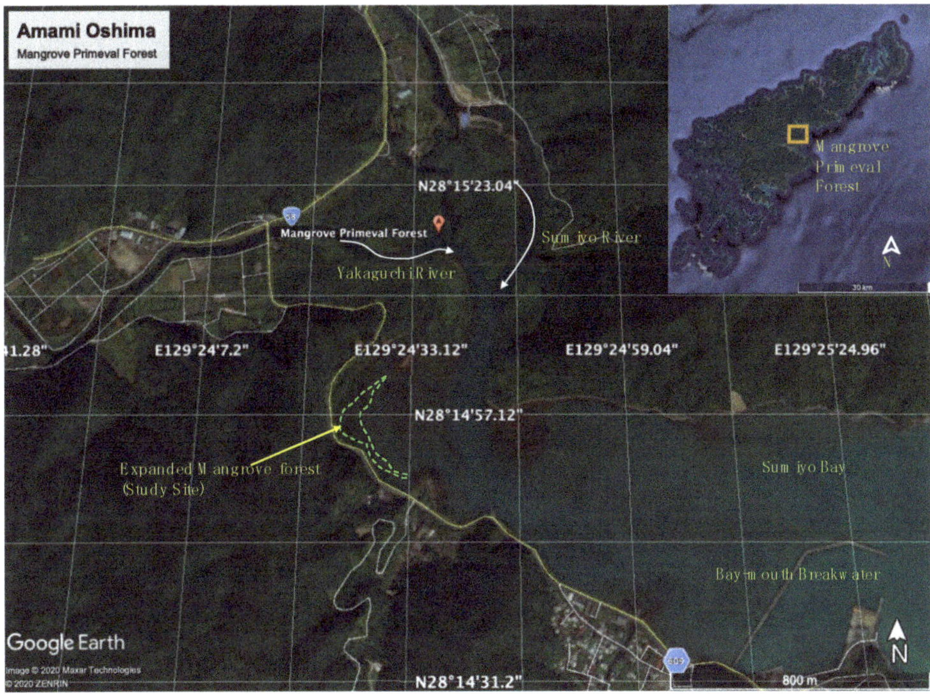


**Figure 2.** Mangrove study site, which is located at the inner-most part of Sumiyo Bay in Amami (study site location: 28°14′57.3″ N 129°24′24.8″ E).

**Figure 3.** Mangrove seeds (propagules of *K. obovata*) collected in the Amami study site.

**Figure 4.** Measurement of the topography in the mangrove forest.

### 2.2. Laboratory Test

Twenty seeds of *K. obovata* were transported to the laboratory in Tokyo and planted in a range of different types of soil: (i) Amami's native soil, (ii) a mixture of sand, silt, and compost, and (iii) coastal sand taken in Tokyo. The growth of these plants was monitored for approximately six months from June to December 2019. The duration of laboratory observations was consistent with that of the in situ plantation test in Amami. The laboratory growing test was conducted to monitor the growth of mangroves in a controlled environment. The test simulated a situation where mangrove plantations are started from seedlings that were originally grown in a pot. Another important idea behind the field survey and laboratory plantation was to investigate the early growing stage of mangrove plants, because the design of the portable reef depends highly on initial plant growth.

## 3. Field Observation Results

### 3.1. Survey in the Study Site

The land slope of a mangrove forest approximately 80 m wide was measured to be as mild as 1/100 on average. This can be considered as a gentle slope but not extremely flat. The density of mangrove plants varied extensively depending on location, i.e., offshore, midshore, and near-shore regions. When approaching the offshore area, it was observed that the number of plants tended to decrease. The lowest mangrove densities were found offshore, where water depths reached half a meter at high tide. Mangrove density was very sparse offshore, whereas 2–4 plants/m$^2$ in the midshore and 10–20 small mangrove plants/m$^2$ around two matured shrubs onshore were observed, as shown in Figure 5a,b.

**Figure 5.** (**a**,**b**) Observed mangrove density at the study site in Amami (photos were taken by a drone).

### 3.2. Mangrove Plantation Test at the Research Site

Figure 6a,b illustrate the plantation of mangroves and their growth from May 2019 to December 2019. Twenty seeds with an initial seedling length of 18.5 cm on average (standard deviation of 3.5 cm) were planted in the offshore zone. The diameter of the seeds varied among the seedlings, with the widest part being 5–10 mm. As shown in Figure 3, some seedlings were completely straight, whereas the others were significantly curved, as observed during the plantation. The color of the seeds varied from green to brownish-green. Of the many plantations, one was made around an existing mangrove plant to clearly identify it during a future survey, as in Figure 6a,b, and the remaining seedlings were planted in nearby locations. In the December 2019 visit, it was found that the survival rate of planted seedlings was 75% (15 out of 20 seeds). Hydrological disturbances, such as high waves, unusual tides, and currents, especially during the typhoon season (July to October), can result in the uprooting of seedlings in soft sediments. Fortunately, however, no strong typhoon approached Amami during the half-year of this survey. Nevertheless, the five seedlings died or washed away for some reason. Various microbial organisms, such as bacteria, fungi, viruses, nematodes, and insects [47], and abiotic factors, such as high salinity levels and low and extremely high temperatures, could have adversely affected the growth of the mangrove plants [2]. The loss may also have been caused by sapling damage by animals, e.g., sea crabs, which selectively eat young seedlings. The water temperature and salinity level in May 2019 were 28.3 °C and 8‰, respectively, while the temperature reduced to 26.1 °C and the salinity increased to 19‰ in December 2019. The rise in salinity may have been because the water discharge from the two rivers is higher in spring than in winter. The precipitation chart is shown in Figure 7, which also supports the observation that salinity levels might have varied because of the difference in precipitation, which is higher in the rainy season (around June) and lower in the winter season (around December).

**Figure 6.** (**a**) Mangrove seeds planted around a prominent existing mangrove in May 2019 and (**b**) mangrove plant growth in December 2019.

### 3.3. Water-Level Observations

Figure 8 depicts water levels and their fluctuations in the study site, recorded using the pressure sensor, on the one-day cycle of the predicted astronomical tide of the bay. The water level fluctuates in correspondence with astronomical tides. It dropped from nearly 0.4 to 0.05 m above the ground level in 2 h, between high to medium tidal ranges. This observation was conducted on a half-moon day during a medium tidal phase. However, the actual water level oscillated with a short period of approximately 20 min on top of the tidal curve. Although it has not been confirmed, this seems to be a sort of seiche that occurs in the bay. As a result, actual currents in mangrove forests may be faster than those induced by pure astronomical tidal forcing. However, the maximum velocity measured in the field was approximately 10 cm/s. This is significantly slower than the velocity generated in a tidal-dominant river mouth where a tidal current of more than 1 m/s often occurs [48].

Because the water depth is very shallow in the study site, frictional effects are believed to be responsible for a significant reduction in flow speed. This level of tidal currents is considered to be less impactful to young mangroves.

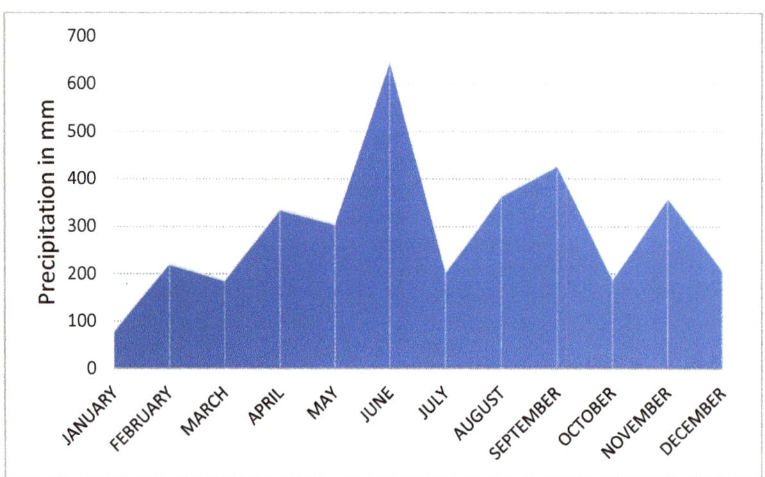


**Figure 7.** Monthly precipitation in 2019, recorded at Naze WMO Station in Amami (Lat 28°22.7′ N Lon 129°29.7′ E)—source: Japan Meteorological Agency.

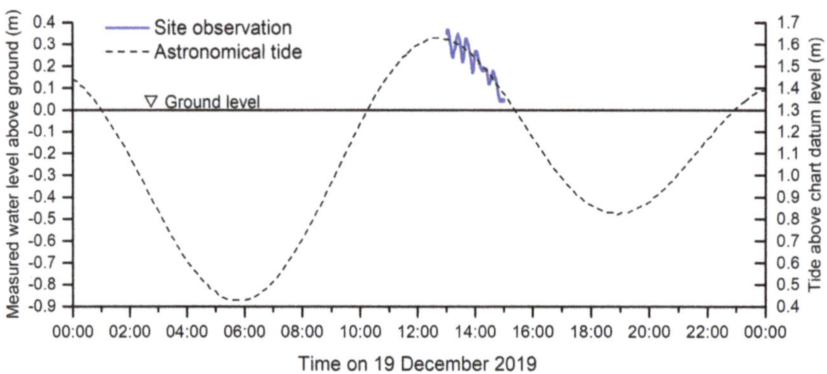


**Figure 8.** Water levels recorded at 1-s intervals for approximately 2 h (19 December 2019) presented in coordination with local astronomical tide data of the day.

### 3.4. In Situ Wave Experiment

In the December survey, an in situ wave experiment was conducted to observe the strength of plants against waves. Waves were manually generated with a paper board for approximately 2 min, which impacted different age groups of plants, such as a month, a half year, and one year. The purpose of this test was to verify the differences in response to waves and critical wave height among the three young mangroves in different growth stages. The maximum wave height during manual wave generation was estimated to be approximately 10 cm through visual analysis with a video image. Figure 9a shows a 17-cm-long one-month-old mangrove that had two leaves, a short main root of approximately 2 cm, and thin subroots. The plant was mostly submerged during the test (Figure 9b). The one-month-old plant was broken entirely and submerged by a 2-min continuous-wave impact, as in Figure 9c,d, whereas a half-year-old and one-year-old mangrove 39 and

80 cm in height, respectively, survived without visible damage, as shown in Figure 9e. The six-month-old mangrove plant shown in Figure 9f developed a stiff root system with a length of one third the total plant length. Hence, it was firmly rooted in the sediment and could withstand the waves. The field investigation revealed that mangroves in the very early stage of growth (a few months) were particularly weak, whereas a half-year-old or older mangrove can sufficiently withstand moderate waves. Thus, special protection is required to protect mangrove seedlings from high waves in the initial two to three months. As a rough estimate, the failure limit of mangroves is considered to be a wave height of 0.1 m, which was used as a basis for the structural design of a portable reef.

**Figure 9.** On-site manual wave generating test: (**a**) one-month mangrove, (**b**) before the wave test, (**c**) during the wave test, (**d**) after the wave test, (**e**) half-year-old and one-year-old mangroves during the wave test, and (**f**) half-year-old mangrove pulled out.

### *3.5. Comparison of Mangrove Growth between Laboratory and Field Experiments*

Figure 10a,b illustrate the mangrove evolution in terms of plant height and number of leaves in the laboratory. Mangrove growth was also measured six months after plantation at the Amami research site, as shown in Figure 6b, and compared with the sixth-month growth of plants in the laboratory. In the laboratory in Tokyo, the growth of the mangrove plant was observed for six months, and measurements were taken for the first, third, and sixth months. Pregerminated mangrove seeds (propagules) brought from the site were planted to observe their growth in different soil states: native soil, a mixture of sand, silt, and compost, and pure sand. The growth of mangroves was observed and measured in terms of the average height and the average number of leaves with standard deviations, as shown in Figure 10a,b. Pictures of propagule growth were taken at zero, one, three, and six months of laboratory plantation, as shown in Figure 11a–d. Although the initial growth was similar among the three soil types, the fastest plant growth was observed in the mixture of sand, silt, and compost after six months. Two plants in the mixture soil had withered at the end of the sixth month, despite showing good growth until three months. The average height of mangrove plants in Amami was 49 cm, with an average of 7.3 leaves per plant, as in Figure 10a,b, while indoor plants demonstrated growth of a height of 38.5 cm with the number of leaves up to 4.3 on average. The laboratory-grown plants were 21% lower in height than the plants grown in Amami. Similarly, the average number of leaves after six months was 41% lower than that of the plants at the Amami research site. The laboratory-grown plants appeared weak, and the stems turned and bent downward, as shown in Figure 11d. The plants grown in native soil looked healthiest among the three soils.

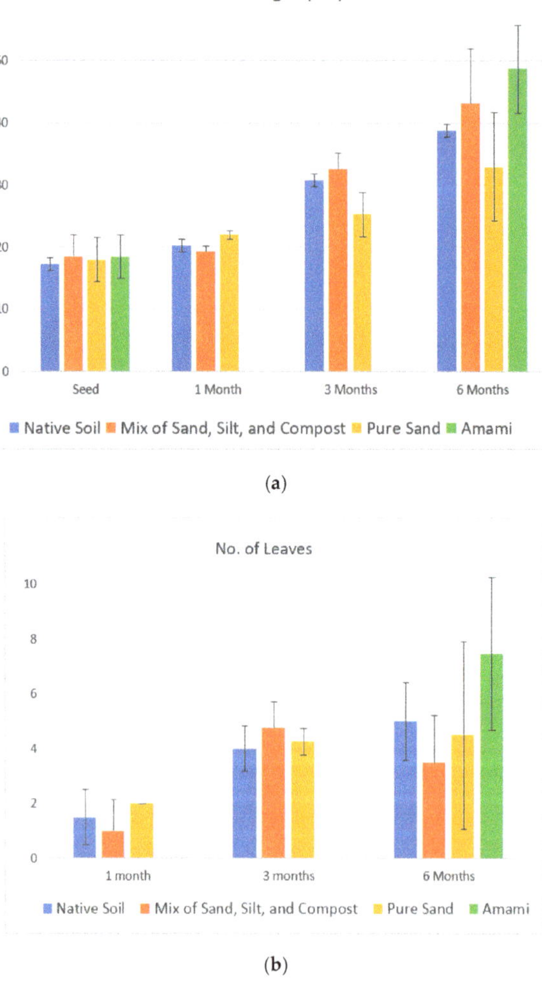


**Figure 10.** (**a**,**b**) Average mangrove growth in terms of height above the soil and the number of leaves recorded during six months in the laboratory: error bars indicate the range of minimum to maximum.

The germination of seedlings began with the development of the stem with a pair of leaves in the first month in all soil media, as shown in Figure 11b. After the third month, the average height of plants in all soil media was 28 cm above the soil surface, with an average number of leaves of 3.5. The average height and number of leaves grown in mangrove plants in all types of soil medium at the end of the sixth month were 38.3 cm and 4.3, respectively. The average increase in plant height and number of leaves in all soil media in the first, third, and sixth months were 3.7, 7.7, and 9.7 cm and 2, 1.5, and 0.8, respectively. With the arrival of new leaves, the plants lost their older leaves, which turned yellow and withered before dropping off see Figure 11c,d. Loss of leaves and decrease in leaf growth indicate lower leaf health because of environmental stresses [43]. This is probably because of the low temperature in November and December, as shown in Figure 12.

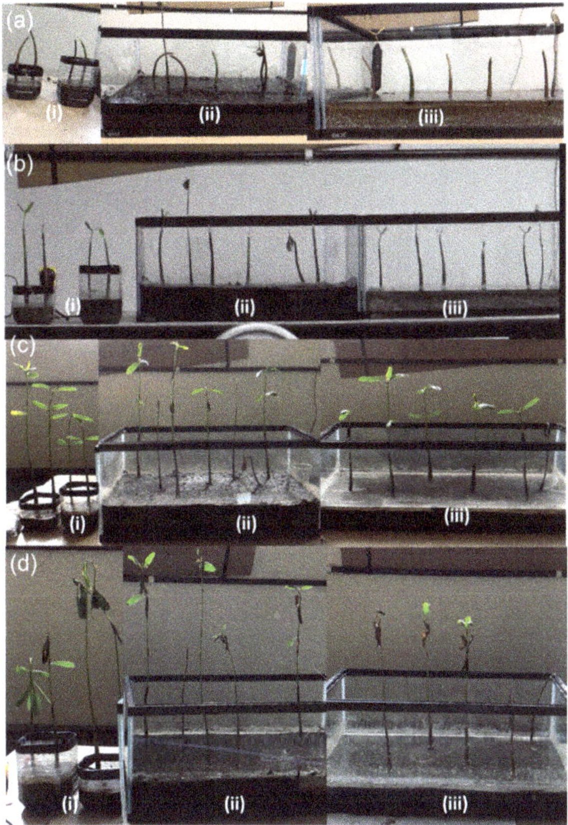


**Figure 11.** Growth of laboratory mangroves in (i) native soil, (ii) mix of sand, silt, and compost, and (iii) pure sand (**a**) on day of the plantation (5 June 2019), (**b**) after one month (10 July 2019), (**c**) after three months (9 September 2019), and (**d**) after sixth months (10 December 2019): an example of the growth record has been uploaded as a time-lapse video at http://www.ide.titech.ac.jp/~takagi/Labmangrove.html.

The air temperature differences in Tokyo and Amami for the period of six months from June to December 2019 are shown in Figure 12. On average, the temperature in Amami was 2.8 °C higher than in Tokyo. However, the maximum temperature in Tokyo (28.8 °C) was 0.4 °C higher than that in Amami (28.4 °C) in August. In December, the average temperature dropped to 8.5 °C in Tokyo, while that in Amami was 12.1 °C. The cold weather during the early winter season probably retarded the growth of mangrove plants and resulted in the withering of the leaves.

The diameter of the laboratory mangroves remained almost unchanged and was less than 4 mm on average, while the average stem diameter of Amami mangroves was 5.4 mm, which is 35% thicker than that of the laboratory mangroves. The Amami mangrove stems appeared tougher and had a deeper green color than those in the laboratory. The leaf surface of the Amami plants was also slightly thicker, whereas that of the laboratory mangroves was thinner and lighter in color.

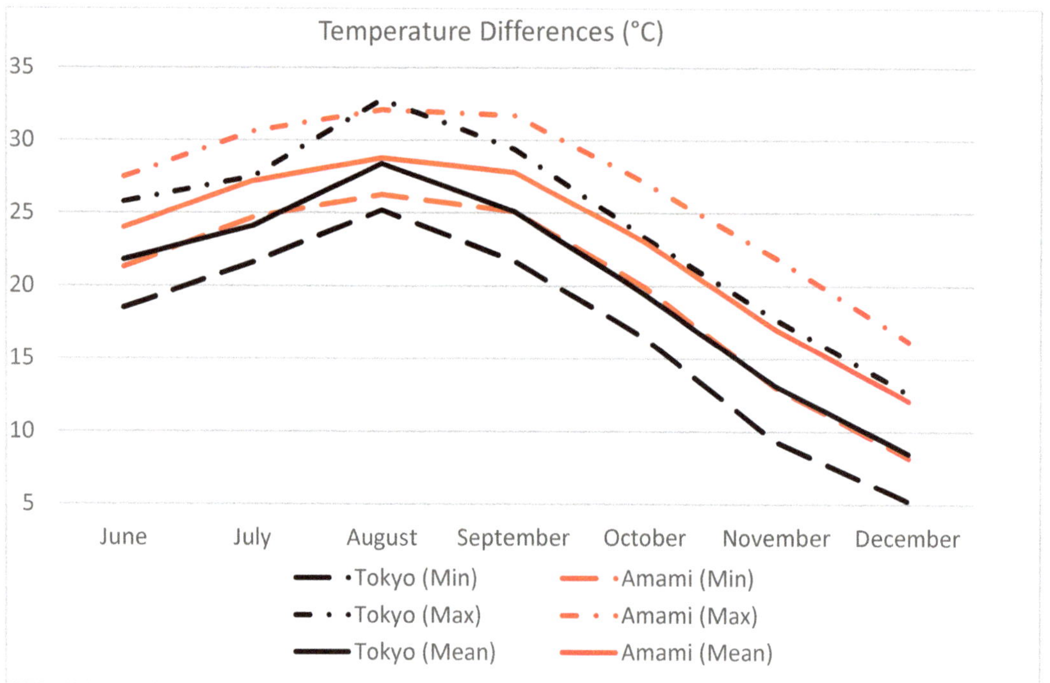


**Figure 12.** Comparison of average minimum, mean, and maximum temperatures between Tokyo and Amami (Kagoshima Prefecture)—source: Japan Meteorological Agency.

The observations of the plants grown in the laboratory suggest that mangrove plants initially grown under a controlled environment do not necessarily show similar growth to that of the plants grown in the field. Laboratory-grown plants were thin, with reduced growth rate and leaf numbers of 20.8% and 41%, respectively, compared with the Amami mangroves, clearly suggesting that the laboratory mangroves were weaker. Seedlings raised in pots are often planted on the coast for mangrove restoration [31,49,50]. Such plants grown in a different location may not acclimatize themselves well and may not easily survive when transported and replanted in the field. In Sungai Haji Dorani, Malaysia, only 30% of the transplanted plants survived [51]. Given all these observations, it is ideal to replant mangroves directly in the required location rather than in nursery plantations. It is expected that portable reefs placed in front of the mangroves can safeguard seeds and young plants from hydrodynamic disturbances and promote the initial growth of mangroves.

## 4. Case Study: Design of Portable Reef as a Community-Based Breakwater

A portable reef was investigated. It was a community-based breakwater composed of a low-crested rubble mound with single-sized stones. In this section, a portable reef designed as a case study based on the wave and topographic conditions in Amami is described. Because many researchers have studied the stability, wave attenuation, and wave transmission of rubble breakwaters for several decades [52–57], the existing formulae were used to address the extent to which the portable reef can be reduced in size while maintaining favorable wave conditions for the growth of young mangroves.

### 4.1. Stability of Rubble Mound

A simple design is desirable to achieve an economical and community-oriented countermeasure. The design of extreme wave conditions would result in heavy stone weights in achieving sufficient structural stability, but such materials cannot be easily handled as community-based activities without the use of heavy equipment. Hence, extreme events, such as tropical cyclones, should be omitted from design considerations. Based on field observations and astronomical tide levels at the study site in Amami, 40 cm is considered the maximum water depth $d$. The breaking wave criteria proposed by Weggel ($H_B < 0.78d$) lead to a maximum wave height of approximately 0.3 m for this depth condition. Hence, 0.3 m was the design wave height in this case analysis (Table 1). The wave period was assumed to vary from 2 to 3.5 s as a short wave in very shallow waters. Although the wave period seems negligible, it has a substantial impact on the portable reef and mangrove plants. Following the convention in coastal engineering, a significant wave (average of the upper one-third) was used as the design wave. In this case study, a rubble mound was designed that can withstand the design waves and confirm how well the transmitted waves can be mitigated.

**Table 1.** Design conditions.

| Variables | Notation | Value |
|---|---|---|
| Wave heights | H | 0.3 m |
| Wave periods | T | 2 to 3.5 s |
| Water depth | d | 0.4 m |

The traditional method of designing rubble breakwaters assumes a stable structure with no damage or statically less than 5% damage levels [57]. A Hudson stability formula was developed from experimental investigations on a permeable breakwater subjected to nonovertopping waves. The equation states the relationship between the armor unit weight and the wave height at the toe of the structure, as shown in Equation (1) [58]:

$$W = \frac{\gamma_s Hs^3}{K_D (s-1)^3 cot\alpha} \tag{1}$$

where $W$ is the weight of a single armor unit, $\gamma_s$ is the specific stone weight, $K_D$ is the dimensionless stability coefficient, $s$ is the specific gravity of the armor unit, $\alpha$ is the structural slope angle, and $H_s$ is the significant wave height. Equation (1) does not consider the damage level, irregular wave conditions, wave period, storm duration, and permeability of stones.

However, it is essential to allow deformation of the system to some extent, particularly in the case of smaller stones. Hence, the formula obtained by Ahrens [59], which designs a low-crested rubble-mound, reef-type breakwater without a multilayer cross section, was applied. Figure 13 illustrates the concept of the low-crested breakwater. Here, $h_c{}'$ is the initial crest height, and $h_c$ is the crest height at the end of the wave impacts, based on an empirical equation from the experiment. In addition, B is the crest width (three median stones wide: $3D_{n50}$). The stability can be examined by considering the crest height that sunk as a result of continuous wave impacts.

Equation (1) is analyzed for the range of significant wave heights ($H_s$), with $K_D$ being 1.2 for quarry stone, smoothly rounded for breaking waves [57]. Here, $\gamma_s$ is taken as 2800 kg/m$^3$, and the slope of the structure is considered to be 1V:2.5H, leading to a slope angle $\alpha$ of 21.6°. The relationship between $H_s$ and the armor unit weight is plotted in

Figure 14. When 0.3 m is used as the design wave height, the lowest stone weight is calculated to be approximately 6 kg.

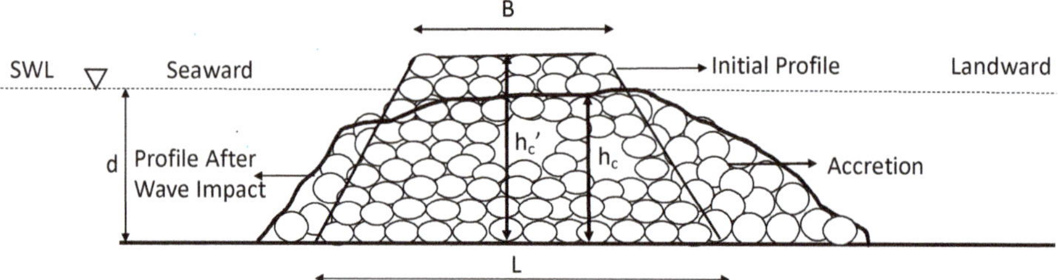


**Figure 13.** Typical reef profile before and after damage, adapted from Ahrens (1989) [58].

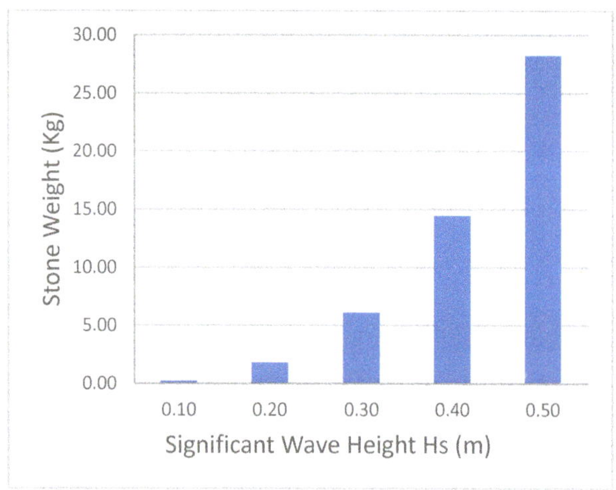


**Figure 14.** Required weight of single stone against a given wave height.

The structural stability with selected stone weights can be further investigated using Equations (2)–(4). The stability number ($N_s$) of the unit is defined as [58]:

$$N_s = \frac{H_s{}^{2/3} L^{1/3}}{\Delta D_{n50}} \tag{2}$$

where $L$ is the wavelength calculated using the wave period (T) and water depth. The reduction in the crest height of the structure was estimated by the Equation (3) which was modified by Van der Meer after reanalyzing the data of Ahrens [58]:

$$h_c = \sqrt{\frac{A_t}{a \exp N_s}} \tag{3}$$

$$a = -0.028 + 0.045\, C' + 0.034 \frac{h_c{}'}{h} - 6 \times 10^{-9} B_n{}^2 \tag{4}$$

Here, $A_t$ is the structural cross-sectional area ($Bh_c{}'$+$C'h_c{}'^2$), C′ is the average structural slope, $N_s$ is the spectral stability number, $B_n$ is the bulk number, and d is still water depth. Short wave periods in the range between T = 2 and 3.5 s are assumed to calculate the wavelength $L = \sqrt{gd}$T and wave steepness $S_{op} = 2\pi H_s/\text{g}\text{T}^2$. The deformation in crest height ($h_c$) can be estimated using Equations (3) and (4). Figure 15 shows a graph of the crest height reduction factor ($h_c/h_c'$) versus wave steepness. If $h_c/h_c'$ exceeds 1, the structure is fully stable, whereas, when $h_c/h_c'$ drops below 1, the structure is less stable [57]. The structural stability increases with the increase in wave steepness. A stone weight more than 15 kg showed $h_c/h_c' > 1$, indicating a stable condition for the design wave. Hence, for the design considerations of stone, a weight of 15 kg can be used as an optimal weight, rather than 6 kg.

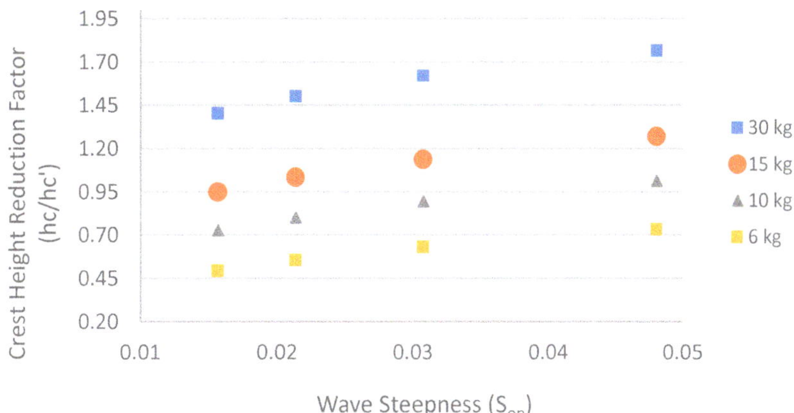


**Figure 15.** Crest height reduction factor ($h_c/h_c'$) versus wave steepness ($S_{op}$).

### 4.2. Wave Transmission and Cross-Sectional Design

Wave breaking and energy dissipation are promoted when waves are transmitted over a reef [60]. The degree of wave transmission is estimated by the coefficient of wave transmission ($K_t$), which is defined by the transmitted wave divided by the incident wave heights ($H_t/H_i$). Wave transmission depends on the geometry of the reef, mainly on crest width and water depth, wave conditions, permeability, freeboard (crest height above water level) ($R_c$), and wave steepness ($S_{op}$) [61]. The prediction of the wave transmission characteristics of breakwaters has been studied, and equations for $K_t$ have been established. The following equation is used for the present analysis of wave transmission over a portable reef [57]:

$$K_t = \left(0.031\,\frac{H_i}{D_{n50}} - 0.24\right)\frac{R_c}{D_{n50}} - 2.6\,S_{op} - 0.05\frac{H_i}{D_{n50}} + 0.85 \tag{5}$$

Here, $K_t$ is derived for the proposed portable reef design (Figure 16. The $K_t$ value markedly varies with the change in the relative crest height ($R_c/D_{n50}$). The minimum $K_t$ and maximum $K_t$ expected of a portable reef breakwater fall between 0.07 and 0.51 for wave periods of 2 to 3.5 s. The lower the relative crest height, the higher the transmission. The lower the height of the breakwater, the smaller the stone volume required. However, a reduction in the freeboard increases the transmission of the waves and adversely affects the growth of mangroves. In the field experiments in Amami, it was found that mangroves of approximately one- or two-month-old plants were washed away by waves, even at a wave height of approximately 10 cm. Hence, $K_t$ needs to be set below 0.3 in the case of a wave height of 30 cm at the portable reef.

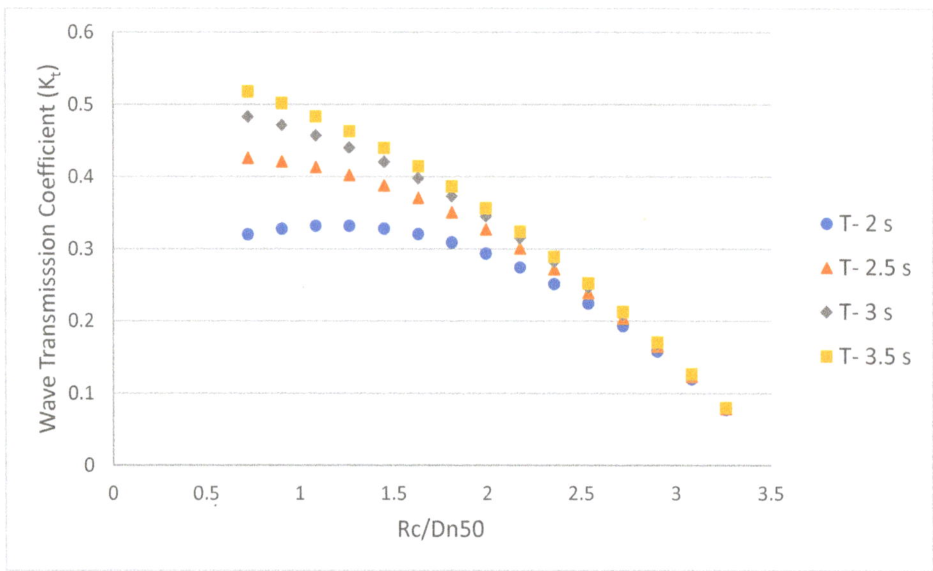


**Figure 16.** Wave transmission versus relative crest height.

As a result of these considerations, the breakwater cross section shown in Figure 17 was selected as one of the candidates to form the entire reef system as simply and feasibly as possible for the local community. The total number of stones needed to form a trapezoid per cubic meter was estimated as 145 (average of 15 kg each). Because of the low weight of the stone, construction can be accomplished without the use of heavy machinery if several workers collaborate. This is an advantage in areas where the ground is loose, such as where mangroves grow. There is a possibility that the reef top may sink a little bit owing to settlement, but it will be easy to replenish. A mangrove plantation is implemented after the reef is installed. However, plants may be washed away because of the disturbance caused by wave overtopping if planted immediately behind the reef. Therefore, it is recommended to maintain a certain distance between the reef and the plantation. It is necessary in future research to investigate how much distance is needed.

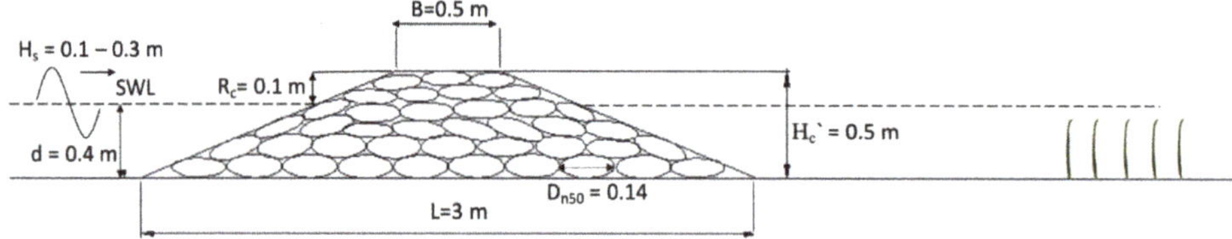


**Figure 17.** Example of a portable reef and mangrove planting.

## 5. Conclusions

The restoration of mangrove forests has not always been successful, despite enormous attempts. In this study, the importance of protecting young mangroves from hydrodynamic disturbances was addressed to improve the implementation of plantations. Early growth of mangroves was studied both in the natural environment (a mangrove forest on Amami) and in a controlled environment (a laboratory in Tokyo). It was observed that plants could grow in any type of soil, even in the indoor environment, but the plants grown in the field looked healthier and stronger than those in the laboratory after six months. The mangrove growth test suggests that direct planting of seedlings in the restoration site is preferable rather than transporting germinated seedlings on a pot. Thus, a portable reef system was proposed to act as an effective wave attenuator to facilitate the growth of young mangrove seedlings. The Amami site was investigated to derive the design wave conditions necessary for protecting early mangroves. Sixth-month old mangrove plants can survive under normal wave conditions; hence, the service period of the portable reef system can be set as short as six months. The examination of the low-crested breakwater proposed by Ahrens was applied for the design of a portable reef system considering the structural stability and wave transmission ratio. As an example, the reef dimensions were designed and it was suggested that a 50-cm-high reef with approximately 15-kg stones is sufficient to protect against waves and effectively promote early mangrove growth. Once the mangroves have grown for about six months, the portable reef is no longer needed and can be dismantled and transported to other locations in the vicinity for reuse. In this way, community-based mangrove plantations should be able to continue in the long term without a high cost.

**Author Contributions:** Conceptualization, H.T. and S.S.; Field Investigation, H.T., S.S., and R.S.; Indoor Test, S.S. and R.S.; Writing—Original Draft Preparation, S.S.; Writing—Review, Modifying, and Editing, H.T.; Funding Acquisition, H.T. All authors have read and agreed to the published version of the manuscript.

**Funding:** This research was funded by a grant awarded to the Tokyo Institute of Technology (Japan Society for the Promotion of Science, 16KK0121 and 19K04964).

**Data Availability Statement:** Not applicable.

**Acknowledgments:** The authors are grateful for the assistance of Fan Jinghui during the series of field surveys in Amami. The study is supported by the Japan Educational Exchanges and Services (JEES), which awarded a T-Banaji Indian student scholarship for one of the authors.

**Conflicts of Interest:** The authors declare no conflict of interest.

## References


1. Ewel, K.C.; Twilley, R.R.; Ong, J. Different kinds of mangrove forests provide different goods and services. *Glob. Ecol. Biogeogr. Lett.* **1998**, *7*, 83–94. [CrossRef]
2. Kathiresan, K.; Bingham, B.L. Biology of mangroves and mangrove ecosystems. In *Advances in Marine Biology*; Academic Press Inc.: Cambridge, MA, USA, 2001; Volume 40, pp. 81–251.
3. Vermaat, J.E.; Thampanya, U. Mangroves mitigate tsunami damage: A further response. *Estuar. Coast. Shelf Sci.* **2006**, *69*, 1–3. [CrossRef]
4. Kathiresan, K.; Rajendran, N. Coastal mangrove forests mitigated tsunami. *Estuar. Coast. Shelf Sci.* **2005**, *65*, 601–606. [CrossRef]
5. Teh, S.Y.; Koh, H.L.; Liu, P.L.F.; Ismail, A.I.M.; Lee, H.L. Analytical and numerical simulation of tsunami mitigation by mangroves in Penang, Malaysia. *J. Asian Earth Sci.* **2009**, *36*, 38–46. [CrossRef]
6. Delfino, R.J.; Carlos, C.; David, L.; Lasco, R.; Juanico, D.E. Perceptions of Typhoon Haiyan-affected communities about the resilience and storm protection function of mangrove ecosystems in Leyte and Eastern Samar, Philippines. *Clim. Disaster Dev. J.* **2015**, *1*, 15–24. [CrossRef]
7. Hu, Z.; Suzuki, T.; Zitman, T.; Uittewaal, W.; Stive, M. Laboratory study on wave dissipation by vegetation in combined current-wave flow. *Coast. Eng.* **2014**, *88*, 131–142. [CrossRef]
8. Mazda, Y.; Magi, M.; Ikeda, Y.; Kurokawa, T.; Asano, T. Wave reduction in a mangrove forest dominated by *Sonneratia* sp. *Wetl. Ecol. Manag.* **2006**, *14*, 365–378. [CrossRef]
9. Parvathy, K.G.; Bhaskaran, P.K. Wave attenuation in presence of mangroves: A sensitivity study for varying bottom slopes. *Int. J. Ocean Clim. Syst.* **2017**. [CrossRef]
10. Mcivor, A.; Möller, I.; Spencer, T.; Spalding, M. Reduction of Wind and Swell Waves by Mangroves. In *Natural Coastal Protection Series: Report 1 Cambridge Coastal Research Unit Working Paper 40*; The Nature Conservancy: Arlington, VA, USA; Wetlands International: Wageningen, The Netherlands, 2012; p. 27.
11. Takagi, H. Long-term design of mangrove landfills as an effective tide attenuator under relative sea-level rise. *Sustainability* **2018**, *10*, 1045. [CrossRef]
12. Takagi, H. "Adapted mangrove on hybrid platform"—Coupling of ecological and engineering principles against coastal hazards. *Results Eng.* **2019**, *4*, 100067. [CrossRef]
13. Thampanya, U.; Vermaat, J.E.; Sinsakul, S.; Panapitukkul, N. Coastal erosion and mangrove progradation of Southern Thailand. *Estuar. Coast. Shelf Sci.* **2006**, *68*, 75–85. [CrossRef]

14. Horstman, E.M.; Dohmen-Janssen, C.M.; Narra, P.M.F.; van den Berg, N.J.F.; Siemerink, M.; Hulscher, S.J.M.H. Wave attenuation in mangroves: A quantitative approach to field observations. *Coast. Eng.* **2014**, *94*, 47–62. [CrossRef]
15. Kathiresan, K. Mangrove forests of India. *Curr. Sci.* **2018**, *114*, 976–981. [CrossRef]
16. Noor, T.; Batool, N.; Mazhar, R.; Ilyas, N. Effects of Siltation, Temperature and Salinity on Mangrove Plants. *Eur. Acad. Res.* **2015**, *2*, 14172–14179.
17. Friess, D.A.; Rogers, K.; Lovelock, C.E.; Krauss, K.W.; Hamilton, S.E.; Lee, S.Y.; Lucas, R.; Primavera, J.; Rajkaran, A.; Shi, S. The State of the World's Mangrove Forests: Past, Present, and Future. *Annu. Rev. Environ. Resour.* **2019**, *44*, 89–115. [CrossRef]
18. Hamilton, S.E.; Casey, D. Creation of a high spatio-temporal resolution global database of continuous mangrove forest cover for the 21st century (CGMFC-21). *Glob. Ecol. Biogeogr.* **2016**, *25*, 729–738. [CrossRef]
19. Rasmeemasmuang, T.; Sasaki, J. Wave Reduction in Mangrove Forests: General Information and Case Study in Thailand. *Handb. Coast. Disaster Mitig. Eng. Plan.* **2015**, *2015*, 511–535. [CrossRef]
20. Sui, L.; Wang, J.; Yang, X.; Wang, Z. Spatial-temporal characteristics of coastline changes in Indonesia from 1990 to 2018. *Sustainability* **2020**, *12*, 3242. [CrossRef]
21. Duke, N.C.; Meynecke, J.-O.; Dittmann, S.; Ellison, A.M.; Anger, K.; Berger, U.; Cannicci, S.; Diele, K.; Ewel, K.C.; Field, C.D.; et al. A world without mangroves? *Science* **2007**, *317*, 41–44. [CrossRef]
22. Primavera, J.H. Development and conservation of Philippine mangroves: Institutional issues. *Ecol. Econ.* **2000**, *35*, 91–106. [CrossRef]
23. Rahman, M.A.A.; Asmawi, M.Z. Local Residents' Awareness towards the Issue of Mangrove Degradation in Kuala Selangor, Malaysia. *Proc. Soc. Behav. Sci.* **2016**, *222*, 659–667. [CrossRef]
24. Macintosh, D.J.; Ashton, E.C.; Havanon, S. Mangrove Rehabilitation and Intertidal Biodiversity: A Study in the Ranong Mangrove Ecosystem, Thailand. *Estuar. Coast. Shelf Sci.* **2002**, *55*, 331–345. [CrossRef]
25. Mcleod, E.; Salm, R.V. *Managing Mangroves for Resilience to Climate Change IUCN Global Marine Programme*; International Union for Conservation of Nature (IUCN): Gland, Switzerland, 2006; 64p.
26. Valiela, I.; Bowen, J.L.; York, J.K. Mangrove Forests: One of the World's Threatened Major Tropical Environments. *Bioscience* **2001**, *51*, 807. [CrossRef]
27. Lee, S.Y.; Hamilton, S.; Barbier, E.B.; Primavera, J.; Lewis, R.R. Better restoration policies are needed to conserve mangrove ecosystems. *Nat. Ecol. Evol.* **2019**, *3*, 870–872. [CrossRef] [PubMed]
28. Van Tao, D.; Ha, N.H. Mangrove replanting, Disaster preparedness and many other benefits. *Clim. Coast. Coop.* **2011**, 176–178.
29. Thompson, B.S. The political ecology of mangrove forest restoration in Thailand: Institutional arrangements and power dynamics. *Land Use Policy* **2018**, *78*, 503–514. [CrossRef]
30. Primavera, J.H.; Esteban, A.J.M.A.; Esteban, J.M.A. A review of mangrove rehabilitation in the Philippines: Successes, failures and future prospects. *Wetl. Ecol Manag.* **2008**, *16*, 345–358. [CrossRef]
31. Primavera, J.H.; Rollon, R.N.; Samson, M.S. The Pressing Challenges of Mangrove Rehabilitation: Pond Reversion and Coastal Protection. In *Treatise on Estuarine and Coastal Science*; Wolanski, E., McLusky, D., Eds.; Elsevier Inc.: Waltham, MA, USA, 2011; Volume 10, pp. 217–244.
32. Kamali, B.; Hashim, R. Mangrove restoration without planting. *Ecol. Eng.* **2011**, *37*, 387–391. [CrossRef]
33. Kodikara, K.A.S.; Mukherjee, N.; Jayatissa, L.P.; Dahdouh-Guebas, F.; Koedam, N. Have mangrove restoration projects worked? An in-depth study in Sri Lanka. *Restor. Ecol.* **2017**, *25*, 705–716. [CrossRef]
34. Teutli-Hernández, C.; Herrera-Silveira, J.A.; Comín, F.A.; López, M.M. Nurse species could facilitate the recruitment of mangrove seedlings after hydrological rehabilitation. *Ecol. Eng.* **2019**, *130*, 263–270. [CrossRef]
35. Elster, C. Reasons for reforestation success and failure with three mangrove species in Colombia. *For. Ecol. Manag.* **2000**, *131*, 201–214. [CrossRef]
36. Le Minor, M.; Bartzke, G.; Zimmer, M.; Gillis, L.; Helfer, V.; Huhn, K. Numerical modelling of hydraulics and sediment dynamics around mangrove seedlings: Implications for mangrove establishment and reforestation. *Estuar. Coast. Shelf Sci.* **2019**, *217*, 81–95. [CrossRef]
37. Tamin, N.M.; Zakaria, R.; Hashim, R.; Yin, Y. Establishment of Avicennia marina mangroves on accreting coastline at Sungai Haji Dorani, Selangor, Malaysia. *Estuar. Coast. Shelf Sci.* **2011**, *94*, 334–342. [CrossRef]
38. Takagi, H.; Sekiguchi, S.; Thao, N.D.; Rasmeemasmuang, T. Do wooden pile breakwaters work for community-based coastal protection? *J. Coast. Conserv.* **2020**, *24*, 1–11. [CrossRef]
39. Stanley, O.D.; Lewis, R.R. Strategies for Mangrove Rehabilitation in an Eroded Coastline of Selangor, Peninsular Malaysia. *J. Coast. Dev.* **2009**, *12*, 142–154.
40. Lewis, R.R. Ecological engineering for successful management and restoration of mangrove forests. *Ecol. Eng.* **2005**, *24*, 403–418. [CrossRef]
41. Lewis, R.R.; Brown, B.M.; Flynn, L.L. Methods and criteria for successful mangrove forest rehabilitation. In *Coastal Wetlands: An Integrated Ecosystem Approach*; Gerardo, M.E., Perillo, E.W., Cahoon, D.R., Hopkinson, C.S., Eds.; Elsevier: Amsterdam, The Netherlands, 2019; pp. 863–887.
42. Quarto, A.; Thiam, I. Community-Based Ecological Mangrove. *Nat. Faune Wetl. Int.* **2018**, *32*, 39–45.
43. Chen, L.; Wang, W.; Li, Q.Q.; Zhang, Y.; Yang, S.; Osland, M.J.; Huang, J.; Peng, C.; Chen, L.; Wang, W.; et al. Mangrove species' responses to winter air temperature extremes in China. *Ecosphere* **2017**, *8*, e01865. [CrossRef]

44. Tai, A.; Hashimoto, A.; Oba, T.; Kawai, K.; Otsuki, K.; Nagasaka, H.; Saita, T. Growth of mangrove forests and the influence on flood disaster at Amami Oshima Island, Japan. *J. Disaster Res.* **2015**, *10*, 486–494. [CrossRef]
45. Nabiul, I.K.M.; Kabir, M.E. Ecology of *Kandelia obovata* (S., L.) Yong: A Fast-Growing Mangrove in Okinawa, Japan. In *Participatory Mangrove Management in a Changing Climate*; DasGupta, R., Shaw, R., Eds.; Springer: Tokyo, Japan, 2017; pp. 287–301.
46. Sheue, C.R.; Liu, H.Y.; Yong, J.W.H. *Kandelia obovata* (Rhizophoraceae), a new mangrove species from Eastern Asia. *Taxon* **2003**, *52*, 287–294. [CrossRef]
47. Das, S.K.; Patra, J.K.; Thatoi, H. Antioxidative response to abiotic and biotic stresses in mangrove plants: A review. *Int. Rev. Hydrobiol.* **2016**, *101*, 3–19. [CrossRef]
48. Takagi, H.; Quan, N.H.; Anh, L.T.; Thao, N.D.; Tri, V.P.D.; Anh, T.T. Practical modelling of tidal propagation under fluvial interaction in the Mekong Delta. *Int. J. River Basin Manag.* **2019**, *17*, 377–387. [CrossRef]
49. Chan, H.T.; Baba, S. *Manual on Guidelines for Rehabilitation of Coastal Forests Damaged by Natural Hazards in the Asia Pacific Region*; International Society of Mangrove Ecosystem (ISME): Nishihara, Japan; International Tropical Timber Organization (ITTO): Yokohama, Japan, 2009; p. 66.
50. Ravishankar, T.; Ramasubramanian, R. *Manual on Mangrove Raising Techniques*; M.S. Swaminathan Research Foundation (MSSRF): Chennai, India, 2004; ISBN 9788578110796.
51. Hashim, R.; Kamali, B.; Tamin, N.M.; Zakaria, R. An integrated approach to coastal rehabilitation: Mangrove restoration in Sungai Haji Dorani, Malaysia. *Estuar. Coast. Shelf Sci.* **2010**, *86*, 118–124. [CrossRef]
52. Dattatri, J.; Raman, H.; Shankar, N.J. Performance Characteristics of Submerged Breakwaters. In Proceedings of the 16th International Conference on Coastal Engineering, Hamburg, Germany, 27 August–3 September 1978; American Society of Civil Engineers: New York, NY, USA, 1978; pp. 2153–2171.
53. Twu, S.-W.; Liu, C.-C.; Hsu, W.-H. Wave damping characteristics of deeply submerged breakwaters. *J. Waterw. Port. Coast. Ocean Eng.* **2001**, *127*, 97–105. [CrossRef]
54. Shirlal, K.G.; Rao, S.; Ganesh, V. Manu Stability of breakwater defenced by a seaward submerged reef. *Ocean Eng.* **2006**, *33*, 829–846. [CrossRef]
55. Sindhu, S.; Shirlal, K.G. Manu Prediction of wave transmission characteristics at submerged reef breakwater. *Proc. Eng.* **2015**, *116*, 262–268. [CrossRef]
56. Calabrese, M.; Vicinanza, D.; Buccino, M. 2D Wave setup behind submerged breakwaters. *Ocean Eng.* **2008**, *35*, 1015–1028. [CrossRef]
57. Van der Meer, J.W.; Pilarczyk, K.W. Stability of Low-Crested and Reef Breakwaters. In Proceedings of the 22nd International Conference on Coastal Engineering, Delft, The Netherlands, 2–6 July 1990; American Society of Civil Engineers: New York, NY, USA, 1991; Volume 2, pp. 1375–1388.
58. Chasten, M.A.; Rosati, J.D.; McCormick, J.W. *Engineering Design Guidance for Detached Breakwaters as Shore Stabilization Structures*; Randall, R.E., Ed.; US Army Corps of Engineers: Washington, DC, USA, 1993.
59. Ahrens, J.P. Stability of Reef Breakwaters. *J. Waterw. Port Coast. Ocean Eng.* **1989**, *115*, 221–234. [CrossRef]
60. Thaha, M.A.; Muhiddin, A.B. The combination of low crested breakwater with mangroves to reduce the vulnerability of the coast due to climate change. In *Asian and Pacific Coasts: Proceedings of the 6th International Conference on Asian and Pacific Coasts*; World Scientific: Hongkong, China, 2011; pp. 541–550.
61. Centre d'Etudes Techniques Maritimes Et Fluviales (CETMEF); Construction Industry Research and Information Association (CIRIA); CUR Building & Infrastructure. *The Rock Manual: The Use of Rock in Hydraulic Engineering*, 2nd ed.; C683 CIRIA: London, UK, 2007; Chapter 5, pp. 487–756.

International Journal of
*Environmental Research and Public Health*

*Article*

# Economic Aspects of a Concrete Floating Offshore Wind Platform in the Atlantic Arc of Europe

**Eugenio Baita-Saavedra [1], David Cordal-Iglesias [1], Almudena Filgueira-Vizoso [2] and Laura Castro-Santos [3,*]**

[1] Escola Politécnica Superior, Universidade da Coruña, Esteiro, 15471 Ferrol, Spain; eugenio.baita@udc.es (E.B.-S.); david.cordal@udc.es (D.C.-I.)
[2] Departamento de Química, Escola Politécnica Superior, Universidade da Coruña, Esteiro, 15471 Ferrol, Spain; almudena.filgueira.vizoso@udc.es
[3] Departamento de Enxeñaría Naval e Industrial, Escola Politécnica Superior, Universidade da Coruña, Esteiro, 15471 Ferrol, Spain
* Correspondence: laura.castro.santos@udc.es

Received: 23 September 2019; Accepted: 18 October 2019; Published: 25 October 2019

**Abstract:** The objective of this paper is to examine the economic aspects of a concrete offshore wind floating platform in the Atlantic Arc of Europe (Portugal and Spain). The life-cycle cost of a concrete floating offshore wind platform is considered to calculate the main economic parameters that will define the economic feasibility of the offshore wind farm. The case of study is the concrete floating offshore wind platform Telwind®, a spar platform with a revolutionary way of installing using a self-erecting telescopic tower of the wind turbine. In addition, the study analyses thirteen locations in Spain and twenty in Portugal, including the Atlantic islands of both countries. Results indicate that the economically feasible location to install a concrete offshore wind farm composed of concrete platforms is the Canary Islands (Spain) and Flores (Portugal).

**Keywords:** floating offshore wind; concrete wind platform; economic feasibility; IRR; NPV; LCOE

## 1. Introduction

The global economy is committed to reducing greenhouse gas emissions [1]. To this end, policies to reduce them are being promoted. The 2015 Paris agreement [2] highlights the importance of cross-border cooperation in environmental matters between member states, including Spain and Portugal. The priority objective is to reduce greenhouse gases by 20% compared to 1990 and reach 80% by 2050 [3]. In recent years, the European Union has made considerable efforts to increase electricity generation through renewable sources. The percentage of renewable energies in the final electrical consumption has doubled from 8.5% in 2004 to 17% in 2016 [4]. Although the EU, as a whole, is on its way to achieve the objectives of H2020, some member states, including Spain and Portugal, will have to make additional efforts to meet their obligations.

In 2008, the first generation of commercial marine energy systems was introduced, with the first platforms located in the United Kingdom and Portugal (Seagen and Pelamis, respectively) [5]. With this, there are currently three main sorts of energy devices from which energy can be produced commercially (wind, waves, and tides) in marine regions. Each of these technologies can be presented individually or as a combination of several, depending on the characteristics of each area, and they are clean energies both in the sense that they do not produce greenhouse gases [6], but also in the visual aspect, which is minimal compared to other energy sources [7].

Within these oceanic energy sources, offshore wind energy can become one of the main sources of the future European energy system. It is one of the basic energies to achieve the objectives of climate change policies [8]. Wind Europe expects 150 GW of wind capacity to be achieved in 2030 that would supply 14% of Europe's electricity demand. Analyzing wind power, it can be seen that offshore power has augmented by 19.7% in the last 10 years [9], being therefore, one of the renewable energies with the greatest potential [10]. In addition to this, as stated by Noori et al., offshore wind turbines produce 48% less greenhouse gas emissions per kWh of produced electricity than onshore wind turbines.

It was also found that the higher the capacity of the wind turbine, the lower the environmental impact.

According to Chipindula et al. and Shifeng Wang, Wang et al., the life-cycle GHG (Greenhouse Gas) emission intensity is 0.082 kg $CO_2$-equivalent (eq)/Mega-joule (MJ) and 0.130 kg $CO_2$-eq/MJ for an onshore and offshore wind turbine, respectively. Both onshore and offshore energy have lower GHG emissions than coal plants (Lei Xu et al. 2018).

Within the marine wind energy, there are two types: Fixed structures (up to 50 m of depth) and floating platforms (over 50 m of depth). Spain and Portugal have very deep continental platforms in areas very close to the coast, which makes it necessary to install floating platforms. Within these, the only commercial ones (Windfloat of Principle Power [11] and Hywind of Statoil [12,13]) are made entirely of steel: It is a very common material in maritime civil engineering, which is resistant to corrosion, does not have high maintenance costs, and is easy to obtain. However, materials such as concrete have a number of characteristics that make it interesting to analyze platforms built with this material. Therefore, this article is focused on the economic analysis of an offshore concrete floating wind platform, a 10 MW Telwind® [14] (mast platform with a revolutionary form of installation), in the Atlantic Arc of Europe. The cost of the life cycle [15–17] will be considered to determine the main economic parameters that will determine the economic viability of the park: Cost of the life cycle, levelized cost of energy (LCOE), net present value (NPV), and internal rate of return (IRR). The study analyzed thirteen locations in Spain [17] (the Iberian Peninsula and the Canary Islands) and twenty in Portugal [18] (the Iberian Peninsula, the Azores Islands, and the Madeira Islands) [19]. The results show which are the most suitable areas, in economic terms, to install a marine wind farm of concrete.

## 2. Materials and Methods

The work methodology of this study is founded on life-cycle cost methodology for floating offshore wind steel platforms [15], but applying the procedure to concrete structures, as defined in the case study.

### 2.1. Determination of Wind Production

The annual produced energy depends on the characteristic power curve of the turbine and the wind speed distribution function of each location. The DTU 10 MW wind turbine with a constant capacity of 10 MW has been selected for the analysis [20]. Figure 1 shows the power curve of this turbine.

In this study, the Weibull distribution is used to characterize wind behavior. The Weibull distribution is a continuous probability distribution that is usually used to define the variation of wind speed at a given location, and thus, describe the behavior of the wind through parameters that define its probability function [21,22].

$$f(v) = \left(\frac{k}{c}\right) \cdot \left(\frac{v}{c}\right)^{k-1} \cdot e^{[-(\frac{v}{c})]^k} \quad (1)$$

where $k$: shape factor Weibull distribution, $c$: scale factor Weibull distribution, $v$: wind speed.

Data of the parameters of each location are necessary. The energy produced is calculated considering Equation (2), where $P_{PC}(v)$ is the power curve of the offshore wind turbine and $p_{Weibull}(v,k,c)$ is the probability function.

$$E = \int_0^{v_{out}} P_{PC}(v) \cdot f(v) \quad (2)$$

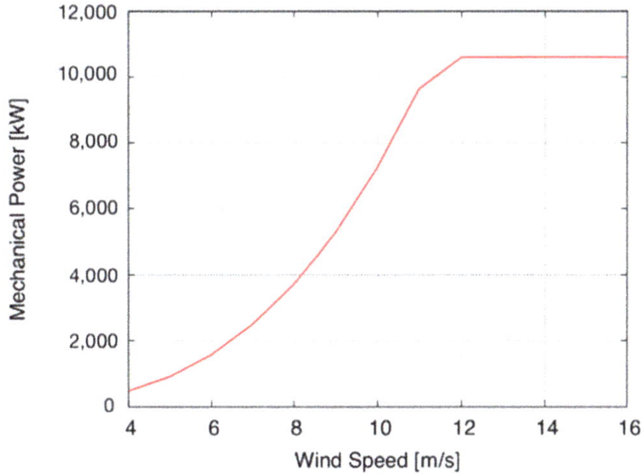


**Figure 1.** Power curve of the DTU 10MW wind turbine.

### 2.2. *Life Cycle Cost Assessment*

Figure 2 shows the phases of the LCS (life cycle system) of a floating offshore wind farm.

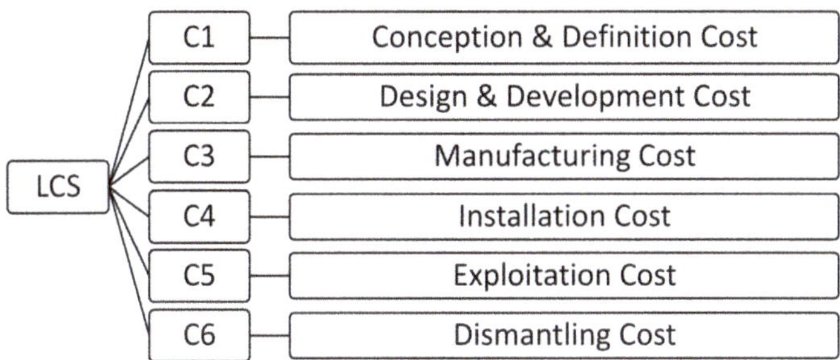


**Figure 2.** Life cycle system (LCS) phases. Source: [15].

In this way, the total cost system of the life cycle of the farm (LCS) is obtained from (3) [15]

$$LCS = C_1 + C_2 + C_3 + C_4 + C_5 + C_6 \quad (3)$$

The manufacturing costs and the installation costs are some the most relevant costs in terms of the differences between the traditional steel floating offshore wind platforms and the concrete one considered in this study.

To analyze manufacturing and installation costs, each of the components that constitute a floating offshore wind farm has been considered and studied separately as a subphase of costs.

The subphases of costs considered for $C_3$ and $C_4$ have been the generator, the floating platform, the mooring, the anchoring, and the electric system. In this way, total manufacturing and installation costs have been obtained as the sum of all subphases.

Equation (4) allows to obtain the transport costs of an offshore wind turbine ($C_{41}$).

$$C_{41} = C_{411} + C_{412} \quad (4)$$

where:

- $C_{411}$: Onshore installation and precommissioning cost (€).
- $C_{412}$: Offshore commissioning cost (€).

The main costs that may affect each operation have been considered. For onshore installation and precommissioning, a self-propelled modular transporter (SPMT) is used to carry out the transport operations of the turbine components.

$$C_{411} = N_{turbine} \cdot (C_{nacelle} + SPMT_{blades} + SPMT_{nacelle} + C_{other}) \tag{5}$$

where:

- $N_{turbine}$: Number of turbines.
- $C_{nacelle}$: Nacelle assembly cost of a 10 MW turbine (€).
- $SPMT_{bladess}$: $SPMT$ for concrete blades cost of a 10 MW turbine (€).
- $SPMT_{nacelle}$: $SPMT$ for nacelle cost of a 10 MW turbine (€).
- $C_{other}$: Other costs of a 10 MW turbine (€).

For offshore commissioning, the costs of technicians and specialized personnel have been considered.

$$C_{412} = N_{turbine} \cdot \left(N_{tech} \cdot r_{daily} \cdot n_{days}\right) \tag{6}$$

where:

- $N_{tech}$: Number of offshore commissioning technicians.
- $r_{daily}$: Offshore commissioning technician daily rate (€/days).
- $n_{days}$: Time required for offshore commissioning (days).

As in the equations shown, many other input data are necessary to complete the entire cost assessment.

Exploitation costs ($C_5$) consist of insurance (obtained from 1% of $C_1 + C_2 + C_3 + C_4$), administration and operations costs (data acquisition cost, SAP (systems, applications, products, and data processing) and maritime coordination cost, meteorological prediction cost), maintenance costs (turbine, export cable, grid connection, and substructure maintenance, inter-array cable survey) and logistics both onshore and offshore.

In order to calculate decommissioning costs ($C_6$), percentages have been used according to the costs of the dismantled material, as shown in Table 1.

**Table 1.** Decommissioning costs (% of $C_4$). Source: [15].

| Description | Percentage of $C_4$ |
|---|---|
| Complete wind turbine—floating | 70% |
| Electric cables | 10% |
| Substation | 90% |
| Mooring | 90% |

### 2.3. Determination of Economic Parameters

The main economic indicators used in this study that will determine the economic feasibility of the farm are: The internal rate of return (IRR), the net present value (NPV), and the levelized cost of energy (LCOE).

The NPV represents the present value of future cash flows that will be created or are the result of a specific investment.

This is a dynamic criterion since it takes into account the updating of cash flows, both their amount and the time when they are obtained, in order to homogenize them over time. The result is obtained in absolute terms of the monetary units.

The NPV was calculated as follows:

$$NPV = -CF_0 + \frac{CF_1}{1+k} + \frac{CF_2}{(1+k)^2} + \ldots + \frac{CF_{n-1}}{(1+k)^{n-1}} + \frac{CF_n}{(1+k)^n} \quad (7)$$

where:

- $CF_0$: Initial investment (€).
- $CF_n$: Cash flow in time n (€).
- $n$: Project lifetime (year).
- $k$: Discount rate.

The criterion of acceptance around the NPV is that it is positive, and, among several projects, it will be more convenient than that with a higher NPV.

The internal rate of return (IRR) measures the expected future returns for a given investment, and implies the supposed case of an opportunity to invest. It is used, together with the NPV, as a criterion for deciding between the acceptance and rejection of an investment project. The higher the IRR, the greater the profitability.

The IRR corresponds to the discount rate ($k$) that the NPV makes zero. Therefore, it is defined with the following expression:

$$0 = -CF_0 + \frac{CF_1}{1+k} + \frac{CF_2}{(1+k)^2} + \ldots + \frac{CF_{n-1}}{(1+k)^{n-1}} + \frac{CF_n}{(1+k)^n} \quad (8)$$

The levelized cost of energy (LCOE) is the most representative indicator to calculate the cost of wind energy production. The cost components that has been estimated to calculate this indicator are the initial investment and the operation and maintenance costs. The annual value of the LCOE is estimated considering (9):

$$LCOE = \frac{\sum_{t=0}^{N_{farm}} \frac{LCS_t}{(1+r)^t}}{\sum_{t=0}^{N_{farm}} \frac{E_t}{(1+r)^t}} \quad (9)$$

where:

- $LCOE$: $LCOE$ of wind energy (€/MWh).
- $E_t$: Energy produced by the farm (MWh/year).
- $N_{farm}$: Life-cycle of the farm (years).

## 3. Configuration of the Offshore wind Farm Platform, TELWIND

The floating offshore wind platform studied in this work is the Telewind®, an evolved concrete spar floating offshore wind platform designed by the Spanish enterprise Esteyco (see Figure 3). The offshore wind farm studied has a capacity of 500 MW.

The main components of the Telewind® platform are (see Figure 4): Telescopic tower, upper tank, tendons, lower tank, and mooring lines.

The Telewind® platform is composed of two parts (see Figure 5): The upper structure (US) and the lower tank (LT). Moreover, the upper structure is composed of the upper tank (UT), tower 0 (T0), tower 1 (T2), and tower 2 (T2). T1, T0, UT, and LT are built in concrete and T2 is built in steel for the Arcwind project. The description of the main parts of the Telewind® platform is as follows (from courtesy of Esteyco):

- Wind turbine generator (WTG): It is worth to note that only the nacelle, hub, and blades are part of WTG.
- Tower (TW): Telescopic concrete tower divided into several sections with variable thickness. The telescopic tower is divided into three sections (T0, T1, T2). It should be noted that the upper

section (T2) of the tower is completely made of steel, while T0 and T1 are made of concrete. The T0 is partially submerged and it is part of the "platform hull", it means it is designed according to accepted offshore industry design criteria taking into account the hydrostatic pressures, waves, and currents plus the contribution due to wind forces.

- Upper tank (UT): Wet part of the platform which comprises the cylindrical base. It provides buoyancy in excess to guarantee the stability of the system. Although it is not really relevant at this stage, the base is internally compartmented to minimize free surface effects, sloshing and to improve the structural strength. Note that the submerged part of the T0 is part of the tower (TW).
- Upper structure (US): In order to simplify the nomenclature, US comprises the WTG + TW + UT.
- Lower tank (LT): Suspended body filled with a combination of solid and water ballast when it is in place. For installation purposes, the lower tank is water ballasted until it gets the final location.
- Tendon suspension system: Cable connections between the UT and the LT. Tendons may be made of steel or synthetic fibers. The final configuration of lines and its connection with UT and LT will be defined in a later stage of the project.
- Mooring system: The mooring system is comprised by anchors, mooring lines, connectors, and links. It is composed of several catenary mooring lines (chain, fibers or mixed systems). The final arrangement will be defined later in detail phases.

**Figure 3.** Telewind® platform. Source: Figure courtesy of Esteyco [15].

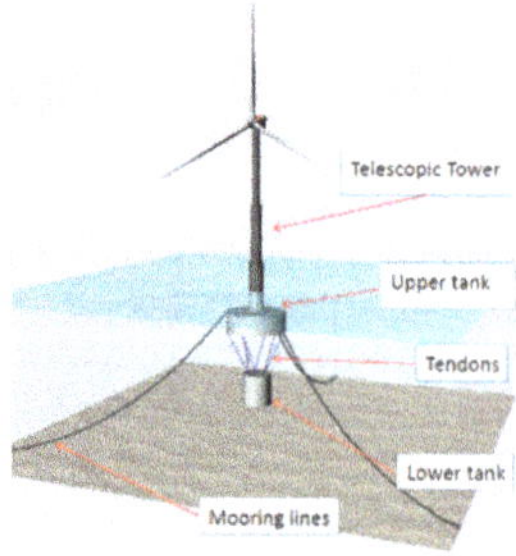


**Figure 4.** Main parts of the Telewind® structure. Source: Figure courtesy of Esteyco [23–25].

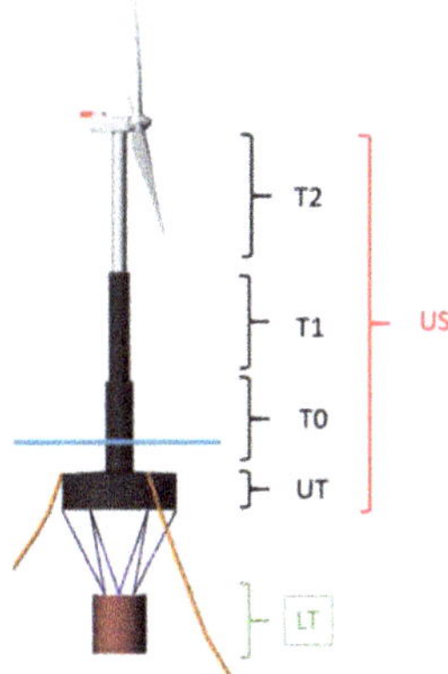


**Figure 5.** Parts of the Telewind® platform. Source: Figure courtesy of Esteyco.

The main characteristics of the Telewind® platform are shown in Table 2.

**Table 2.** Characteristics of the Telewind® platform. Source: Data courtesy of Esteyco.

| Characteristics | Value | Units |
|---|---|---|
| Water depth | 110 | m |
| Hub height above MSL | 119 | m |
| Wind turbine power | 10 | MW |
| Metacentric height in place (GM) | >3 | m |
| Metacentric height transport (GM) | >2 | m |
| Tilt static angle | <10° | ° |
| Overall heave period (T3) | >30 | s |
| Overall pitch period (T3) | >30 | s |

## 4. Study Locations

The case of the study is focused on two countries of the Atlantic Arc of Europe (see Figure 6): Spain and Portugal. The offshore wind sector will have great importance in these countries, which have been pioneers in two different aspects of the offshore wind: Floating offshore wind building using shipyards and floating offshore installation, respectively. Portugal installed the second floating platform in the world, the WindFloat semisubmersible platform, years ago. Nowadays the WindFloat Atlantic project (Windplus) is being carried out as the first floating offshore wind farm in the Iberian Peninsula. Additionally, Navantia Fene (in A Coruña province, Galicia region of NW of Spain) built five Hywind spar platforms for the Hywind Scotland farm, the first commercial floating offshore wind farm in the world, and it is building five platforms for the Kinkardine farm, which will be installed in the UK [26–28].

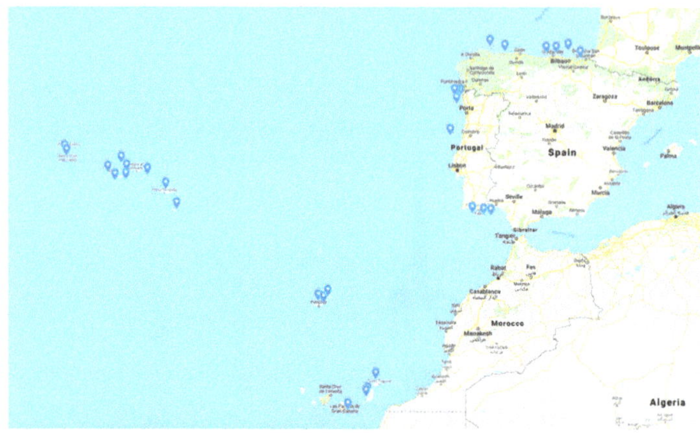


**Figure 6.** Locations of the study. Source: Own elaboration.

In this context, thirteen locations have been considered in Spain: Nine in the Iberian Peninsula (A Guarda-Baiona 1, A Guarda-Baiona 2, Ribadeo, Navia, San Vicente de la Barquera, Santander, Bilbao, Mutriku, Huelva) and four in the Canary Islands (Lanzarote, Fuerteventura 1, Fuerteventura 2, Gran Canaria). In addition, twenty locations have been considered in Portugal: Seven in the Iberian Peninsula (Vianao do Castelo 1, Viana do Castelo 2, Póvoa do Varzim, Porto, Figueira da Foz, Algarve-Albufeira, Algarve-Faro), ten in the Azores Islands (Flores 1, Flores 2, Faial, Pico 1, Pico 2, Sao Jorge, Graciosa, Terceira, Sao Miguel, Santa María) and three in the Madeira Islands (Sao Vicente-Santana, Porto da Cruz-Caniçal, Porto Santo). Their coordinates are in Table 3.

**Table 3.** Coordinates of the locations.

| Location | | Coordinates | |
|---|---|---|---|
| SPAIN | A Guarda-Baiona 1 | 41.86 N | 9.18 W |
| | A Guarda-Baiona 2 | 41.86 N | 9.32 W |
| | Ribadeo | 43.83 N | 7.33 W |
| | Navia | 43.63 N | 6.53 W |
| | San Vicente de la Barquera | 43.56 N | 4.22 W |
| | Santander | 43.57 N | 3.66 W |
| | Bilbao | 43.67 N | 3.00 W |
| | Mutriku | 43.39 N | 2.33 W |
| | Huelva | 36.76 N | 7.30 W |
| CANARY ISLANDS | Lanzarote | 29.22 N | 13.74 W |
| | Fuerteventura 1 | 28.54 N | 14.19 W |
| | Fuerteventura 2 | 28.40 N | 14.29 W |
| | Gran Canaria | 27.77 N | 15.31 W |

**Table 3.** *Cont.*

| Location | | Coordinates | |
|---|---|---|---|
| PORTUGAL | Viana do Castelo 1 | 41.86 N | 9.00 W |
| | Viana do Castelo 2 | 41.82 N | 9.31 W |
| | Póvoa do Varzim | 41.52 N | 9.20 W |
| | Porto | 41.05 N | 9.27 W |
| | Figueira da Foz | 40.21 N | 9.56 W |
| | Algarve-Albufeira | 36.86 N | 8.32 W |
| | Algarve-Faro | 36.79 N | 7.68 W |
| AZORES | Flores 1 | 39.48 N | 31.35 W |
| | Flores 2 | 39.31 N | 31.25 W |
| | Faial | 38.62 N | 28.92 W |
| | Pico 1 | 38.27 N | 28.49 W |
| | Pico 2 | 38.310 N | 27.89 W |
| | Sao Jorge | 38.67 N | 27.85 W |
| | Graciosa | 39.01 N | 28.14 W |
| | Terceira | 38.51 N | 26.66 W |
| | Sao Miguel | 37.90 N | 25.63 W |
| | Santa María | 37.04 N | 25.00 W |
| MADEIRA ISLANDS | Sao Vicente-Santana | 32.89 N | 16.99 W |
| | Porto da Cruz-Caniçal | 32.82 N | 16.68 W |
| | Porto Santo | 33.11 N | 16.44 W |

On the other hand, it is important to consider the distance farm–shore (m), the distance farm–onshore facilities (m) and the bathymetry (m) of the location, which are shown in Table 4, and the separation between wind turbines (Figure 7).

**Table 4.** Main characteristics of the locations selected.

| Location | | Distance Farm-Shore (m) | Distance Farm-Onshore Facilities (m) | Depth (m) |
|---|---|---|---|---|
| SPAIN | A Guarda-Baiona 1 | 8 | 35.2 | 150 |
| | A Guarda-Baiona 2 | 35.6 | 55.7 | 500 |
| | Ribadeo | 17.6 | 89.5 | 150 |
| | Navia | 10.3 | 48.7 | 150 |
| | San Vicente de la Barquera | 14.1 | 40.7 | 500 |
| | Santander | 8 | 15.8 | 400 |
| | Bilbao | 18.5 | 25.6 | 700 |
| | Mutriku | 9.7 | 60.7 | 400 |
| | Huelva | 29.5 | 93 | 500 |

**Table 4.** *Cont.*

| Location | | Distance Farm-Shore (m) | Distance Farm-Onshore Facilities (m) | Depth (m) |
|---|---|---|---|---|
| CANARY ISLANDS | Lanzarote | 9.6 | 76 | 800 |
| | Fuerteventura 1 | 12.8 | 73.2 | 800 |
| | Fuerteventura 2 | 13.6 | 87 | 500 |
| | Gran Canaria | 8.2 | 9.7 | 400 |
| PORTUGAL | Viana do Castelo 1 | 12 | 12.7 | 100 |
| | Viana do Castelo 2 | 27.8 | 28.5 | 150 |
| | Póvoa do Varzim | 20 | 30.4 | 100 |
| | Porto | 33.1 | 33.1 | 150 |
| | Figueira da Foz | 51 | 62.1 | 150 |
| | Algarve-Albufeira | 20.3 | 36.3 | 100 |
| | Algarve-Faro | 20.3 | 26.1 | 600 |
| AZORES | Flores 1 | 2 | 17 | 250 |
| | Flores 2 | 2.3 | 5 | 300 |
| | Faial | 1.8 | 28 | 1000 |
| | Pico 1 | 12.2 | 19 | 1000 |
| | Pico 2 | 7 | 42 | 500 |
| | Sao Jorge | 1 | 38 | 700 |
| | Graciosa | 2.3 | 11 | 500 |
| | Terceira | 16 | 15 | 700 |
| | Sao Miguel | 1.7 | 47 | 700 |
| | Santa María | 3 | 19 | 500 |
| MADEIRA ISLANDS | Sao Vicente-Santana | 1.8 | 38 | 500 |
| | Porto da Cruz-Caniçal | 2.8 | 17.4 | 200 |
| | Porto Santo | 1 | 13 | 80 |

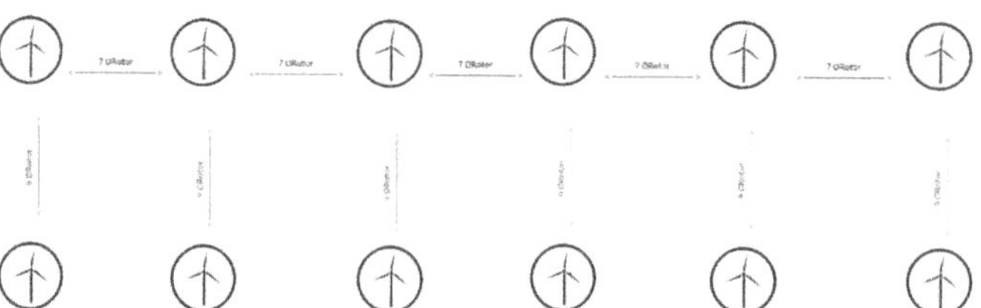

**Figure 7.** Structure of a wind farm. Source: own elaboration.

The distance between wind turbines is seven times the diameter of the rotor of the wind turbine and the distance between lines of wind turbines is nine times the diameter of the rotor of the wind turbine, as shown in Figure 7.

On the other hand, three alternatives have been considered regarding the electric tariff considered (50, 100, and 150 €/MWh) (see Table 5). It is important to notice that the electric tariff changes depending on the country selected. Nowadays, there is no specific electric tariff regulation for floating offshore wind in Spain. While the electricity tariffs in Spain and Portugal are different, the range of tariffs in each country should be still stated. This is the reason why these alternatives have been taken into account.

**Table 5.** Alternatives based on the electric tariff considered (€/MWh).

| ALTERNATIVE | Electric tariff (€/MWh) |
|---|---|
| ALTERNATIVE 1 | 50 |
| ALTERNATIVE 2 | 100 |
| ALTERNATIVE 3 | 150 |

## 5. Results

Once the case study and the different alternatives have been defined, the calculation methodology explained above has been applied to obtain the life cycle cost and the economic parameters. The economic parameters LCOE, NPV, and IRR have been obtained in order to compare the locations proposed.

Figure 8 shows the life cycle costs of each proposed alternative. The most important phases of the life cycle of a floating offshore wind farm are the manufacturing cost and the operation and maintenance costs. The alternative with a lower life cycle cost is Sao Jorge, which is due to the strategic conditions of the location (1 km distance from shore).

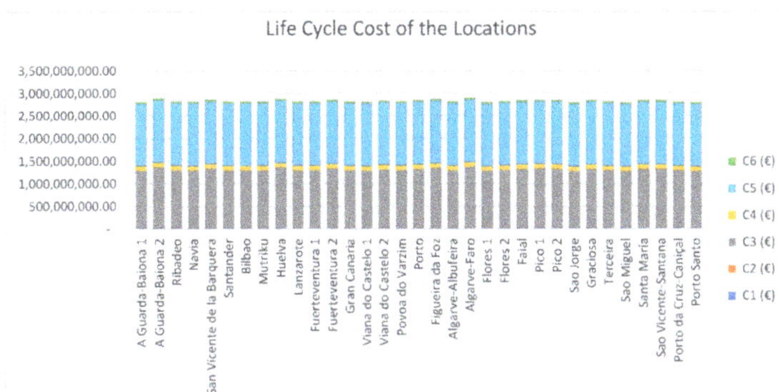


**Figure 8.** Life-cycle cost of the locations.

The manufacturing costs considered are almost the same, since the size of the farm and the number of turbines is constant. The maintenance costs may vary depending on the distance to the coast, but the differences between the locations studied are not very large, ranging between 8 and 35 km.

The relevance of each life cycle phase is shown in Figure 9 for the São Jorge location. The most important costs of a floating offshore windfarm are exploitation cost and manufacturing cost, reaching 50% and 45%, respectively.

To obtain the LCOE, the energy cost obtained by the wind turbines in each location by different alternatives has been calculated. The energy produced has been estimated as a function of the wind distribution and the characteristic power curve of the selected 10 MW turbine.

The LCOE comparison offers information on what might be the best locations for an offshore wind farm without taking into account the parameter of the electricity tariff, only the annual electricity production. As Figure 10 shows, Huelva (Spain) and Algarve-Faro (Portugal), have been the locations with the highest LCOE, exceeding 160 €/MWh.

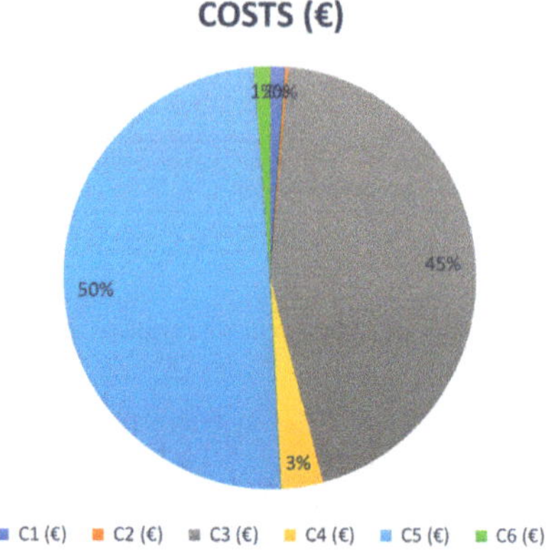


**Figure 9.** Life-cycle cost of São Jorge.

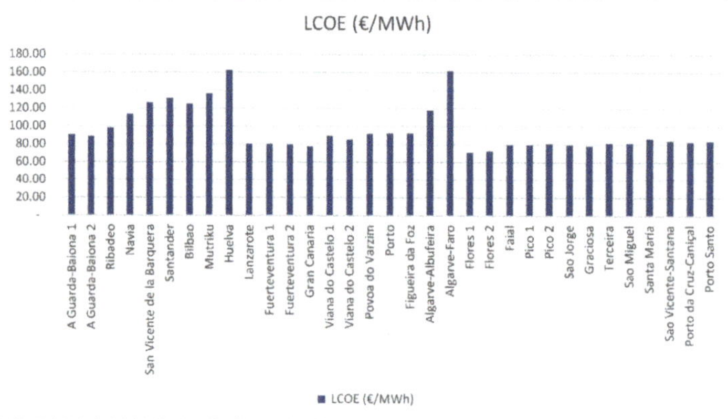


**Figure 10.** LCOE results.

The locations that have shown the best results have been the Canary Islands, Azores and the Madeira Islands. The lowest LCOE of all is Flores 1 (Azores) with 71.04 €/MWh. Given that the LCOE does not depend on the tariff, these results are maintained for all alternatives proposed.

Figure 11 shows how the IRR increases when using a higher electricity tariff. Alternative 2 (e_rate = 100 €/MWh) implies an increase in the IRR by approximately 0.1 points with respect to the values obtained in Alternative 1 (e_rate = 50 €/MWh). Alternative 3 (e_rate = 150 €/MWh) implies an increase in the IRR by approximately 0.07 points with respect to the values obtained in Alternative 2 (e_rate = 100 €/MWh). Comparing these results with the opportunity cost considered in this study (4.92%), Alternative 1 does not offer viability to any location because the results of IRR are lower than 4.92% of the capital cost considered. Alternative 2 would be accepted in terms of IRR for almost all locations, except some in Spain: Santander, Bilbao, Huelva, Mutriku. Projects that consider Alternative 3 give a higher return than the minimum required for all locations.

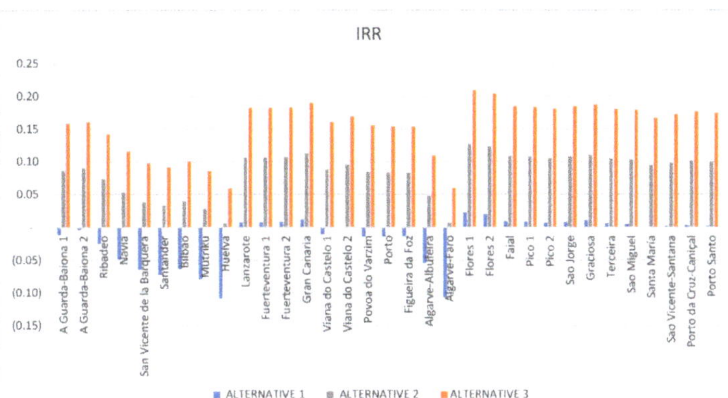


**Figure 11.** Internal rate of return (IRR) results.

For the pessimistic scenario, the IRR only reaches positive values in the locations mentioned above that obtained lower LCOE values.

The influence of the variation of the electricity rate on the NPV has also been taken into account in this study and for the locations analyzed. The farm will be economically feasible if the NPV is higher than zero. All the results obtained from NPV have been negative for the pessimistic scenario (see Figure 12). However, the increase in the rate considered for alternative 2 with respect to alternative 1 allows many locations to reach positive values: A Guarda-Baiona 1 and 2, Ribadeo, Canary Islands, Azores and Madeira Islands, etc.

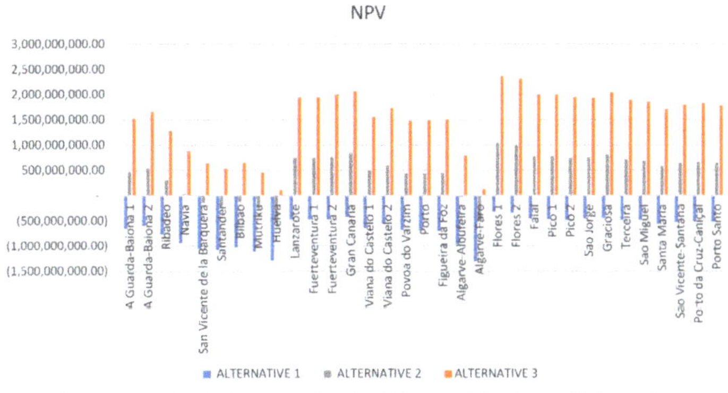


**Figure 12.** Net present value (NPV) results.

When a higher rate is considered for alternative 3 (e_rate = 150 €/MWh), NPV values increase significantly. All locations are positives now, and some of them exceed €2 billion, such as Gran Canaria (Canary Islands), Graciosa, Flores 1 and 2 (Azores).

## 6. Conclusions

This paper has analyzed economically a concrete offshore wind floating platform in the Atlantic Arc of Europe. The life-cycle cost of a concrete floating platform has been considered to calculate the main economic parameters. It is based on the life-cycle phases of the farm: Conception and definition, design and development, manufacturing, installation, exploitation and dismantling. Moreover, the economic parameters calculated are: Levelized cost of energy (LCOE), internal rate of return (IRR) and net present value (NPV). Their analysis determines the feasibility of the offshore farm.

The study considers the concrete floating offshore wind platform Telewind®, a spar platform with a 10 MW offshore wind generator.

In addition, the study analyzes thirteen locations in Spain (the Iberian Peninsula and the Canary Islands) and twenty in Portugal (the Iberian Peninsula, the Azores Islands, and the Madeira Islands).

The lowest LCOE of all is Flores 1 (Azores) with 71.04 €/MWh. This location is also the best IRR result for the pessimistic scenario (with an electric tariff of 50 €/MWh), giving 0.02%. In general, island locations have obtained better economic parameters. For alternative 1, NPV in Bilbao is 1,025,288,577 €. Nevertheless, Fuerteventura 2 (Canary Islands) and Porto da Cruz-Caniçal (Madeira) is 464,581,509 € and 497,607,426 € respectively.

Therefore, results show the most suitable economic area to install a floating offshore wind farm composed of concrete platforms in the South Atlantic area of Europe (Spain and Portugal) that are the Canary Islands for Spain and Flores for Portugal. These places can represent the future developments in offshore wind industry.

**Author Contributions:** Introduction: L.C.-S., A.F.-V.; Materials and methods: E.B.-S., D.C.-I.; Configuration of the offshore wind farm platform, TELWIND: L.C.-S., A.F.-V.; Study Locations: L.C.-S., A.F.-V.: Results: E.B.-S., D.C.-I., L.C.-S., A.F.-V.; Conclusions: E.B.-S., D.C.-I., L.C.-S, A.F.-V.; Supervision: L.C.-S.

**Funding:** This research was funded by EAPA_344/2016 ARCWIND project, co-financed by the European Regional Development Fund though the Interreg Atlantic Area Programme.

**Acknowledgments:** This work was performed in the scope of the ARCWIND project (EAPA_344/2016), co-financed by the European Regional Development Fund though the Interreg Atlantic Area Programme. The authors would like to thank the enterprise Esteyco for the information provided.

**Conflicts of Interest:** The authors declare no conflicts of interest

## References

1. Enkvist, P.-A.; Naucler, T.; Rosander, J. A cost curve for greenhouse gas reduction. *McKinsey Q.* **2007**, *1*, 35–45.
2. European Union. Acuerdo de París—Acción por el Clima.
3. European Union. Directive 2009/28/EC of the European Parliament and of the Council of 23 April 2009 on the Promotion of the Use of Energy from Renewable Sources and Amending and Subsequently Repealing Directives 2001/77/EC and 2003/30/EC (Text with EEA Relevance). *Off. J. Eur. Union* **2002**, *5*, 16–62.
4. Eurosat Statistics Explained Estadísticas de energía renovable 2018.
5. Esteban, M.; Leary, D. Current developments and future prospects of offshore wind and ocean energy. *Appl. Energy* **2012**, *90*, 128–136. [CrossRef]
6. Reimers, B.; Ozdirik, B.; Kaltschmitt, M. Greenhouse gas emissions from electricity generated by offshore wind farms. *Renew. Energy* **2014**, *72*, 428–438. [CrossRef]
7. Ladenburg, J. Visual impact assessment of offshore wind farms and prior experience. *Appl. Energy* **2009**, *86*, 380–387. [CrossRef]
8. Wieczorek, A.J.; Negro, S.O.; Harmsen, R.; Heimeriks, G.J.; Luo, L.; Hekkert, M.P. A review of the European offshore wind innovation system. *Renew. Sustain. Energy Rev.* **2013**, *26*, 294–306. [CrossRef]
9. Wind Europe. *Wind Energy in Europe: Scenarios For 2030*; Wind Europe: Brussels, Belgium, 2007.
10. Sun, X.; Huang, D.; Wu, G. The current state of offshore wind energy technology development. *Energy* **2012**, *41*, 298–312. [CrossRef]
11. Cermelli, C.; Roddier, D.; Aubault, A. WindFloat: A Floating Foundation for Offshore Wind Turbines—Part II: Hydrodynamics Analysis. In *Volume 4: Ocean Engineering; Ocean Renewable Energy; Ocean Space Utilization, Parts A and B*; ASME: Two Park Avenue, NY, USA, 2009; pp. 135–143.
12. Skaare, B. Development of the Hywind Concept. In *Volume 9: Offshore Geotechnics; Torgeir Moan Honoring Symposium*; ASME: Two Park Avenue, NY, USA, 2017; p. V009T12A050.
13. Driscoll, F.; Jonkman, J.; Robertson, A.; Sirnivas, S.; Skaare, B.; Nielsen, F.G. Validation of a FAST Model of the Statoil-hywind Demo Floating Wind Turbine. *Energy Procedia* **2016**, *94*, 3–19. [CrossRef]

14. Dankelmann, S.; Visser, B.; Gupta, N.; Serna, J.; Counago, B.; Urruchi, A.; Fernandez, J.L.; Cortes, C.; Garcia, R.G.; Jurado, A. *TELWIND–Integrated Telescopic Tower Combined with an Evolved Spar Floating Substructurefor Low-Cost Deep Water Offshore Wind and next Generation of 10 MW+ Wind Turbines*; Esteyco Energia: Barcelona, Spain, 2016.
15. Castro-Santos, L.; Diaz-Casas, V. Life-cycle cost analysis of floating offshore wind farms. *Renew. Energy* **2014**, *66*, 41–48.
16. Scheu, M.; Matha, D.; Hofmann, M.; Muskulus, M. Maintenance Strategies for Large Offshore Wind Farms. *Energy Procedia* **2012**, *24*, 281–288. [CrossRef]
17. Topham, E.; McMillan, D. Sustainable decommissioning of an offshore wind farm. *Renew. Energy* **2017**, *102*, 470–480. [CrossRef]
18. Suarez de Vivero, J.L.; Rodriguez Mateos, J.C. The Spanish approach to marine spatial planning. Marine Strategy Framework Directive vs. EU Integrated Maritime Policy. *Mar. Policy* **2012**, *36*, 18–27. [CrossRef]
19. Pacheco, A.; Gorbena, E.; Sequeira, C.; Jerez, S. An evaluation of offshore wind power production by floatable systems: A case study from SW Portugal. *Energy* **2017**, *131*, 239–250. [CrossRef]
20. Mytilinou, V.; Kolios, A.J. Techno-economic optimisation of offshore wind farms based on life cycle cost analysis on the UK. *Renew. Energy* **2019**, *132*, 439–454. [CrossRef]
21. Bak, C.; Zahle, F.; Bitsche, R.; Yde, A.; Henriksen, L.C.; Nata, A.; Hansen, M.H. *Description of the DTU 10 MW Reference Wind Turbine*; DTU Wind Energy Report-I-0092; DTU Library: Lyngby, Denmark, 2013; pp. 1–138.
22. Celik, A.N. A statistical analysis of wind power density based on the Weibull and Rayleigh models at the southern region of Turkey. *Renew. Energy* **2004**, *29*, 593–604. [CrossRef]
23. Myhr, A.; Bjerkseter, C.; Agotnes, A.; Nygaard, T.A. Levelised cost of energy for offshore floating wind turbines in a lifecycle perspective. *Renew. Energy* **2014**, *66*, 714–728. [CrossRef]
24. Esteyco. Available online: https://www.esteyco.com/ (accessed on 23 July 2019).
25. Castro, R. *Uma Introdução às Energias Renováveis: Eólica, Fotovoltaica, e Mini-Hídrica*; Instituto Superior Técnico: Lisbon, Portugal, 2011; Volume 13.
26. Yu, W.; Lemmer, F.; Schlipf, D.; Cheng, P.W.; Visser, B.; Links, H.; Gupta, N.; Dankemann, S.; Counago, B.; Serna, J. Evaluation of control methods for floating offshore wind turbines. *J. Phys. Conf. Ser.* **2018**, *1104*. [CrossRef]
27. Armesto, J.A.; Jurado, A.; Guanche, R.; Counago, B.; Urbano, J.; Serna, J. Telwind: Numerical analysis of a floating wind turbine supported by a two bodies platform. In Proceedings of the International Conference on Offshore Mechanics and Arctic Engineering (OMAE), Madrid, Spain, 17–22 June 2018; Volume 10.
28. Navantia. Available online: www.navantia.es (accessed on 22 July 2019).

International Journal of
*Environmental Research and Public Health*

MDPI

*Article*

# Evaluation of Formal and Informal Spatial Coastal Area Planning Process in Baltic Sea Region

**Edgars Pudzis [1], Sanda Geipele [1,*], Armands Auzins [1], Andrejs Lazdins [1], Jevgenija Butnicka [2], Krista Krumina [2], Indra Ciuksa [2], Maris Kalinka [1], Una Krutova [1], Mark Grimitliht [3], Marii Prii-Pärn [3], Charlotta Björklund [4], Susanne Vävare [4], Johanna Hagström [5], Ingela Granqvist [6] and Malin Josefina Hallor [6]**

1 Institute of the Civil Engineering and Real Estate Economics, Riga Technical University, 1048 Riga, Latvia; edgars.pudzis@rtu.lv (E.P.); armands.auzins@rtu.lv (A.A.); andrejs.lazdins@rtu.lv (A.L.); maris.kalinka@rtu.lv (M.K.); una.krutova@rtu.lv (U.K.)
2 The Ministry of Environmental Protection and Regional Development, 1494 Riga, Latvia; jevgenija.butnicka@varam.gov.lv (J.B.); krista.krumina@varam.gov.lv (K.K.); indra.ciuksa@varam.gov.lv (I.C.)
3 Saaremaa Municipality, 93819 Kuressaare, Estonia; mark.grimitliht@saaremaavald.ee (M.G.); marii.prii-parn@saaremaavald.ee (M.P.-P.)
4 The Government of Åland, 22 100 Mariehamn, Finland; charlotta.bjorklund@regeringen.ax (C.B.); susanne.vavare@regeringen.ax (S.V.)
5 Mariehamn Municipality, 22 101 Mariehamn, Finland; johanna.hagstrom@mariehamn.ax
6 Norrköping Municipality, 601 81 Norrköping, Sweden; ingela.granqvist@norrkoping.se (I.G.); malin.hallor@norrkoping.se (M.J.H.)
* Correspondence: sanda.geipele@rtu.lv

**Citation:** Pudzis, E.; Geipele, S.; Auzins, A.; Lazdins, A.; Butnicka, J.; Krumina, K.; Ciuksa, I.; Kalinka, M.; Krutova, U.; Grimitliht, M.; et al. Evaluation of Formal and Informal Spatial Coastal Area Planning Process in Baltic Sea Region. *Int. J. Environ. Res. Public Health* **2021**, *18*, 4895. https://doi.org/10.3390/ijerph18094895

Academic Editor: M. Dolores Esteban

Received: 16 March 2021
Accepted: 28 April 2021
Published: 4 May 2021

**Publisher's Note:** MDPI stays neutral with regard to jurisdictional claims in published maps and institutional affiliations.



**Abstract:** Many shared views of both scholars and practitioners reflect spatial planning as a place-creating process that must be understood from a multi-level perspective. Formal and informal planning modes have variations in planning practices in different countries. In this study, we aimed to evaluate the interaction of formal and informal spatial planning in the frame of the spatial planning system in the Baltic Sea region. We were searching to highlight the involvement possibilities of territorial communities in the spatial planning process around the Baltic Sea region, focusing on coastal areas and their specific features in Latvia, Estonia, the Åland Islands of Finland, and Sweden. Involved experts expressed views based on a pre-developed model to identify how institutionalized formal spatial planning relates with informal interventions. This allowed the development and proposal of a model for coastal area spatial planning and implementation. We concluded that in the spatial planning approach, the governance works differently in different countries, and coastal area spatial planning differs from regular spatial planning. The information base is sufficient to initiate spatial planning at the municipal level, but municipalities should be more active, involving territorial communities in the planning, implementation, and control of municipal spatial planning, as this ensures a greater interest in the use of planning outcome.

**Keywords:** formal planning; informal planning; spatial planning process; coastal area spatial planning; planning levels; community involvement; territorial community; coastal communities

## 1. Introduction

With increasing urbanization, particularly in the coastal areas of the Baltic Sea region, there are several sustainable development problems identified, e.g., environmental pollution, community living, and preservation of cultural and natural heritage. A comprehensive and local needs-based planning approach to the sustainable development of maritime and coastal areas in the Baltic Sea region is relevant and consequently better for decision-making by local and regional governments [1]. Therefore, the formal and informal spatial planning process of the coastal zone needs to be evaluated. To understand the importance and interrelationship of formal and informal planning, it is necessary to assess the stages of mobilization, planning, implementation, and monitoring (analysis structure).

An increasing proportion of the world's population is based in coastal cities because of the beneficial circumstances and services they provide [2]. The Baltic Sea coast is attractive in many aspects. Interests from different sectors all have claims on the coastal areas resulting in major pressures on valuable marine habitats, natural resources, and ecosystem services. There is a joint requirement for improved management strategies to work with spatial planning in a more holistic, sustainable, and efficient way. In the frame of the Interreg Central Baltic project Coast4us in all participating countries (Latvia, Estonia, the Åland Islands of Finland, and Sweden), the problem is that many relevant interests, e.g., environmental, economic, social, or cultural, are not sufficiently recognized in the planning process. This non-holistic approach results in actions that are less viable and are less adapted to local interests, and thus they become less cost-efficient. The idea of the project Coast4us is therefore to work with a holistic approach in the planning process, involving stakeholders of different interests in a participative way, and create sustainable marine and coastal zone plans. The main objective of the project Coast4us is to create sustainable plans for marine and coastal areas [1].

To achieve faster territorial growth and to provide positive changes in regions and local municipalities, it is essential to involve wider society, including entrepreneurs, local communities, and citizen groups, in addressing territorial development issues. A formal planning approach refers to the provision of a formal planning process which is regulated by established institutional settings and implemented through a set hierarchy of competences, e.g., planning and governance levels. Formal planning also provides formal planning tools, e.g., a long-term comprehensive plan for the local government territory (municipality). In the context of planning, Syssner and Meijer [3] assume that formal planning processes contain elements of informality. The concept of formal planning can be used to refer to the kind of planning that is government-led and shaped mainly by formal structures and through formal negotiation. However, beyond the formal planning system, we may recognize some attempts of citizens to "adapt space" according to their needs.

Informal planning approaches, such as community planning and participatory budgeting, might help local governance better understand the needs of local society, especially in large municipalities also addressing the needs of peripheral areas. It might help to provide the best tailored solutions for different communities/areas. In addition, it might improve better societal trust in public governance. Furthermore, an informal planning approach might provide an opportunity for the society/active citizens to present their needs to the local governance, providing more fruitful dialogue with local governance and engaging in a public decision-making process. It might help them to better contribute to the development of living environment attracting necessary support from local governments and other parties, promote local patriotism and a sense of belonging to their place of life, as well as to improve their knowledge about the development planning and implementation processes. Local initiatives are of great importance as often the local community might address the challenges in the most efficient way. These assumptions have been discussed among participants of the project Coast4us during organized networking and workshops. However, it is necessary to explore how formal and informal spatial planning processes in the specific coastal conditions support the development and implementation of long-term strategies and plans, as well as how these processes interact with and influence the sustainability of local communities.

The scientific literature on the territorial community emphasized five core elements. First, locus, a sense of place, referred to a geographic entity ranging from neighbourhood to city size, or a particular milieu around which people gathered, such as, a church or recreation centre. Second, sharing, common interests and perspectives, referred to common interests and values that could cross geographic boundaries. Third, joint action, a sense of coherence and identity, included informal common activities such as sharing tasks and helping neighbours, but these were not necessarily intentionally designed to create community cohesion. Fourth, social ties involved relationships that created the ongoing sense of cohesion. Finally, a diversity referred not primarily to ethnic groupings, but to the social complexity within communities in which a multiplicity of communities co-existed [4]. We may see the territorial community as the common territory of existence, the presence of common interests of local importance, social interaction of community members in the process of ensuring these interests, psychological self-identification of each member with the community, common communal property, and payment of utility taxes [2]. Therefore, in the present study, the territorial community in Latvia is a village, in Estonia, an island, in Sweden, a village, camping site, and unexplored island, and in the Åland Islands of Finland, urban and rural areas [1].

The experts of participating countries in the project Coast4us addressed their views about the spatial planning system and process in their territorial community. **In Latvia**, the villagers have a better understanding of the situation on the ground and a better understanding of the needs of the inhabitants of the village than the people from outside of the village. The involvement of the population allows for better control of the result to be achieved. The villagers better maintain their objects because they are more interested. **In Estonia,** continuously involving and engaging different social groups in the early visioning and planning process should be a binding part of every development. **In Sweden**, it is important to listen to local experts and actors on site. Therefore, it is necessary to involve and engage them in an early stage for development. **In Åland**, it is important to get local people involved in an early stage in the planning process, to ensure people understand what needs to be implemented. The purpose of the whole process should also be described based on sustainability principles, because planning and a future sustainable society with a working green infrastructure must go hand in hand. We need to prevent the effects of climate change and we need to increase and strengthen biodiversity because it gives society a greater resilience.

The **hypothesis** of this study is that spatial planning systems in the Baltic Sea region are in the process of changing in order to involve society in the spatial development of their territory.

For this study, several **research questions** were developed. First, how can the spatial development planning process improve the life of local communities in the specific coastal conditions? Second, what is the significance of both formal and informal planning approaches, and the impact of their interaction? Third, what are the possibilities to involve in the coastal area spatial planning process for the population groups in territorial communities?

In order to answer the research questions, a multi-element study is needed, which includes: (1) theoretical aspects on the features of formal and informal planning; (2) an overview of special needs of coastal communities, comparison, and evaluation of the spatial planning process in specific coastal areas (considering that one coastal area consists of different countries with different planning systems); and, (3) the impact assessment of the stages of the spatial planning on meeting the special needs of the coast.

The multi-element study is being developed for the Baltic Sea coast, based on information from national experts, as well as the study of specific pilot areas around the Baltic Sea.

**The pilot areas of Latvia** are two small village communities from two different coastal local municipalities: (1) Tuja village in Salacgriva municipality, and (2) Garupe village in Carnikava municipality. **The pilot area in Estonia** is the largest island in Estonia-Saaremaa. This is the main island of Saare county and it belongs to the West Estonian Archipelago. Since the end of 2017, based on all twelve former municipalities, Saaremaa has been governed as one municipality. **The pilot areas on mainland Åland** are: (1) an urban area in the city of Mariehamn, and (2) a rural area in Sund municipality. **The pilot areas in Sweden** are from three different coastal local municipalities: (1) a small village in Arkösund/Norrköping municipality, (2) a camping site in Ekön/Valdemarsvik municipality, and (3) the unexplored island Bergön in Söderköping municipality.

**The purpose of the research** is to evaluate the interaction of formal and informal spatial planning processes in the coastal area through comparative study, considering the hierarchies of spatial planning systems in selected and differently experienced country cases around the Baltic Sea region. Accordingly, **the objectives** are set as follows: (1) to describe the theoretical aspects of the formal and informal planning; (2) to compare and evaluate the spatial planning process of Latvia, Estonia, the Åland Islands of Finland, and Sweden by levels of the spatial planning system; (3) to conduct an expert survey on the impact of formal and informal planning processes by using the stages of cooperation, i.e., mobilisation, planning, implementation, and monitoring; and (4) to propose a model supporting coastal area spatial planning and implementation for community involvement.

The research uses: (1) a literature review method for an overview of theoretical aspects of the formal and informal planning and a comparative evaluation of the planning process; (2) an expert assessment method employed by the experts from Latvia, Estonia, Finland, and Sweden for a description of the situation, according to the objectives of the project and the study, as well as to conduct an expert survey; and (3) discourse analysis and synthesis as well as graphical methods for designing main research results, including a proposed model. The collection and analysis of secondary and primary data are performed by the authors of the study. The expert survey in (2) is based on the methodology previously developed by researchers at Riga Technical University [5].

## 2. Theoretical Background

The theoretical aspects of formal and informal spatial planning are closely related to settlements, i.e., villages and towns. Historically, developed cities and other places are the largest complex adaptive systems in human culture and have always been changing over time according to largely unplanned patterns of development [6]. To a large extent, the development of processes is shaped by the potential for co-optation. Erikson [7] outlined four characteristics of collaboration: (1) the bonding between the parties, incorporating the user representatives in the organizations and their institutional logic; (2) the organizational framing of the user involvement activities, setting the initial rule for how to act/speak, where to act/speak, when to act/speak, as well as what to speak about; (3) the organizational control exercised as the activities took place, directing the discussions and interactions to align with the interests of the welfare organizations; and (4) the resistance exercised by user representatives, enabling them to influence the organizations and contribute to change. Successful collaboration is based on available resources. Two resource groups are important in the planning process, namely, human resources and organizational resources. Their significance primarily consists of two facets: first, that they utilize existing knowledge; and second, that they create legitimacy and feelings of pride and belonging in the local community [3,8]. Collaboration can take place in a formal, semi-formal, and informal way, which means that spatial planning also has these forms of collaboration. From the best-practice perspective, decision-making in spatial planning must be decentralized, and the tools of spatial planning must be less binding (which has been broadly practised in Switzerland, for instance) [9].

The idea of the informal organization was first introduced by Barnard [10]. He compared the informal organization to a clique or an exclusive group of people that naturally forms over time. According to Barnard, this can be achieved by linking Maslow A's theory of needs and central ideas to new concepts of leadership [11]. Many scholars have studied the involvement of informal social groups in the planning processes, including spatial planning. For example, Certoma [12] defined five tools, evidence, knowledge, encouragement, evaluation, and assistance, which are related to the aim of the study. However, we must first distinguish between spatial planning systems and planning practices, the latter of which reflects the planning culture. Reimer, et al. [13], interpreted planning systems as "dynamic institutional technologies, which define corridors of action for planning practice, which may, however, nonetheless display a good deal of variability". Fürst [14] equated the planning culture to the values, attitudes, mind-sets, and routines shared by those taking part in the planning process. Reimer, et al. [13] provided some arguments that planning practices inherent to the system cannot be drawn only from a comparison of legal–administrative framework conditions.

Comprehensive literature analysis allows for identifying the characteristics of informal spatial planning. Based on such an analysis, Mishra [9] formulated that informal planning should: facilitate the formal process; adds flexibility; contributes with matured results by discourses; is used as an ad hoc system when required; implies a degree of innovation, continuously meeting the planning challenges; is a non-traditional planning mode not influenced by the hierarchy culture of planning; is a multi-level collaboration process-oriented to consensus in decision-making; requires governance to avoid potential conflict and progress towards a legitimate solution; requires the inclusion of different stakeholders' unbiased results or planning direction; and vitally needs transparency as it keeps the stakeholders of different levels well informed and avoids potential conflicts.

Following a "socio-institutionalism perspective", this article refers to the assumptions guiding spatial planning and development. Therefore, the legal and administrative structures and competencies that shape possible spatial development or changes in land use (formal institutional settings) are introduced first to search for governance relations in planning and implementation processes. Institutions are established to organize a spatial planning process and provide measures for public involvement. Thus, the institutional performance refers to administrative structures, policy styles, institutional and social settings, collective actions, and social learning [15]. The planning process is concerned with deliberative plan-making, applying various planning modes of different scales, e.g., national, regional, local, and more detailed. However, the planning process may differ in terms of the extent of public involvement. This may be assessed, for instance, by analyzing the activities of stakeholders' deliberation and informal population groups.

Formal and informal (complementary) spatial planning tools provide the necessary support to improve planning practices, but positively-influenced practices substantiate discourses (e.g., desirable dominating ideas) in spatial planning [15]. Informal planning tools often are developed as a result of using project-oriented techniques and integrated assessment tools, e.g., nature protection plans, management plans of water bodies, or assessments of ecosystem services. Relevant processes, e.g., formal and informal spatial planning, local development, and protection of valuable landscapes and related consequent decision-making, strive for collaborative learning by understanding the values of land-related resources and their most efficient usage. Auzins [15] provided main objectives for introducing a values-led planning approach as it promotes improved, more supportive, and collaborative territorial governance as well as informal institutions and organisational forms, as they significantly support formal spatial planning and social settings driven by common and local place-based interests.

At the same time, it is clear that spatial planning approaches and governance work differently in different countries. Indeed, even in Europe you can find varieties of approaches starting from integrated, regional-economic to just land-use management approaches. Coastal area spatial planning differs from regular spatial planning because it is connected with specific water objects (Baltic Sea in the case of this study). The water object is placed on territories of different countries with different legislation, history, and governance. With these circumstances, it is important to analyze planning conditions, and share knowledge to, from one perspective, sustainably use and develop a common resource (Baltic Sea in case of this study), and from another perspective, make equal opportunities for the development of coastal communities, regardless of political circumstances.

Spatial planning, implementation, and monitoring need to have a certain order to ensure common processes and results [3]. At the same time, for instance, the latest research in village and community planning in Latvia proposed that the methods of involvement must be as diverse as possible [16]. This clearly indicates that, as a result of societal growth, the planning process must be changeable and modern, which forces the research community to constantly look for new and appropriate opportunities for formal and informal societal involvement and motivation.

Community involvement has been shown to make a positive contribution to planning and development processes. At its best, community involvement can enable the following: processes to be sped up; resources to be used more effectively; product quality and feelings of local ownership to improve; added value to emerge; confidence and skills to increase for all; and conflicts to be more readily resolved. Public participation should be an indispensable element in human settlements, especially in planning strategies and in their formulation, implementation, and management. It should influence all levels of government in the decision-making process to further the political and economic growth of human settlements [17].

Recently, Geipele, et al. [18] highlighted several specific spatial planning problems of coastal communities in the Baltic Sea region by evaluating community involvement in participatory processes: (a) lack of communication, regardless of country and region; (b) the weak involvement of different social groups; (c) insufficient coastal and environmental management; (d) excessive regulatory enactments; and (e) summer-year-round population conflict. These problems must be taken into account, but have not been analysed in this study.

## 3. Materials and Methods

In general, qualitative research methodology (descriptive, logical, comparative expert assessment) and quantitative research methodology (factual comparison methods) are used for the study.

The experts, who have been involved for the development of the study, represent the local (municipal) and governmental authorities as well as scientific sectors to assess the formal and informal territorial coastal spatial planning process in Latvia, Estonia, the Finnish Åland Islands, and Sweden. An expert group consisting of four subgroups and representing particular countries has been established for the study: (1) in Latvia, three scientific experts and three experts from national-level authority; (2) in Estonia, two experts from municipal authority; (3) in the Åland Islands of Finland, two experts from the government and one expert from municipal authority; and in (4) Sweden, two experts from local, municipal authorities. The size of the expert subgroups was determined by the number of participants involved in the project (determined by the team leader).

For the overview, analysis, and discussion of spatial planning systems, the experts explored particular institutional settings, legal instruments, and policy planning documentation. The results are summarized in Table A1 (see Appendix A). The expert survey was conducted to assess the impact of formal and informal spatial planning on community development by using four stages of cooperation. From the preliminary networking in the frame of the Coast4us project, it was acknowledged that selected experts were professionally competent enough to deliver knowledge that is accumulated from formal and informal population groups of communities, and thus, represents the dominating opinion of local communities in specific pilot areas of the coast. Therefore, they gave some discursive influence on research, as they were largely in charge of relevant spatial planning and implementation processes. The designed structured expert survey consisted of two parts to particularly analyze the significance of both formal and informal planning. Each part included 12 questions to the respondents. The responses to these questions allowed scoring from 1 to 10 (from low significance to high significance). The average rating data demonstrated some interpreted trend and ground synergistic planning models. The Section 6 of the study displays summarized results and proposes the model for cooperation as a synthesized outcome.

## 4. Pilot Areas for Research

Figure 1 shows marked pilot areas of this study in the Baltic Sea region.

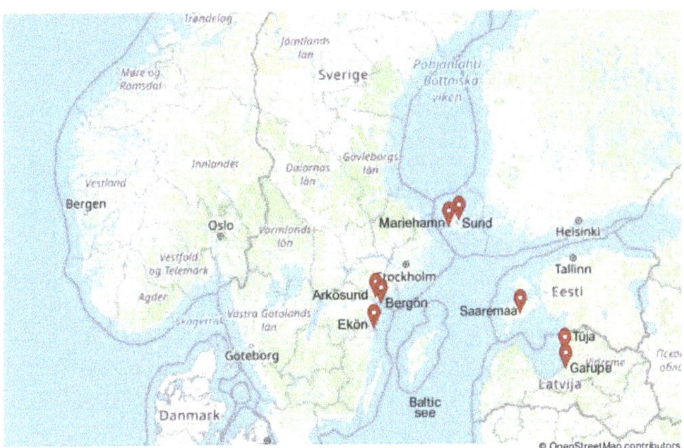


**Figure 1.** Pilot areas of the present study on the Baltic Sea region map.

### *4.1. The Case Study of Latvia: Description of the Villages*

Garupe village is in the Piejura lowland, Carnikava municipality (Latvia) about 4 km to the southwest from the administrative centre of Carnikava, and 26 km from the centre of Riga. The village is located in the surrounding area of capital Riga. The environmental conditions in Garupe are largely determined by the fact that the village is in the territory of the Eimura-Mangali polder. The spatial structure of Garupe village is characterised by dense summer house/dwelling house construction, a drain system, and a dense street network. Garupe is a residential village. The population of Garupe is gradually transforming from a seasonal to a permanent population. On average, 405 people live in Garupe. During the summer, the total population reaches over 1000 people. People work in nearby areas, especially in Riga. Experts concluded that Garupe is a typical urban extension of the Riga city and the local community is forming at its stage of development.

Tuja village is a populated place in Liepupe parish, Salacgriva municipality, Latvia. It is located on the coast of the Gulf of Riga, on the banks of Zakupite, 33 km from the centre of Salacgriva district and 75 km from Riga. On average, 276 people live in Tuja. In Tuja, the territory of detached houses occupies 44%, recreation area—11%, territory of garden plots—2%, and mixed business area—5% (approximate distribution). There are two campsites and a beach in Tuja. There is a long fishing tradition and public services are available.

### *4.2. The Case Study of Estonia: Description of the Island*

Saaremaa municipality is located on the largest island in Estonia-Saaremaa. It is located in the Baltic Sea, south of Hiiumaa island and west of Muhu island, and belongs to the West Estonian Archipelago. It is the fourth largest island in the Baltic Sea. Its territory is 2718 km$^2$ and the shoreline is 874 km long. There are 4 large islands in Saaremaa municipality: Saaremaa, Abruka, Vilsandi, and Kõinastu and a number of smaller islands. In Saaremaa municipality, there is 1 town, Kuressaare, 9 boroughs, and 427 villages. There are 31,466 inhabitants (as of 1 April 2019), and the population is descending and ageing. Inhabitants work mostly in electronics, boat building, food, or the plastic industry, but also in agriculture, fishing, and tourism. Many people are self-employed, such as little holiday homeowners, craftsmen, small producers or small farmers, and many people do telework.

Because of its mild maritime climate and a variety of soils, Saaremaa has a rich flora, including rare orchid species, but also a wide variety of rare wildlife species, ranging from insects to seals. There are also a number of semi-natural communities (wooded meadows, alvars, floodplain meadows, and coastal meadows). All this biological richness is the reason why almost 18% of the Saaremaa community is under nature conservation.

Beautiful nature, interesting geology, safe environment, unique cultural heritage, and many created facilities (spas, hotels, tourism farms) are the reasons why Saaremaa is a quite popular tourism destination. In addition, it is common to own a summer house there, to be used seasonally. Saaremaa can be visited by ferry or by plane, but there are also many small marinas for yachts or other small boats.

The aforementioned components need a comprehensive understanding and vision. In the project, the main aim is to generate valuable input material for the Saaremaa municipality, including a comprehensive plan regarding coastal area region.

### 4.3. The Case Study of the Åland Islands of Finland: Description of the City and Rural Area

Mariehamn city is the capital of the autonomous Åland Islands, where about 11,000 people (of the total 30,000 Åland population) live. Here, one may find shops, cafés, camping grounds, public beaches, green areas, hotels, conference centres, marinas, and football fields. Mariehamn city also has a big harbour and is an important tourist destination during the summer months. Mariehamn is situated in the south of the mainland on a peninsula, almost completely surrounded by water. It is important to make sure that the activities on land do not harm nature on land or at sea. Therefore, in the project, the aim is to try to incorporate the importance of a functioning green infrastructure in the planning of the city.

Sund municipality is an area where the agriculture dominates, and the challenge is to bring back natural biodiversity and stop nutrient run-off from land to sea.

### 4.4. The Case Study of Sweden: Description of the Village, Camping Site, and Unexplored Island

Arkösund is in the eastern part of the Vikbolandet peninsula, 50 km from Norrköping city and 200 km south from Stockholm. The district is located in the Östergötland region territory.

Östergötland's northern archipelago, with the bay of Bråviken, shores along the Vikbolandet peninsula, and the skerries and islands outside, is called Arkösund Archipelago. The Gränsö-Birkö-Aspö-Arkö island chain outside the channel by Arkösund is composed of larger islands with a fair amount of settlement, both in the form of permanent residences and weekend cottages. The large shallow areas here have proven to have great biodiversity and different types of environments. That is one reason that the county's largest nature reserve, Bråviken Nature Reserve, with more than 9000 hectares of protected land and water, has been accorded the status of a marine nature reserve. Many people who visit Arkösund come by boat since the north–south shipping line from Stockholm passes right by here. Arkösund is a natural junction. The community itself is an old fishing village and a seaside resort. At the end of the 19th century and beginning of the 20th, Arkösund was a summer paradise for wealthy Norrköping residents, which is abundantly clear from the many large villas. A special "resort train" ran on a narrow-gauge railway to the area. There were picturesque small walking paths with footbridges out to the islets, the so-called "Bathing Islands". During the period from March to October, it is much frequented and a centre of activities. Here, one may find marinas, a boatyard, a petrol station, shops, camping grounds, conference centres, pubs, and hotels with entertainment. Although the activities vary from year to year, Arkösund is a place where there is always life and movement.

## 5. Assessment of Different Spatial Planning Levels, Including Formal and Informal Spatial Planning in Latvia, Estonia, Åland Islands of Finland, and Sweden

### 5.1. Overview of Spatial Planning in Latvia

In Latvia, the overall development planning system is determined by the Development Planning System Law, while the Spatial Development Planning Law focuses on spatial development planning at all governance levels. To pay more attention to the involvement of various groups of the inhabitants in the development planning, regulation on the Procedure for Involvement of the General Public in the Development Planning Process was adopted in 2008 [19]. The purpose of the regulation is to foster an effective, open, inclusive, timely, and responsible involvement of the general public into the development planning process, thus increasing the quality of the planning process and ensuring the compliance of planning results with the needs and interests of the population. The regulations on municipal spatial development planning documents and on planning region spatial development planning documents stipulate the involvement of the public in the preparation of spatial development plans [20,21].

**National level**. According to the legislation, at the national level, the following spatial development planning documents shall be elaborated: the Sustainable Development Strategy for Latvia and the National Development Plan. The Sustainable Development Strategy for Latvia is hierarchically the highest national level long-term planning document, which sets out the long-term development priorities and the spatial development perspective of the state. It is followed by the medium-term planning document, the National Development Plan, which defines priorities for sectoral policies, territory development and tasks and activities to be implemented thereof, as well as sources of financing. [22].

Following the Sustainable Development Strategy and National Development Plan, sectoral policy development planning documents shall be elaborated for a medium-term. The Regional Policy Guidelines 2021–2027 is the main medium-term policy planning document, which determines the regional policy of Latvia and elaborates on the National Development Plan. It provides directions of action and tasks in the field of regional development [23]. Development of a maritime spatial plan and national long-term thematic plan for public infrastructure development in the coastal area was introduced in 2014 [24].

**Regional level**. At the regional level, the sustainable development strategy and the development program of a planning region (for each of the 5 planning regions in Latvia) shall be elaborated. The sustainable development strategy of a planning region is a long-term spatial development planning document, specifying the vision of the long-term development, strategic objectives, priorities of the planning region, and the spatial development perspective in written and graphic form. The development program of a planning region (medium-term) shall contain the current situation analysis, tendencies, and forecasts, as well as information regarding the developing process of the development program, and shall define mid-term priorities, the set of measures for the implementation thereof and the procedures for monitoring thereof [25].

**Local (municipal) level**. At the local governmental level, the following planning documents shall be elaborated: the sustainable development strategy, the development program, the spatial development plan (a comprehensive plan), a local plan (for particular areas), and a detailed plan (for specific development areas/sites). The sustainable development strategy of a local government defines the vision of the local government's long-term development, strategic objectives, development priorities, and the spatial development perspective in written and graphic form.

A local government development program (medium-term) shall include the analysis of the current situation, tendencies and forecasts, as well as information regarding the developing process of the development program, and shall define mid-term priorities, the action and investment plan, the number of resources necessary for the implementation of the development program, and particular procedures for monitoring of the development program. A local government development program, which is developed in an integrated manner, is a prerequisite to attract support from the EU funds.

The spatial development plan of a local government may be detailed in a local plan. After the sustainable development strategy comes into effect in a local government, the spatial development plan may be amended in the local plan, insofar as the local plan is not in contradiction with the sustainable development strategy of the local government [26].

**Village (community) level**. The formal development planning system does not provide a framework for planning at the village/community level in Latvia. It is a voluntary-based process, which can be initiated by informal groups of citizens. Although there are several examples of elaborated village plans, their implementation so far was not successful, due to the lack of support, including from public governance. Nevertheless, to attract the funding for local initiatives from the EU Agriculture Fund for rural development (LEADER approach), local action groups (LAG) shall elaborate local action plans, which define and justify priorities and necessary changes in the territory covered by the LAGs. This process is regulated by the regulations of EU funds attraction.

### 5.2. Overview of Spatial Planning in Estonia

**National level**. The hierarchically highest national level long-term planning document is the national spatial plan Estonia 2030+. The national spatial plan is a strategic schematic document aiming to achieve the expedient utilisation of space on the scale of Estonia as a whole. The national spatial plan is being prepared for the entire territory of the country. It defines the policies and trends for sustainable and balanced national spatial development. The purpose of the plan is to obtain spatial bases, informed by the specific character of the environment, for shaping settlement, mobility, national engineering infrastructure, and regional development [27].

In 2017 the government of Estonia initiated a thematic national spatial plan for maritime areas. The Estonian maritime spatial plan is a tool for the long-term planning of the use of the sea, which balances the social, economic, cultural, and environmental interests and needs. Maritime spatial planning enables the determination of where and under what conditions the implementation of different human activities in the marine area is most reasonable. This is to ensure the economic benefits resulting from the exploitation of marine resources, as well as the value of the sea and coastal areas as socially and culturally important areas, keeping in mind that any human activity must be based on the achievement or maintenance of the good status of the marine environment [27,28].

**Regional level**. Regional development planning is directed by the Regional Development Strategy 2014–2020. The key place in the regional development in Estonia is held by the development of centres and making better use of regional differences [29]. Following the Regional Development Strategy, the state is working to ensure consistent growth in all areas, applying the unique potential available due to each area's peculiarities. The Estonian government ratified this strategy and its implementation plan for 2014–2020 in 2014. The strategy focuses on the development needs of all Estonia's regions. The government is investing more than previously in the improvement of work availability and services in areas, which have been adversely affected by urbanization, by emphasizing the strengths and unique aspects of each region.

Main strategic goals are divided into four major groups:

- An environment for households and enterprises in the active regions, which supports their wholeness and competitiveness. To shape a balance regarding the draw of larger urban centres with stronger active regions across Estonia, having improved environments for living and entrepreneurship as well as diverse work service, and activity opportunities.
- An environment in major cities that promotes competitiveness in the international economy. To increase the importance of urban areas as centres of growth for an innovative and science-intensive economy with the help of an increasingly attractive living environment.
- Exploiting region-specific resources with greater skill. This promotes specialization in growing areas of competence and enterprise according to region-specific conditions, and increases clarity in the uniqueness of different areas.
- Greater connectedness and ability to grow. For regions to achieve a stronger ability to develop by greater inter-regional connectivity and increasing efficiency in regional cooperation and capacity for growth.

Regional (national and local as well) development planning is regulated by the Planning Act (PLA) [30]. PLA aims to create, through spatial planning, by promoting environmentally sound and economically, culturally, and socially sustainable development. These are the preconditions that are necessary for democratic, long-term, and balanced spatial development that takes into account the needs and interests of all members of the Estonian society to occur.

For each county, a regional development plan is implemented [31]. A county-wide spatial plan aims to define the principles and directions of the spatial development of the entire county, or a part thereof, or another region. A county-wide spatial plan is prepared primarily to express cross-border interests and to balance national and local needs and interests regarding spatial development. County-wide spatial plans are the basis for the preparation of comprehensive plans.

For example, Saaremaa County Plan 2030+ was implemented by the Minister of Public Administration in 2018. The main purpose of it is to balance the national and local interests, consider the local situation as well as to support the county's spatial development, and ensure balanced and sustainable settlement structure and quality of life in the situation where the population of the county is shrinking and ageing [31].

**Local (municipal) level**. The functions and competence of a local authority include the organisation of spatial planning in the rural municipality or city. The instrument that directs spatial strategic planning and spatial development in the local municipality is a comprehensive plan. It is mandatory for every municipality. A comprehensive plan aims to define the principles of and directions in the spatial development of the entire territory of a rural municipality or city or a part of such territory.

The PLA sets some functions of a comprehensive plan, such as: to specify the conditions directing the development of human settlement; to define the boundaries of areas of repeated flooding on the coastline; to set the high water marks of internal bodies of water with an extensive flooding area; to specify the conditions to ensure the functioning of the green network and to determine the restrictions resulting from such network; to state the conditions of public access to shore paths; and to extend or reduce the building exclusion zone of the shore or bank. The functions to be fulfilled by a comprehensive plan are decided following the spatial needs of the local authority and the purpose of the plan. For example, Saaremaa municipality currently has 22 comprehensive plans in force. In 2018, Saaremaa's local council initiated by a resolution the preparation of a new comprehensive plan of the entire territory of the new municipality [32].

To implement the comprehensive plan and to create an inclusive spatial solution for the planning area, a detailed spatial plan is prepared to plan construction works. The detailed spatial plan grounds initialization of construction works.

**Village (community) level**. In Estonia, there is no formal village and community planning stage or practice. The level of a municipality (local government level) involves village planning. The comprehensive planning process gives possibilities to plan and work on different levels, including village visions. This is a good way to connect visions of different levels. Formally possible ways may not be the best solution, because informal ways may be more attractive and efficient for locals to negotiate and agree on village visions.

### 5.3. *Overview of Spatial Planning in the Åland Islands of Finland*

The Åland Islands have an autonomous governance model, which means that they have their own legal rules on some issues and all planning processes are formal.

**National level**. Åland has a law called "Plan-och bygglagen" [33] (the Planning and Building Act). The purpose of this law is to regulate land use and construction, so that: the conditions for a good living environment are created and preserved; an ecologically, economically, socially, and culturally sustainable development is promoted; and cultural-historical values are preserved.

With planning and building permits matters, the following must be observed: the provisions of the Act on nature conservation; the Act on forest management; the Act on environmental impact assessment and environmental assessment; the Act on environmental protection; the Act on the protection of culturally valuable historical buildings; the Act on ancient monuments; the Act on the protection of the maritime cultural heritage; the Act on the application to the Åland Health Protection; and the Water Act for the Åland.

**Local (municipal) level**. It is a municipal matter following the Planning and Building Act to decide the planning of the use of land and water. Each municipality must have a current municipal overview that covers the entire territory of the municipality. The municipal overview shall indicate the direction for the long-term development of the physical environment and guide decisions on the use of land and water areas and how these will be changed and preserved.

The regulation on land use and development within the municipality is provided through general and detail plans. The general plan sets out the main features of land use in the entire municipality or part of it. The detailed plan specifies how a limited area of land in the municipality should be used and built.

When the draft plan is drawn up, Åland municipality and other bodies, legal entities, and persons affected by the planning process are able to consult and express themselves in writing or orally. The purpose of the hearing is to improve the decision basis and provide opportunities for transparency and influence. The submitted proposals are considered, and the results are communicated to the public (stakeholders). Before a plan is adopted, the municipality must "exhibit" the proposal for at least 30 days. Municipal members and others have the right to submit comments on the plan proposal in writing during the exhibition period.

The municipality handles the planning, controls, and supervision of construction in its territory. If necessary, the government of Åland can decide on land use for certain important social functions or for certain purposes that are of great importance to the society, e.g., traffic networks, harbours and airports, energy production, and waste management.

The level of a municipality involves village planning. There is no expert opinion on the level of village planning. Through the Coast4us project, we have learned how important it is to involve local people and their knowledge as well as municipalities and other local operators at an early stage in various planning processes and decisions. This is to increase the understanding of the population on how to plan the coastal zone in the long-term and in a sustainable way.

### 5.4. Overview of Spatial Planning in Sweden

**National level**. Swedish planning processes are influenced by EU directives, especially legislation concerning environmental issues. There are several laws and regulations on how one can build and shape the Swedish environment (in all aspects). There are plans for municipalities and regions, but there is no national planning for the entire territory of Sweden. However, the state can affect the plans of municipalities and regions with national goals and interests. Based on the Planning and Building Act, the county administrative boards should make sure that national targets are realised and that everyone considers the national interests which exist. The actions of county administrative boards can have a positive or negative impact, including on the health and safety of people, and the risk of accidents or floods.

**Regional level**. Regional planning works with larger areas than municipalities. Regions coordinate planning across municipal borders. These plans exist so that each region can develop based on its circumstances.

**Local (municipal) level**. The municipality works with physical planning. Physical planning is about how to use land and water areas, where buildings and roads should be located and how they should be designed. The municipalities follow the Planning and Building Act. There are three types of municipal physical plans: structure plans, detailed development plans, and special area regulations. The structure plan should cover the entire municipality's area. It displays how the municipality would like the city and land to be in the future and which areas the municipality thinks should and should not be used for building purposes. Detailed development plans have rules for where new buildings may be located and how they should appear. Special area regulations are based on the structure plans and detailed development plans. For instance, they may ground the decisions about the territories of holiday houses [34]. Through the networking in Coust4us project, cooperation with the colleagues from Norrköping municipality was established. They contacted local associations (informal community groups). In cooperation, it was possible to listen and learn how to comply with the Planning and Construction Law, and to use the planning process to create better products in the long run.

When analyzing the spatial planning systems of different countries, it can be concluded that the governance of spatial planning in the explored countries of the Baltic Sea region each work differently. However, spatial planning in general and the planning systems are similar, as they include the national, regional, and local government levels. No country has a formal level of planning in the village or community. In all countries, the most direct and closest public involvement in the spatial planning process is at the local government level. Local governments are responsible for both sustainable spatial development planning and land use planning.

## 6. Comparison and Evaluation of Formal and Informal Spatial Planning Process in Latvia, Estonia, Åland Islands of Finland, and Sweden

In this section, the authors have used information and conclusions from parallel and previous research conducted by individual study authors on the phases [5] and methods [18] of public involvement in spatial development.

According to the assessment of different spatial planning levels in Latvia, Estonia, Åland Islands of Finland, and Sweden, the involved experts prepared the reports. They summarized the results on mobilization, planning, implementation, and monitoring ranked by the municipal role in planning, strengths, and weaknesses (see Table A1 in Appendix A).

Comparing the reports by the Latvian, Estonian, Finnish, and Swedish experts, there is evidence that the spatial planning process is associated with a hierarchical structure. Long-term development documents at national, regional, and municipal levels (keywords: sustainability, efficiency, resources) provide the main guidelines for site maintenance and use.

When analysing the involvement of local community (specific coastal territories in each country case) at lowest and closest level for an individual, it has been concluded that at the beginning of the planning process (mobilization), municipalities in all participating countries invite citizens and stakeholders (informal groups) to get involved in the planning process due to the dissemination of information and discussions. All countries' experts emphasized that the information is sufficient for the initial planning process. The weakness of the initial planning phase lies in the lack of communication with informal population groups.

In the spatial planning process, the local authorities have an information base (legislation, statistics, reports, opinions) to carry out the planning work. Experts emphasized that there is insufficient information about specific places and objects, and their functionality. There is information that in Åland, the autonomy function allows for extensive use of information. Informal groups do not participate in the planning process (document preparation), but they participate in the discussion of plans that have already been developed before they enter into force.

The plan is implemented by the municipality, following the developed plan, and granted funding. Both the plan as a planning tool and the implementation process are public. As a weakness, experts mentioned the impact of external factors that can change the course of the project, including various communication barriers that can cause controversy. Experts did not mention the role of informal groups in the implementation phase of the plan.

During the control phase, the municipalities monitor the implementation of the plan and provide reports following the established regulatory framework. At this stage, the availability and operability of the information are important factors. Experts mentioned the process was difficult to monitor, but did not mention the role of informal groups in the control process.

To better understand the benefits and challenging issues of the project Coast4us implementation process, an expert survey of responsible persons involved in the project was conducted. The essence of this survey was to assess the impact of the formal and informal spatial planning process in the specific coastal conditions, considering community involvement. The significance scale from 1 (insignificant) to 10 (significant) was used in the survey.

Significance averages range from 4.7 to 7.7, which is quite wide. In the mobilization phase, the significance indicators are closer (6.0–7.3), which can be explained by the great importance of village development and sustainability. At the planning stage, there is a larger range of indicators of significance (5.4–7.7), which can be explained by the ambiguity of the goal definition and planning process (the coordination of opinions). The averages of the significance of the implementation phase are slightly scattered (5.6–6.8), which can be explained by the compromise reached in the planning process, but more in-depth research would be needed. In the monitoring phase, which is closely related to the implementation phase, the significance indicators of the obtained results are scattered (4.7–6.9), which can be explained by the evaluation of the process and the result (see Figure 2).

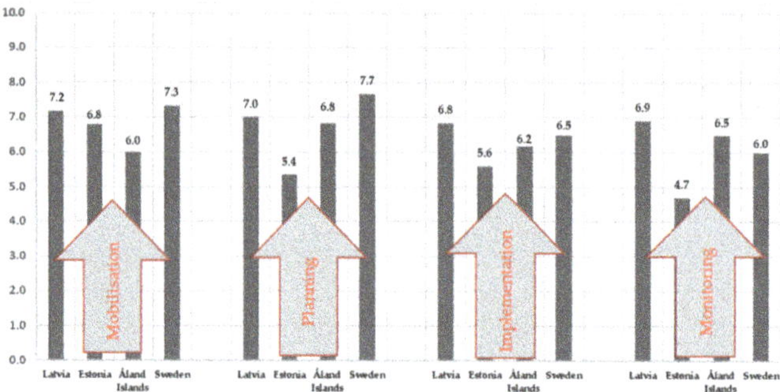


**Figure 2.** The summary of survey significance indicators (formal and informal spatial planning process).

Based on the expert evaluation summary of the spatial planning process (mobilization, planning, implementation, and monitoring) and the answers provided by experts, the theoretical models of the functioning of informal organizations, as well as the results of expert discussions, a model has been developed by all research authors.

The results of the study show that mobilisation, planning, implementation, and monitoring binds to Eriksson's model of collaboration [7]. A successful project is based on a concerted goal that is in the interest of the parties involved (central government, local government, and informal groups). Collaboration in planning is essential to achieve the goal, as citizens are often more aware of the situation and they will be the ones who will use the results. Stakeholders' participation in project implementation should be ensured as it avoids conflict situations, but monitoring ensures more efficient use of resources and a result of much higher quality. The members of local communities can be mobilized if their goals are clear, and these goals meet their interests. The spatial planning process should be open, involving and listening to the local community, which in this study were coastal communities. In the process of implementing the plan, the local community must be interested, and it will bring good results (and aid in future cooperation). A good result is achieved by monitoring the progress of the project and efficient use of resources, and all of this is confirmed by the informal groups (see Figure 3).

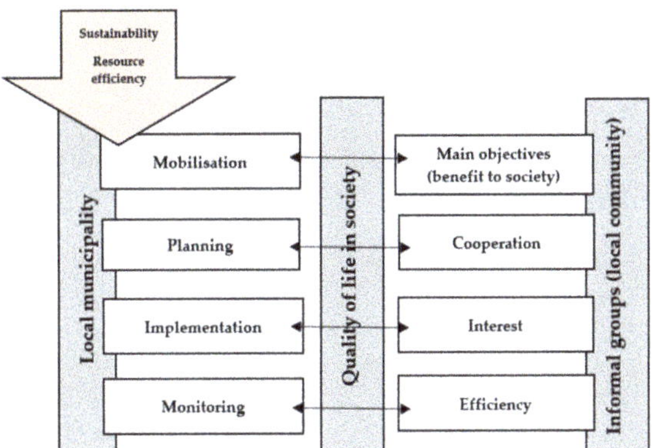


**Figure 3.** Model of cooperation between municipalities and territorial communities in the spatial development planning process.

In addition to the aforementioned and conducted comparative research, there is a clear need to comply with the national regulatory framework in the process of community engagement while creating new and modern solutions for informal and motivated public involvement in the spatial planning process. At the same time, the study exploring different countries with the unifying object of the Baltic Sea, highlights the specifics of coastal area spatial planning. Therefore, regarding the usual sustainability factors, particular attention should be paid to natural and environmental values, conservation, and development of ecosystems, etc. These specific features, as well as the changing and rapid growth of society in the direction of community development, offer new directions of research for both academics and scientists.

## 7. Conclusions

The hypothesis of this study is confirmed because there is evidence that spatial planning systems are in the process of transformation and in fact "approach" the local population, as the focus of development shifts to the needs of a particular person in a specific place (local community) using new informal methods.

This study is a result of an extensive and in-depth collaboration between participants from different countries and spatial planning traditions of the Baltic Sea region, but with a common interest in formal and informal spatial planning processes in coastal areas.

According to research questions mentioned in the introduction of this study and the multi-element research conducted, it can be concluded:

- both similarities and differences of spatial planning approaches have been detected in different countries that are placed around one water object, namely the Baltic Sea. All countries that have been analyzed have hierarchical planning systems and historically have used formal "top-down" spatial planning approaches. In recent years, spatial planning systems are changing to more "bottom-up" systems, but each country is conducting these processes in different ways. This makes risks to sustainable governance of the Baltic Sea and coastal communities;
- in all countries standard spatial planning process steps (mobilization, planning, implementation, and monitoring) are done, but it is clear that when a new "bottom-up" system is implemented, informal and more community-involved activities are done in step mobilization and planning. However, steps implementation and monitoring is mostly done by municipalities and civil servants. This, by opinion of the researcher group, leads to situations in which the community loses interest in being active inhabitants, because local community can not affect real implementation of their ideas and needs. Besides, they are not a part of change management of sustainable development.

This creates a large risk of local conflicts as well as causing loss of interest of local development.

It is important to take into account the condition that coastal territory planning has very specific circumstances, and is connected with one water object that does not have physical borders, but sustainable development of these territories is very connected with this main resource (Baltic Sea). Mostly, development problems are similar, but they can be solved only when actions around the water object are equivalent. From our point of view, when different countries have mostly hierarchical planning structures, it is important that there are international agreements about main sustainable actions that are supplemented with "bottom up" solutions at the local community level, and at the same time experiences of local solution best practices are shared around coastal areas of one water object with the goal of sustainability and harmonization of actions.

The authors of this study propose to authorities of the spatial planning process for coastal areas around the Baltic Sea region to create a system, model, or even regulation, to make local (coastal) communities a part of the planning system at all phases (including implementation phase). This would reach such goals as: sustainable development, a "bottom-up" approach, and active citizenship. In addition, there is a proposal for better communication between countries, municipalities, and local communities around one water object to share best practices and harmonize actions.

**Author Contributions:** E.P., Conceptualization, Formal Analysis, Investigation, Methodology, Resources, Supervision, Validation, Roles/Writing—Original Draft, Writing—Review & Editing. S.G., Conceptualization, Formal Analysis, Funding Acquisition, Investigation, Methodology, Resources, Project Administration, Supervision, Validation, Roles/Writing—Original Draft, Writing—Review & Editing. A.A., Formal Analysis, Investigation, Methodology, Resources, Writing—Review & Editing. A.L., Conceptualization, Data Curation, Formal Analysis, Investigation, Methodology, Resources, Software, Validation, Visualization, Roles/Writing—Original Draft, Writing—Review & Editing. J.B., Formal Analysis, Resources, Roles/Writing—Original Draft, Writing—Review & Editing. K.K., Formal Analysis, Resources, Roles/Writing—Original Draft, Writing—Review & Editing. I.C., Formal Analysis, Resources, Roles/Writing—Original Draft, Writing—Review & Editing. M.K., Resources, Software. U.K., Resources, Software. M.G., Formal Analysis, Resources, Roles/Writing—Original draft, Writing—Review & Editing. M.P.-P., Formal Analysis, Resources, Roles/Writing—Original Draft, Writing—Review & Editing. C.B., Formal Analysis, Resources, Roles/Writing—Original Draft, Writing—Review & Editing. S.V., Formal Analysis, Resources, Roles/Writing—Original Draft, Writing—Review & Editing. J.H., Formal Analysis, Resources, Roles/Writing—Original Draft, Writing—Review & Editing. I.G., Formal Analysis, Resources, Roles/Writing—Original Draft, Writing—Review & Editing. M.J.H., Formal Analysis, Resources, Roles/Writing—Original Draft, Writing—Review & Editing. All authors have read and agreed to the published version of the manuscript.

**Funding:** This research was funded by the INTERREG Central Baltic program project "Coast4us" within the P2 project priority—Sustainable use of the common resources, project number 627.

**Institutional Review Board Statement:** Not applicable.

**Informed Consent Statement:** Not applicable.

**Data Availability Statement:** Not applicable.

**Conflicts of Interest:** The authors declare no conflict of interest.

## Appendix A

**Table A1.** Spatial planning process steps (mobilization, planning, implementation, and monitoring).

| Country | Stage | Strengths | Weaknesses |
|---|---|---|---|
| | | **Mobilisation** | |
| **Latvia** | The local government invites the village representatives to participate in the planning process, disseminate information in a formal way (website, newspaper, etc.). | Widely available resources for the dissemination of information. Available communication contacts with citizens. | There is no direct daily contact with potential participants. No traditions yet. |
| **Saaremaa municipality Estonia** | Invites interest groups to participate in the planning process. Information is disseminated in a formal way (website, newspaper, exchange of formal letters, social media, etc). | Widely available resources for the dissemination of information. | The report does not identify any weaknesses. |
| **Åland Islands of Finland** | All interested parties shall be allowed to consult and make their views known in writing or orally where this is appropriate for the purpose and relevant to the planning process. | There is communication with citizens and local organizations. | The locals should be involved at an earlier stage before a proposal is made, communication should begin at an earlier stage. |
| **Sweden** | All stakeholders shall be allowed to consult and make their views known, in writing or orally, as appropriate to the purpose and relevant to the planning process. There is no strict national strategy for spatial planning. | There is communication with citizens and local organizations, and national interests are respected. | The report does not identify any weaknesses. |
| | | **Planning** | |
| **Latvia** | The local government, based on statistical research, modern solution identification and clustering of working groups, prepares the documents that are publicly discussed and approved within the framework of laws and regulations. | Resources are available for extensive statistical research. Resources are available for research in various industries, attracting both local and foreign specialists. Possible exchange of experience by using gathered contacts. | The use of formalised procedures. The lack of knowledge about specific problems in a specific place. Formal discussion of documents, as well as reliance on the laws and regulations that in the case of Latvia are often fragmented, or there is no available funding or other resources to introduce the regulations. |
| **Saaremaa municipality Estonia** | The local government, based on research, knowledge from previous plans, strategies, and information from different involved parties (interest groups, governmental institutions, etc.) prepares the documents that are publicly discussed and approved within the framework of laws and regulations. | Resources are available for extensive statistical research. Possible exchange of experience by using gathered contacts. | Insufficient knowledge about specific problems or possibilities in a specific place. |
| **Åland Islands of Finland** | Planning is formal, the island has its legislation, but the planning process follows general national legal norms. | The Government of Åland and other municipalities, authorities, legal persons, and persons shall be allowed to consult and make their views known, in writing or orally, as appropriate to the purpose and relevance of the plan. The purpose of the hearing is to improve the basis for decisions and to ensure transparency and influence. | Municipal members and others have the right to submit written comments on the plan proposal during the exhibition. There is limited time for collaboration in planning. |
| **Sweden** | Planning must respect national interests, EU law (special environmental issues). | All stakeholders are involved in the planning. | There is no national planning for the entirety of Sweden. |

**Table A1.** *Cont.*

| Country | Stage | Strengths | Weaknesses |
|---|---|---|---|
| | | **Implementation** | |
| **Latvia** | Following the approved action plan and investment plan, the local government ensures the implementation of activities. | There are public resources available for the implementation of activities. There are tangible and intangible assets available to ensure the place for the implementation of activities. | Restrictions on the activities directed towards the regulation of statutory acts. Not being "on the site" ** makes it difficult to identify the most effective solution. |
| **Saaremaa municipality Estonia** | Following the approved strategic and funding plan, the local government ensures the implementation of the planning document and its principles. | The adopted planning document is public. Principles directing spatial planning are available online. | Planning document may not be flexible enough as there may occur changes in laws, unpredictable developments, etc. (not being up to date). Adopted principles may be understood differently (e.g., because of the wording). |
| **Åland Islands of Finland** | According to plan. | About the results of the hearing and proposals based on the different opinions that have been made shall be reported when the plan proposal is presented for the public. | The report does not identify any weaknesses. |
| **Sweden** | The planning process and implementation are regional and local in nature. | A better understanding of local conditions, fewer errors in implementation. | The report does not identify any weaknesses. |
| | | **Monitoring** | |
| **Latvia** | Following the regulations, the local government draws up regular reports on the progress, informs the Council about the reports, and publishes the reports online. | The data are made available, and there are restricted access databases. There are specialists to perform the work, and the necessary capacity is ensured. | A difficult-to-manage and slow (change) management process. Formal reports based on data collection. |
| **Saaremaa municipality Estonia** | Following the Planning Act, the local municipality needs to follow, review, and implement adopted spatial plans. | The planning process is public. There are specialists to perform the work, and the necessary capacity is ensured. | The report does not identify any weaknesses. |
| **Åland Islands of Finland** | The municipality handles the planning, controls, and supervision of construction within the municipality. | The report does not identify any strengths. | The municipality handles the planning, controls, and supervision of construction within the municipality. |
| **Sweden** | Monitoring shall be provided at all levels, following national interests and regulatory requirements. | The planning process is public. There are specialists to perform the work, and the necessary capacity is ensured. | It is indistinct who has the legal right to appeal the detailed development plans. This confuses and prolongs the process. |

** Not being "on the site" – not living in a community on a daily basis, lack of knowledge of the community daily processes.

## References


1. Coast4us: Coast for, us. INTERREG CB Programme Project Home Page. Available online: http://coast4us.com (accessed on 20 January 2020).
2. Moroz, O. Territorial Community: Essence, Formation and Contemporary Ukrainian Realities. *Democr. Gov.* **2008**, *2*, 18–26.
3. Syssner, J.; Meijer, M. Informal Planning in Depopulating Rural Areas. *Eur. Countrys.* **2017**, *9*, 458–472. [CrossRef]
4. MacQueen, K.M.; McLellan, E.; Metzger, D.S.; Kegeles, S.; Strauss, R.P.; Scotti, R.; Blanchard, L.; Trotter, R.T., 2nd. What is Community? An Evidence-based Definition for Participatory Public Health. *Am. J. Public Health* **2001**, *91*, 1929–1938. [CrossRef] [PubMed]
5. Pudzis, E.; Geipele, S.; Geipele, I. Community Participation in Village Development: The Scale of Latvia. *Balt. J. Real Estate Econ. Constr. Manag.* **2016**, *4*, 84–99. [CrossRef]
6. McGranahan, G.; Balk, D.; Anderson, B. The Rising Tide: Assessing the Risks of Climate Change and Human Settlements in Low Elevation Coastal Zones. *Environ. Urban.* **2007**, *19*, 17–37. [CrossRef]
7. Eriksson, E. Four Features of Cooptation: User Involvement as Sanctioned Resistance. *Nord. Welf. Res.* **2018**, *3*, 7–17. [CrossRef]

8. Meijer, M.; Syssner, J. Getting Ahead in Depopulating Areas—How Linking Social Capital is Used for Informal Planning Practices in Sweden and The Netherlands. *J. Rural Stud.* **2017**, *55*, 59–70. [CrossRef]
9. Mishra, A. The Process of Informal Spatial Planning: A Literature Overview. *Balt. J. Real Estate Econ. Constr. Manag.* **2019**, *7*, 216–227. [CrossRef]
10. Barnard, C. *The Functions of the Executive*; Harvard University Press: Cambridge, MA, USA, 1938.
11. Nikezić, S.; Dželetović, M.; Vučinić, D. Chester Barnard: Organizational-Management Code for the 21st Century. *Procedia Soc. Behav. Sci.* **2016**, *221*, 126–134. [CrossRef]
12. Carmona, M. The formal and informal tools of design governance. *J. Urban Des.* **2017**, *22*, 1–36. [CrossRef]
13. Reimer, M.; Getimis, P.; Blotevogel, H. *Spatial Planning Systems and Practices in Europe: A Comparative Perspective on Continuity and Changes*; Routledge: New York, NY, USA, 2014. [CrossRef]
14. Fürst, D. Planning cultures en route to a better comprehension of "planning processes". In *Planning Cultures in Europe. Decoding Cultural Phenomena in Urban and Regional Planning*; Knieling, J., Othengrafen, F., Eds.; Ashgate: Farnham, UK, 2009; pp. 23–38. ISBN 978-0-7546-7565-5.
15. Auzins, A. Capitalising on the European Research Outcome for Improved Spatial Planning Practices and Territorial Governance. *Land* **2019**, *8*, 163. [CrossRef]
16. Zamarina, M. Ciemu plānošana viedai attīstībai. Rokasgrāmata. 2020. Available online: https://www.academia.edu/43376031/Ciemu_pl%C4%81no%C5%A1ana_viedai_att%C4%ABst%C4%ABbai_rokasgr%C4%81mata (accessed on 14 July 2020).
17. Wates, N. The Community Planning Handbook: How People Can Shape Their Cities, Towns and Villages in Any Part of the World. Available online: http://library.uniteddiversity.coop/REconomy_Resource_Pack/Community_Assets_and_Development/The_Community_Planning_Handbook-How_People_Can_Shape_Their_C.pdf (accessed on 12 August 2020).
18. Geipele, S.; Kundziņa, A.; Pudzis, E.; Lazdiņš, A. Evaluation of Community Involvement in Participatory Process—Lessons Learned in the Baltic Sea Region. *Archit. Urban Plan.* **2020**, *16*, 56–65. [CrossRef]
19. Procedures for the Public Participation in the Development Planning Process. Republic of Latvia Cabinet Regulation No. 970, adopted 25 August 2009. *Legal Acts of the Republic of Latvia*. Available online: https://likumi.lv/ta/en/en/id/197033 (accessed on 12 February 2020).
20. Municipal Spatial Development Planning Documents. Republic of Latvia Cabinet Regulation No. 628, adopted 14 October 2014. *Legal Acts of the Republic of Latvia*. Available online: https://likumi.lv/ta/id/269842-noteikumi-par-pasvaldibu-teritorijas-attistibas-planosanas-dokumentiem (accessed on 12 February 2020).
21. Rules on Planning Region Spatial Development Planning Documents. Republic of Latvia Cabinet Regulation No. 402, adopted 16 July 2013. *Legal Acts of the Republic of Latvia*. Available online: https://likumi.lv/doc.php?id=258626 (accessed on 25 January 2020).
22. Development Planning System Law. *Legal Acts of the Republic of Latvia*. Available online: https://likumi.lv/ta/en/en/id/175748 (accessed on 23 March 2020).
23. On the Regional Policy Guidelines for 2021–2027. Republic of Latvia Cabinet Regulation No. 587, adopted 26 November 2019. *Legal Acts of the Republic of Latvia*. Available online: https://likumi.lv/ta/id/310954-par-regionalas-politikas-pamatnostadnem-2021-2027-gadam (accessed on 20 January 2020).
24. Maritime Spatial Plan 2030: The Maritime Spatial Plan for the Marine Inland Waters, Territorial Sea and Exclusive Economic Zone Waters of the Republic of Latvia. Republic of Latvia Cabinet Order No. 232, of 21 of May 2019. *National Level Long-Term Spatial Development Planning Document*. Available online: https://drive.google.com/file/d/1mKigVjv6N03cjgPkwR5RSItcQezsn5zY/view (accessed on 20 March 2020).
25. Regional Development Law. *Legal Acts of the Republic of Latvia*. Available online: https://likumi.lv/ta/en/en/id/61002 (accessed on 27 January 2020).
26. Spatial Development Planning Law. *Legal Acts of the Republic of Latvia*. Available online: https://likumi.lv/ta/en/id/238807-spatial-development-planning-law (accessed on 20 January 2020).
27. National Spatial Plan Estonia 2030+. *Regionaalministri Valitsemisala*. Available online: https://eesti2030.files.wordpress.com/2014/02/estonia-2030.pdf (accessed on 8 January 2020).
28. Saaremaa valla üldplaneering. Available online: https://www.saaremaavald.ee/et/saaremaa-valla-uldplaneering (accessed on 27 January 2020).
29. Regional Development and Policy. *Rahandusministeeruim*. Available online: https://www.rahandusministeerium.ee/en/regional-development-and-policy (accessed on 27 January 2020).
30. Planning Act. *Riigi Teataja*. Available online: https://www.riigiteataja.ee/en/eli/ee/505042019003/consolide/current (accessed on 20 May 2020).
31. Saare Maakonnaplaneering 2030+. *Rahandusministeeruim*. Available online: https://maakonnaplaneering.ee/saare-maakonnaplaneering (accessed on 20 May 2020).
32. Maakonnaplaneeringud. *Rahandusministeeruim*. Available online: https://maakonnaplaneering.ee/maakonna-planeeringud (accessed on 18 May 2020).
33. Planning and Building Act (2008: 102) for the Province of Åland. *Ålands Landskapsregering*. Available online: https://www.regeringen.ax/alandsk-lagstiftning/alex/2008102 (accessed on 4 April 2020).
34. Swedish National Board of Housing, Building and Planning. In *Government Offices of Sweden*. Available online: https://www.government.se/government-agencies/swedish-national-board-of-housing-building-and-planning/ (accessed on 16 May 2020).

International Journal of
*Environmental Research and Public Health*

*Review*

# Life Cycle Assessment on Wave and Tidal Energy Systems: A Review of Current Methodological Practice

**Xizhuo Zhang, Longfei Zhang, Yujun Yuan and Qiang Zhai ***

Department of Mechanical Engineering, School of Mechanical, Electrical & Information Engineering, Shandong University, Weihai 264209, China; 201700800328@mail.sdu.edu.cn (X.Z.); 201836529@mail.sdu.edu.cn (L.Z.); 201700800324@mail.sdu.edu.cn (Y.Y.)

* Correspondence: zhaiqiang@sdu.edu.cn; Tel.: +86-631-568-83-38

Received: 13 February 2020; Accepted: 27 February 2020; Published: 2 March 2020

**Abstract:** Recent decades have witnessed wave and tidal energy technology receiving considerable attention because of their low carbon emissions during electricity production. However, indirect emissions from their entire life cycle should not be ignored. Therefore, life cycle assessment (LCA) has been widely applied as a useful approach to systematically evaluate the environmental performance of wave and tidal energy technologies. This study reviews recent LCA studies on wave and tidal energy systems for stakeholders to understand current status of methodological practice and associated inherent limitations and reveal future research needs for application of LCA on wave and tidal technologies. The conformance of the selected LCAs to ISO 14040 (2006) and 14044 (2006) are critically analyzed in strict accordance with the ISO stepwise methodologies, namely, goal and scope definition, life cycle inventory (LCI) analysis, as well as life cycle impact assessment (LCIA). Our systematically screening of these studies indicates that few of the selected studies are of strict conformance with ISO 14040 and 14044 standards, which makes the results unreliable and thus further reduces the confidence of interested stakeholders. Further, our review indicates that current LCA practice on wave and tidal energies is lacking consideration of temporal variations, which should be addressed in future research, as it causes inaccuracy and uncertainties.

**Keywords:** wave energy; tidal current energy; life cycle assessment; ISO

---

## 1. Introduction

Life cycle assessment (LCA) has been widely recognized as an efficient approach to evaluate the life cycle environmental impacts of a product or service by comprehensively encompassing all processes and environmental releases for specific environmental impact categories. As the most significant contributor to climate change, life cycle greenhouse gas (GHG) emission has been adopted by laws and regulations as an indicator to evaluate the environmental performance of clean energies. For instance, U.S. Energy Independence and Security Act of 2007, Section 526 [1] requires that life cycle GHGs for nonconventional petroleum sources must be less than or equal to such emissions from the equivalent conventional fuel produced from fossil sources. The early incorporation of environmental issues has been requested by the EU Strategic Environmental Assessment procedure (Directive 2001/42/EC) [2]. Declaration of life cycle GHG emissions has been required by current and future environmental regulations [3,4]. Among viable techniques for environmental assessment, LCA is a comprehensive stepwise method, including goal and scope definition, life cycle inventory (LCI) analysis, life cycle impact assessment (LCIA) and interpretation. As a technique of environmental management, the principles, framework, requirements and guidelines are suggested in ISO 14040

(2006) [5] and ISO 14044 [6]. As for wave and tidal energy, although recent decades have witnessed the emergence of new technologies, only a few wave and tidal systems are studied by LCA methods, because the technologies are still at such an early stage that limited funding is available for supporting research beyond technology development [7].

In recent decades, a few researchers have overviewed the LCAs on ocean energy systems. Banerjee et al. discuss the emission characteristics and energy accounting of wave and tidal energy systems for LCAs [8]. A comprehensive review of the current state of the art of research in the field of ocean energy systems, with an emphasis on research beyond technology or technological improvements is presented in [7]. Paredes et al. systematically evaluated the LCA studies of ocean energy technologies and presented a summary of the LCA results [9]. To our knowledge, there is no review work has been reported on a comprehensive and in-depth analysis of the methods adopted by the published LCAs on wave and tidal energy systems. In this study, we conduct an extensive review of recent LCAs on wave and tidal energy systems, with the following purposes: (a) summarizing the current status of the methodological practice; (b) identifying the limitations of methods of LCAs on wave and tidal energies; (c) revealing future research needs for wave and tidal LCAs from the methodological perspective.

## 2. Methodology

### 2.1. Literature Search Strategy

The selected international databases included: Web of Science Core Collection, BIOSIS Previews, Chinese Science Citation Database, Inspec, KCI-Korean Journal Database, MEDLINE, Russian Science Citation Index and SciELO Citation Index. Search keywords for topics and titles included: "ocean energy", or "marine energy", or "marine current energy", or "ocean thermal energy", or "salinity gradient energy", or "wave energy", or "wave power", or "tidal energy", or "tidal power", or "tidal current", or "tidal stream", or "sea turbine", or "wave energy conversion", or "wave energy converter", or "WEC", or "tidal stream/barrage device", or "tidal current turbine"; and "life cycle assessment", or "LCA", or "life cycle analysis", or "environmental assessment", or "environmental impact", or "global warming", or "greenhouse gas", or "GHG", or "carbon footprint", or "carbon dioxide", or "$CO_2$", or "embodied carbon", or "carbon intensity", or "$CO_2$ intensity", or "carbon audit", or "carbon emission", or "energy audit", or "energy accounting", or "energy intensity", or "embodied energy".

### 2.2. Case Studies Refining

Through the application of the above-mentioned keywords in Section 2.1, the search returned 1214 references. The found research literature was then further screened by applying such criteria as:

a) the studies are put together in English;
b) the studies are peer-reviewed journal articles, full conference papers excluding abstracts and posters, theses or dissertations, or official governmental reports;
c) the wave or tidal technologies are designed for production of electricity.

This finally left 18 studies, amongst which 4 cases were about tidal current energies [9–13], 12 cases were about wave energies [3,4,14–23] and 2 cases were about both tidal and wave energies [24,25].

#### 2.2.1. Spatial Distribution of Studied Wave and Tidal Energy Systems

Geographical distribution of the selected LCA studies, as shown in Table 1, indicates that 11 systems are installed in European seas [3,9,11,13–19,22,23], 1 in New Zealand [12], 1 in China [20] and 1 in multicontinental locations [22]; three are located at hypothetical offshore locations [4,21,24].

As shown in Figure 1, the selected literature studied two oscillating surge WECs [22,25], one oscillating water column [21], three attenuators [14–17], two overtopping [3,23], four point absorbers [4,19,20,22], one vertical axis tidal [9], eight horizontal axis tidal [11–13,25], one Archimedes [13] and one tidal range [25].

**Table 1.** General info of the studied wave and tidal systems studied by life cycle assessments (LCAs).

| Reference | Device | Region | Installed Capacity | Capacity Factor | ISO Smartcards | |
|---|---|---|---|---|---|---|
| | | | | | 14040 | 14044 |
| Cavallaro et al. (2007) | Kobold | Italy | 160 kW | N.A. | - | - |
| Douglas et al. (2008) | Seagen | UK | 1.2 MW | 48% | √ | - |
| Rule et al. (2009) | Kaipara Harbor | New Zealand | 200 MW | 37% | - | √ |
| Howell et al. (2013) | DeepGen; OpenHydro; ScotRenewables SR2000; Flumill | UK | 1 MW; 2 MW; 2 MW; 2 MW | N.A. | √ | √ |
| Hans et al. (2007) | Wave Dragon | Denmark | 7 MW | N.A. | - | - |
| Parker et al. (2007) | Pelamis | UK | 750 kW | N.A. | √ | - |
| Thomson et al. (2011a) | Pelamis | UK | 750 kW | N.A. | √ | - |
| Thomson et al. (2011b) | Pelamis | UK | 750 kW | N.A. | √ | - |
| Thomson et al. (2019) | Pelamis | UK | 750 kW | 45% | - | √ |
| Dahlsten et al. (2009) | Seabased | Hypothetical | 20 MW | N.A. | √ | √ |
| Walker et al. (2011) | Oyster | UK | 315 kW | N.A. | - | - |
| Ombach et al. (2014) | Wave Star | Denmark | 1000 kW | N.A. | 14000 | - |
| Zhai et al. (2018) | BRD | China | 10 kW | 50% | √ | - |
| Elginoz et al. (2017) | MUP farm | Hypothetical | 265.5 MW | Various | √ | - |
| Curto et al. (2018) | DEIM | Italy | 30 kW | N.A. | √ | √ |
| Patrizi et al. (2019) | OBREC | Italy | 3 kW | N.A. | - | - |
| Uihlein et al. (2016) | Various | Unspecified | 500kW-1 MW | 34%; 0% | - | - |
| Douziech et al. (2016) | Annapolis; SeaGen; HS1000; HydraTidal; Oyster800 | Canada; Ireland; UK; Norway; UK | 20 MW; 1200 kW; 1000 kW; 1500 kW; 800 kW | N.A. | - | - |

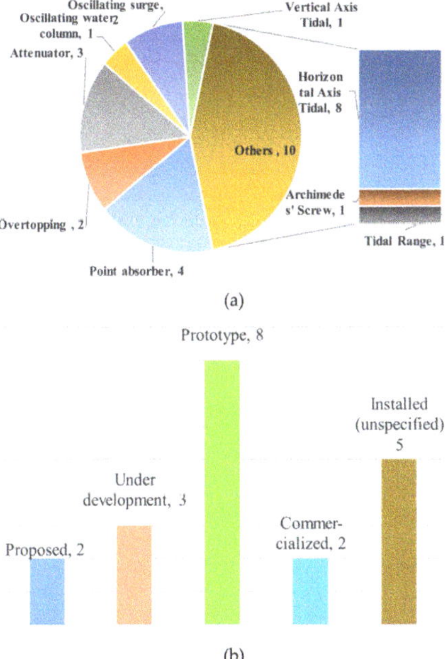


**Figure 1.** Technological coverage and development status of studied wave and tidal power systems. (a) Technological coverage; (b) Development status. Note: the number following each item in (a) indicates the number of the systems by specific type of technology; the number following each item in (b) indicates the number of the systems under the specific development status.

#### 2.2.2. Installed Capacities and Technological Development Status

Installed capacities of the studied systems range from 3 kW to 265.5 MW, as shown in Table 1. Only five studies take the capacity factors of the power generation systems for consideration [11,12,17,20,24]. As Figure 1b demonstrates, two systems are under proposal [12,24]; three are under development [4,21,22]; eight have prototypes installed and tested [3,13,14,19,20,23]; two are commercialized [15,17]; five are claimed being installed as of writing of the papers or reports, however with no further details provided regarding the installation, e.g., it is unknown whether they are pioneer plants or full-scale commercialized systems [9,11,13,18,25].

#### 2.2.3. ISO 14040 and 14044 Conformance Declarations

Currently, ideal practice of LCA is to follow the principles, framework, requirements and guidelines by international standards ISO 14040 (2006a) and 14044 (2006). Our review of the selected literature shows that only five studies claim that they followed ISO 14040 [11,14,15,20,21], two studies claimed that they followed ISO 14044 [12,17], and two claimed that they followed both ISO 14040 and 14044 [4,13].

## 3. Results

### 3.1. Goal and Scope

Definition of goal and scope is the first step of an LCA study, as it defines the purpose or application of the study, as well as the scope of assessment to be conducted. As per our analysis of the selected studies, energy and carbon are the most considered environmental indicators, so some LCA studies merely investigate the life cycle primary energy and carbon of the wave and tidal energy systems [11–14,18,19,23]. In this case, these studies are merely LCI studies, as shown in Table 1, since no life cycle impacts are assessed and discussed.

LCI results are normally calculated and interpreted through such indicators as energy intensity and carbon intensity, as well as energy and carbon payback. Inclusion of LCIA in the LCA studies can comprehensively illustrate the life cycle environmental performance of the studied systems, thus various levels of LCIA were conducted to further reveal the environmental impacts of the wave and tidal systems beyond energy consumption and carbon emission [3,4,9,15–17,20–22,24,25].

#### 3.1.1. Internal and External Application

Internal application of LCA is to identify the most significant contributors such as materials, processes and life cycle stages of ocean energy devices in terms of environmental impacts [11,24] by investigating the associated with life cycle stages [9]. The results can be a reference in identifying system improvement potentials [4,15,20] through choice of substitute materials and processes [4]. LCA results are widely applied for comparison between different wave and tidal energy systems as well as other renewable energy technologies [12,19,20,24].

#### 3.1.2. System Boundary

System boundary determines physical inclusion of materials and processes, temporal inclusion of long- and short-term releasements, as well as geographical factors into the product system for the LCA study. Wave and tidal energy systems are usually divided into such stages as material extraction, manufacturing, installation, operation and maintenance, decommission and disposal. As shown in Table 2, most of the analyzed studies adopt cradle-to-grave boundaries, i.e., include all the above-mentioned life cycle stages in the power system for LCA studies [3,4,9–21,24,25]. Only two studies apply cradle-to-gate system boundaries and exclude the stages of installation, operation and maintenance, decommission and disposal [22,23].

**Table 2.** Goal and scope definition of studied LCAs.

| Reference | Study Type | | | Goal | | Scope | | Cut-off Criteria | | | | | | Allocation | | ICD | CRD |
|---|---|---|---|---|---|---|---|---|---|---|---|---|---|---|---|---|---|
| | | | | | | | | Inputs | | | Outputs | | | | | | |
| | LCA | LCI | CS | Int. | Ext. | System Boundary | Functional Unit | M | E | ES | M | E | ES | Open Loop | Closed Loop | | |
| **Cavallaro et al. (2007)** | √ | - | - | √ | - | **X-grave** | **System** | - | - | - | - | - | - | - | √ | √ | - |
| Douglas et al. (2008) | - | √ | √ | √ | - | X-grave | Undefined[1] | - | √ | √ | - | √ | √ | - | √ | N.A. | - |
| Rule et al. (2009) | - | √ | √ | - | √ | X-grave | 1 kWh | - | √ | √ | - | - | - | - | √[2] | N.A. | - |
| Howell et al. (2013) | | √ | √ | √ | | X-grave | System | √ | - | - | √ | - | - | √[3] | | N.A. | - |
| Hans et al. (2007) | √ | | √ | - | √ | X-grave | 1 kWh | - | - | - | - | - | - | √[4] | √[4] | √ | - |
| Parker et al. (2007) | | √ | √ | √ | | X-grave | 1 kWh | - | √ | √ | - | - | - | √[3] | - | N.A. | - |
| Thomson et al. (2011a) | √ | - | √ | √ | √ | X-grave | 1 kWh | - | √ | √ | - | - | - | √[5] | - | √ | - |
| Thomson et al. (2011b) | √ | - | - | √ | √ | X-grave | 1 kWh | - | √ | √ | - | - | - | √[5] | - | √ | - |
| Thomson et al. (2019) | √ | - | √ | √ | - | X-grave | 1 kWh | - | √ | √ | - | √ | √ | √[5] | - | √ | - |
| Dahlsten et al. (2009) | √ | - | - | √ | - | X-grave | 1 kWh | √ | √ | √ | √ | √ | √ | √[6] | - | √ | - |
| Walker et al. (2011) | - | √ | - | √ | - | X-grave | Undefined | - | - | - | - | - | - | - | √[7] | N.A. | - |
| Ombach et al. (2014) | - | √ | √ | √ | - | X-grave | Undefined | - | - | - | - | - | - | - | √[8] | N.A. | - |
| Zhai et al. (2018) | √ | - | √ | √ | - | X-grave | System | √ | - | - | √ | - | - | - | √ | √ | - |
| Elginoz et al. (2017) | √ | - | √ | √ | √ | X-grave | 1 kWh | - | - | - | - | - | - | - | √[9] | √ | - |
| Curto et al. (2018) | √ | - | √ | √ | - | X-gate | System | - | - | - | - | - | - | - | - | √ | - |
| Patrizi et al. (2019) | - | √ | - | √ | - | X-gate | System | - | - | √ | - | - | √ | - | - | √ | - |
| Uihlein et al. (2016) | √ | - | √ | √ | - | X-grave | 1 kWh | - | - | - | - | - | - | - | √[10] | √ | - |
| Douziech et al. (2016) | √ | - | √ | - | √ | X-grave | 1 kWh | - | - | - | - | - | - | - | √[11] | √ | - |

Abbreviations: CS, comparative study; Int., internal; Ext., external; X, cradle; ICD, impact category definition; CRD, critical review definition; M, materials; E, Energy; ES, environmental significance; √, yes; -, not reported; N.A., not applicable. Notes: [1]/kWh used for energy and carbon intensities; [2]No recycling for disposal of field equip., reuse of half of turbines. [3]Recycling only for steel. [4]Steel, copper, aluminum, bronze, plastics used for other processes; concrete reused for road construction. [5]Recycled content approach for both foreground and background processes. [6]Polluter pays (EPD); no allocation for foreground data. [7]Replacing primary material with recycled material in future. [8]Recycling rate for metals is 90%, some materials not recycled, and instead incinerated or taken to land fill. [9]Ninty percent recycling for metals, otherwise incineration or landfill. [10]Various recycling rate for ferrous and nonferrous metals. [11]ISO/TS 14067 closed-loop procedure.

#### 3.1.3. Functional Unit

Functional unit enables that results from different LCA studies can be compared for product systems with similar functions [5]. As illustrated in Table 2, for wave and tidal energy systems, the main function is electricity production; therefore, the majority of the selected LCA studies define the functional unit as 1 kWh electricity generated [3,4,12,14–17,21,24,25]. Some studies further specify that the 1 kWh electricity is generated and fed to the national grid [4,24,25]. Few studies [11,18,19] do not claim functional unit definitions, however, they use per-kWh electricity for calculation of energy and carbon intensities. On this point then, 1 kWh electricity is the virtual functional unit of these studies. Other studies define the functional units as the entire power systems [9,13,20,23], as they intend to conduct the LCAs for merely internal purpose.

#### 3.1.4. Cut-off Criteria

Exhaustive inclusion of inputs and outputs of the system is neither possible, since a product life cycle contains too many materials and too much energy consumption associated with unit processes and emissions, nor necessary, as the goal of an LCA is normally defined to identify the most significant contributors to the environmental impacts. Definition of appropriate cut-off criteria is therefore necessary to exclude less important inputs and outputs, with setting up percentages of mass, energy or environmental significance [6].

For wave and tidal systems, Table 2 shows that for inputs, energy flow and environmental significance are the most commonly used cut-off criteria [11,12,14–17], as embodied energy and carbon are the most relevant indicators. Only two of the selected LCA studies considered mass as the cut-off criteria [13,20]. Mass, energy and environmental significance are taken into consideration for cutting off by Dahlsten et al., 2009. As for outputs, two studies use mass and energy flow as the cut-off criteria [11,17], two use mass [13,20], and one uses mass, energy and environmental significance [4]. Among the selected LCA studies, only one of them defined environmental significance as the single cut-off criterion [23]. None of the above-mentioned criteria were described within eight of the selected studies [3,9,18,19,21,22,24,25].

#### 3.1.5. Allocation

Often different systems share inputs and outputs, thus, dividing and assigning them in between these systems is critical to ensure the accuracy of LCA studies [5]. There are different approaches for allocation in LCA practice, such as partitioning approach and substitutional approach. Partitioning approach is also called allocation in the sense of the word, which is based on the physical characteristics such as mass, volume and energy content. Through the application of substitutional approach, the burden of some byproducts of the product system is included into the system boundary, which means the burden of these byproducts is avoided when they enter the boundaries of other systems. Partitioning approach is adopted by most of the reviewed articles. As shown in Table 2, open-loop method is applied by some wave and tidal LCA studies, which defines recycling rate for materials (e.g., metallic materials) for foreground and background data [4,13–17]. Other studies described various recycling and reuse rates of materials, which are defined as closed-loop procedure [9–12,18–21,24,25]. In fact, mass and energy flow is so complicated that single open- or closed-loop does not always sufficiently describe the actual product system. Within this context, a combined or hybrid open and closed loop is more appropriate for the allocation modeling [13]. Two of the selected LCA studies did not describe their allocation procedure for their system modeling [22,23].

#### 3.1.6. Impact Categories Definition

The definition of impact categories depends on the goal and scope definition and provides the range of interested environmental issues either from midpoint or endpoint perspective. The selected studies containing LCIAs all discuss the selection of impact categories in the scope definition [3,4,9,15–17,20–22].

#### 3.1.7. Critical Review

A critical review by experts certifies the validation of the LCA method, data collection and calculation and rationality of the interpretation [6]. None of the selected studies provides information regarding expert reviews [3,4,9–25].

### 3.2. *Life Cycle Inventory Analysis*

#### 3.2.1. Data Collection and Data Quality

As shown in Table 3, as per analyzed results of the selected LCA studies, the data for LCI are divided into three groups: primary data, which concern the foreground system; secondary data, which concern the background system; and unavailable data. For the reviewed studies, primary data are mainly collected from designer, developer and manufacturer [3,9–13,15,17–21,24,25]. The Inventory of Carbon and Energy (ICE), a database developed by the University of Bath is adopted as an important primary data source by some researchers [11,14,19]. It is also the case that primary data are based on calculation by the researchers [4,23].

**Table 3.** Data collection sources for studied LCAs.

| Reference | Data collection | | |
|---|---|---|---|
| | Primary (Specific) Data | Secondary (Generic) Data | Unavailable Data |
| Cavallaro et al. (2007) | Designer | ETH-ESU 1996, IDEMAT 2001, BUWAL 1996, ETH-ESU 1996, IDEMAT 2001, and ETH 1996 | Assumptions |
| Douglas et al. (2008) | MCT (designer/manufacturer); Inventory of Carbon and Energy (ICE, a database by the University of Bath); | Literature (existing LCAs, journals, and textbooks) | Assumptions |
| Rule et al. (2009) | Reports regarding the studied systems | Literature | Assumptions |
| Howell et al. (2013) | Manufacturers, brochures and presentations | Literature | Assumptions |
| Hans et al. (2007) | Designer and Manufacturer | EDIP database; literature (existing LCAs and reports) | Assumptions |
| Parker et al. (2007) | Inventory of Carbon and Energy (ICE) | Ecoinvent database; literature (journals, conference papers and previous LCA studies) | Assumptions |
| Thomson et al. (2011a) | Manufacturer | Ecoinvent database | Assumptions |
| Thomson et al. (2011b) | Manufacturer | Ecoinvent database | Assumptions |
| Thomson et al. (2019) | PWP's own records by Parker et al. | Ecoinvent database | Assumptions |
| Dahlsten et al. (2009) | Calculation based on drawing, product sheets, product specific processes | Ecoinvent; | Assumptions |
| Walker et al. (2011) | Company website, device patent, installation contractor, and EMEC | Literature | Assumptions |
| Ombach et al. (2014) | Designer, ICE database, compiled by the University of Bath | Unspecified | Assumptions |
| Zhai et al. (2018) | Designer | Ecoinvent database | Assumptions |
| Elginoz et al. (2017) | Designer | Ecoinvent; literature (reports, thesis, scientific papers) | Assumptions |
| Curto et al. (2018) | Unspecified | Unspecified | Assumptions |
| Patrizi et al. (2019) | Metric computations | Ecoinvent database | Assumptions |
| Uihlein et al. (2016) | JRC ocean energy database | GaBi database | Assumptions |
| Douziech et al. (2016) | Plant developers | Ecoinvent; literature | Assumptions |

Databases are widely used as a main source for secondary data collection. Commonly used databases include Ecoinvent [4,14,15,17,20,21,23,25], Gabi [24], EDIP [3], ETH-ESU 1996, ETH 1996, IDEMAT 2001 and BUWAL 1996 [9]. Literature is another important source for secondary data collection, which includes journals, conference papers, theses and previous LCA studies [3,11–14,18,21,25]. However, two of the studies do not describe the source of secondary data [19,22]. For unavailable data gaps existing for almost all wave and tidal LCA studies, reasonable assumptions are usually made for inputs and outputs of LCI [3,4,9–25].

#### 3.2.2. Data Calculation and Energy Flows

Our analysis shows that sources of fuels and electricity are considered by some studies [3,4,11–17, 20,23–25]. A few studies and do not claim the sources of fuels and electricity [9,18,19,21,22]. None of the selected LCA studies discuss efficiency of conversion and distribution of energy flow. Among the studies considering the different fuels and electricity sources, most of them also describe the inputs and outputs associated with generation and use of that energy flow, except for three cases [11,12,17].

3.2.3. Validation of Data

Validation of data can be performed by establishing balances of mass and energy or by analyzing release factors. Figure 2 indicates that only two of the selected LCA studies [19,21] describe the conservation of mass flow for the collected data. However, no further details about the data validation process were provided. Other studies do not provide information about the data validation procedure via either mass or energy conservation.

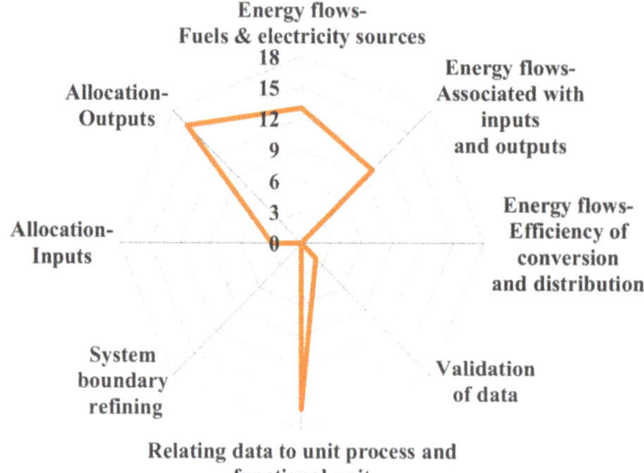


**Figure 2.** Life cycle inventory analysis of studied wave and tidal power systems. Note: the value of each item indicates the number of LCA studies that applied the specific method required in ISOs.

3.2.4. Relating of Data to Unit Process and Reference Flow of the Functional Unit

The relating of data to unit process and reference flow of the functional unit is an optional step of LCI. As shown in Figure 2, most of the selected LCA studies performed the relating of data to unit process and functional unit [3,4,11–25]. Two studies did not provide such information [3,9]. None of the selected LCA studies performed or provided information for the system refining based on the data processing.

3.2.5. Allocation of Inputs and Outputs

An allocation procedure is suggested by ISO since most industrial processes yield multiple, rather than single, outputs and are based on complicated material and energy inputs. As shown in Figure 2, the allocation procedures for reuse and recycling are described by the selected LCA studies except for two [22,23]. Allocation procedures for inputs were considered by few studies [12,17,25].

*3.3. Life Cycle Impact Assessment*

LCIA was conducted on the basis of inventory analysis results by means that the LCI results are assigned, characterized, normalized and weighted with application of given impact categories. Midpoint and endpoint methods look at different stages of the environmental impacts. LCI results are identified and assigned into appropriate impact categories per their environmental relevance and then characterized with specific category indicators so that the specific impact categories can be quantitatively interpreted. A midpoint impact category refers to an impact that contributes to specific aspects of human heath, natural environment or resources. [5]. It is mostly the case that the intended audience decide whether midpoint or endpoint level of environmental impacts should be assessed. As shown in Table 4, midpoint method is commonly adopted for current LCA studies on wave and tidal energy systems [3,4,9,15–17,20–25] except for one study [9].

**Table 4.** Life cycle impact assessment (LCIA) of studied LCAs.

| Reference | Method, Midpoint/End-Point | Classification brk (Assignment of LCI Results) | Characterization brk (Calculation of Indicator Results) | Normalization/Reference (Optional) | Weighting (Optional) |
|---|---|---|---|---|---|
| **Cavallaro et al. (2007)** | **Eco-indicator 99, midpoint and endpoint** | **√** | **Unspecified, no details presented** | **Yes/Europe** | - |
| Hans et al. (2007) | EDIP1997, midpoint | √ | Unspecified, no details presented | Yes/Unspecified | - |
| Thomson et al. (2011a) | EDIP 2003, midpoint | √ | √ | - | - |
| Thomson et al. (2011b) | EDIP 2003, midpoint | √ | √ | - | - |
| Thomson et al. (2019) | ReCiPe and CED, midpoint | √ | √ | - | - |
| Dahlsten et al. (2009) | PCR, midpoint | √ | √ | - | - |
| Zhai et al. (2018) | ReCiPe, midpoint | √ | √ | - | - |
| Elginoz et al. (2017) | CML 2001, midpoint | √ | √ | Yes/Atlantic base scenario | - |
| Curto et al. (2018) | Unspecified, midpoint | √ | √ | - | - |
| Patrizi et al. (2019) | Unspecified, midpoint | √ | √ | - | - |
| Uihlein et al. (2016) | ILCD, midpoint | √ | √ | - | - |
| Douziech et al. (2016) | ReCiPe, midpoint | √ | √ | - | - |

Selected LCAs on wave and tidal energy systems conducted the classification and characterization requirements by ISO 14040 and 14044 via different methods such as Eco-indicator 99 [9], EDIP1997 [3], EDIP 2003 [15,16], ReCiPe [17,20,25], CED [17], PCR [4], CML [21] and ILCD [24]. Normalization is applied to illustrate environmental impact scores by comparison with reference scenarios, such as Europe scenario [9] and Atlantic base scenario [21]. As shown in Table 5, the most relevant impact category is climate change impact, followed by ecotoxicity, resource depletion, human toxicity, eutrophication, ozone layer and acidification. Other impacts such as radiation, particular matter formation, photochemical oxidant formation, bulk waste, land use, slags/ashes and hazardous waste are studied by few case studies. The most significant contributor to each investigated impact category is listed in Table 5.

**Table 5.** Life cycle impact categories and association with most significant contributors.

| Reference | Impact categories | | | | | | | | | | | | | |
|---|---|---|---|---|---|---|---|---|---|---|---|---|---|---|
| | CC | OD | EXT | ACD | EUT | HT | HW | RD | POF | SA | RD | LU | PMF | BW |
| [3] | √[1] | √[1] | √[1] | √[1] | √[1] | √[1] | √[1] | - | √[1] | √[1] | √[1] | - | - | √[1] |
| [4] | M | M | - | M | M | - | - | M | M | - | - | - | - | - |
| [10] | √[1] | √[1] | M | √[1] | √[1] | - | - | M | - | - | - | √[1] | √[1] | - |
| [15] | M | M | O&M | M | O&M | M | M | M | - | M | M | - | - | M |
| [16] | M | M | O&M | O&M | O&M | M | M | M | - | M | M | - | - | M |
| [17] | M | O&M | M | O&M | O&M | M | - | M | O&M | - | M | M | M | - |
| [20] | M | M | EOL | M | EOL | M | - | M | M | - | M | M | M | - |
| [21] | M | M | M | M | M | M | - | M | M | - | - | - | - | - |
| [22] | √[1] | - | √[1] | - | - | √[1] | - | √[1] | - | - | - | - | - | - |
| [23] | M | - | - | - | - | - | - | - | - | - | - | - | - | - |
| [24] | M | M | M | M | M | M | M | M | M | - | M | - | M | - |
| [25] | M | - | EOL | - | - | EOL | - | M | - | - | - | - | M | - |

Notes and Abbreviations: √, investigated; [1] with no contributor specified; -, not investigated; M, manufacturing; O&M, operation and maintenance; EOL, end of life; CC, climate change; OD, ozone depletion; EXT, ecotoxicity; ACD, acidification; EUT, eutrophication; HT, human toxicity; HW, hazardous waste; RD, resource depletion; POF, photochemical oxidant formation; SA, slags/ashes; RD, radiation; LU, land use; PMF, particular matter formation; BW, bulk waste.

## 4. Critical Discussion

Our analyses show that most of current LCA practices on wave and tidal systems are not commendably following the framework, guidelines and requirements established by ISO 14040 and 14044, although ISO standards make the results more convincing to the intended audience. Also, LCA results can be used for formal or official legal disclaimers only if ISO standards are well followed, as regulators and governments commonly count on them for safety insurance in most cases. Finally, results of different LCA studies can only be effectively compared if conducted with same or similar methodological standards.

It is noticed that conventional ISO LCA does not take into account the temporal variability of the inventory data, which is one of the recognized limitations [26–33]. Current LCA practice treats energy, materials, resources and emissions by means of linear summation [33–36]. Thus, various emissions of a material generated at different time periods are treated as a single aggregated emission generated at one time during the life cycle [37]. Apparently, this result is not accurate, as there never exists aggregated emission amount in real world cases [37]. Within this context, recent years have seen development of dynamic life cycle assessment (DLCA) by considering temporal dimension and applying different mathematical models [38]. As pointed out by Müller et al., DLCA shows different results of environmental impacts compared with conventional LCA, especially in climate change and toxicity [39].

Due to absence of commonly recognized mathematical method, the development of DLCA is still at its early stage [40]. Current DLCA application focuses on buildings [41–46] (Batouli and Mostafavi, 2017; Negishi et al., 2018; Hu, 2018; Su et al., 2019; Keiron et al., 2018; Bixler et al., 2019; Cardellini et al., 2018), transportation systems [47,48] and energy systems [49–52] because of their longevities.

Thus, for wave and tidal energy systems, as systems with long lifespans (normally ≥20 years), application of DLCA will help reduce the inaccuracy and uncertainties of environmental impact results. Development of appropriate mathematical methods is encouraged for the conduction of DLCA.

## 5. Conclusions

The scope of this review includes a stepwise check of the selected LCAs on wave and tidal energy systems, from the perspective of their conformance with the ISO 14040 and 14044. The results show that the reviewed studies are carried out in accordance with the ISO standards at different levels. Non-strict conformance with the ISO standards weakens the reliability of the assessment results, whether they are purposed for internal or external applications. This further decreases the comparability between different wave and tidal energy systems, as well as with other energy technologies. Finally, the performed review illustrates that ignorance of temporal variation caused inaccuracy and uncertainty which should be addressed in future research.

**Author Contributions:** Conceptualization and methodology, Q.Z.; validation of data and analysis, X.Z.; writing—original draft preparation, X.Z.; writing—review and editing, L.Z.; project administration, Y.Y. All authors have read and agreed to the published version of the manuscript.

**Funding:** Financial support from Shandong Key Research Program (2019GSF109073) is gratefully acknowledged.

**Conflicts of Interest:** The authors declare no conflict of interest.

## References

1. US Public Law 110–140, 110 Congress, Energy Independence and Security Act of 2007. Available online: https://www.congress.gov/110/plaws/publ140/PLAW-110publ140.pdf (accessed on 18 November 2019).
2. Azzellino, A.; Lanfredi, C.; Contestabile, P.; Ferrante, V.; Vicinanza, D. Strategic environmental assessment to evaluate WEC projects in the perspective of the environmental cost-benefit analysis. In Proceedings of the Twenty-First International Offshore and Polar Engineering Conference, Maui, HI, USA, 19–24 June 2011.
3. Hans, C.S.; Stefan, N.; Stefan, A.; Hauschild, M.Z. Life Cycle Assessment of the Wave Energy Converter: Wave Dragon. 2007. Poster session presented at Conference in Bremerhaven. Available online: https://backend.orbit.dtu.dk/ws/portalfiles/portal/3711218/WaveDragon.pdf (accessed on 11 November 2019).
4. Dahlsten, H. Life Cycle Assessment of Electricity from Wave Power. Swedish University of Agricultural Sciences, 2009. Available online: http://urn.kb.se/resolve?urn=urn:nbn:se:slu:epsilon-s-2115 (accessed on 19 November 2019).
5. ISO14040:2006. *Environmental Management Life Cycle Assessment: Principle and Framework*; CEN: Brussels, Belgium, 2006.
6. ISO14044:2006. *Environmental Management-Life Cycle Assessment: Requirements and Guidelines*; CEN: Brussels, Belgium, 2006.
7. Uihlein, A.; Magagna, D. Wave and tidal current energy-A review of the current state of research beyond technology. *Renew. Sustain. Energy Rev.* **2016**, *58*, 1070–1081. [CrossRef]
8. Banerjee, S.; Duckers, L.; Blanchard, R.E. An overview on green house gas emission characteristics and energy evaluation of ocean energy systems from life cycle assessment and energy accounting studies. *J. Appl. Nat. Sci.* **2013**, *5*, 535–540. [CrossRef]
9. Paredes, M.G.; Padilla-Rivera, A.; Güereca, L.P. Life Cycle Assessment of Ocean Energy Technologies: A Systematic Review. *J. Mar. Sci. Eng.* **2019**, *7*, 322. [CrossRef]
10. Cavallaro, F.; Coiro, D. Life Cycle Assessment (LCA) of a marine current turbine for cleaner energy production. In Proceedings of the 3rd International Conference on Life Cycle Management, Zurich, Switzerland, 27–29 August 2007.
11. Douglas, C.A.; Harrison, G.P.; Chick, J.P. Life cycle assessment of the Seagen marine current turbine. *Proc. Inst. Mech. Eng. Part M J. Eng. Marit. Environ.* **2008**, *222*, 1–12. [CrossRef]
12. Rule, B.M.; Worth, Z.J.; Boyle, C. Comparison of Life Cycle Carbon Dioxide Emissions and Embodied Energy in Four Renewable Electricity Generation Technologies in New Zealand. *Environ. Sci. Technol.* **2009**, *43*, 6406–6413. [CrossRef] [PubMed]

13. Howell, R.J.; Walker, S.; Hodgson, P.; Griffin, A. Tidal energy machines: A comparative life cycle assessment. *Proc. Inst. Mech. Eng. Part M J. Eng. Marit. Environ.* **2013**. [CrossRef]
14. Parker, R.P.M.; Harrison, G.P.; Chick, J.P. Energy and Carbon Audit of an Offshore Wave Energy Converter. *Proc. Inst. Mech. Eng. Part A J. Power Energy* **2007**, *221*, 1119–1130. [CrossRef]
15. Thomson, C.; Harrison, G.; Chick, J. Full Life Cycle Assessment of a Wave Energy Converter. In Proceedings of the IET Renewable Power Generation Conference, Edinburgh, UK, 6–8 September 2011. [CrossRef]
16. Thomson, R.C.; Harrison, G.P.; Chick, J. Life Cycle Assessment in the Marine Renewable Energy Sector. In Proceedings of the LCA XI International Conference, Chicago, IL, USA, 4–6 October 2011.
17. Thomson, R.C.; Chick, J.; Harrison, G. An LCA of the Pelamis wave energy converter. *Int. J. Life Cycle Assess.* **2019**, *24*, 51–63. [CrossRef]
18. Walker, S.; Howell, R. Life cycle comparison of a wave and tidal energy device. *Proc. Inst. Mech. Eng. Part M J. Eng. Marit. Environ.* **2011**, *225*, 325–337. [CrossRef]
19. Ombach, G. Design and safety considerations of interoperable wireless charging system for automotive. *Int. Conf. Ecol. Veh. Renew. Energ.* **2014**, 1–4. [CrossRef]
20. Zhai, Q.; Zhu, L.; Lu, S. Life Cycle Assessment of a Buoy-Rope-Drum Wave Energy Converter. *Energies* **2018**, *11*, 2432. [CrossRef]
21. Elginoz, N.; Bas, B. Life Cycle Assessment of a multi-use offshore platform: Combining wind and wave energy production. *Ocean Eng.* **2017**, 430–443. [CrossRef]
22. Curto, D.; Neugebauer, S.; Viola ATraverso, M.; Franzitta, V.; Trapanese, M. First Life Cycle Impact Considerations of Two Wave Energy Converters. In Proceedings of the Oceans conference, New York, NY, USA, 8 June 2018.
23. Patrizi, N.; Pulselli, R.M.; Neri, E.; Niccolucci, V.; Vicinanza, D.; Contestabile, P.; Bastianoni, S. Lifecycle Environmental Impact Assessment of an Overtopping Wave Energy Converter Embedded in Breakwater Systems. *Front. Energy Res.* **2019**. [CrossRef]
24. Uihlein, A. Life cycle assessment of ocean energy technologies. *Int. J. Life Cycle Assess.* **2016**, *21*, 1425–1437. [CrossRef]
25. Douziech, M.; Hellweg, S.; Verones, F. Are Wave and Tidal Energy Plants New Green Technologies? *Environ. Sci. Technol.* **2016**, *50*, 7870–7878. [CrossRef]
26. Yuan, C.; Wang, E.; Zhai, Q.; Yang, F. Temporal discounting in life cycle assessment: A critical review and theoretical framework. *Environ. Impact Assess. Rev.* **2015**, *51*, 23–31. [CrossRef]
27. Hellweg, S.; Frischknecht, R. Evaluation of long-Term impacts in LCA. *Int. J. Life Cycle Assess.* **2004**, *9*, 339–341. [CrossRef]
28. Pinsonnault, A.; Lesage, P.; Levasseur, A.; Samson, R. Temporal differentiation of background systems in LCA: Relevance of adding temporal information in LCI databases. *Int. J. Life Cycle Assess.* **2014**, *19*, 1843–1853. [CrossRef]
29. Potting, J.; Hauschild, M. Background for Spatial Differentiation in Life Cycle Impact Assessment—The EDIP2003 Methodology. Environmental Project No. 996. Danish Environmental Protection Agency: Copenhagen, Denmark, 2005. Available online: http://www2.mst.dk/Udgiv/publications/2005/87--7614--581--6/html/indhold_eng.htm (accessed on 19 November 2019).
30. Riva, A.; D'Angelosante, S.; Trebeschi, C. Natural gas and the environmental results of life cycle assessment. *Energy* **2006**, *31*, 138–148. [CrossRef]
31. Shah, V.P.; Ries, R.J. A characterization model with spatial and temporal resolution for life cycle impact assessment of photochemical precursors in the United States. *Int. J. Life Cycle Assess.* **2009**, *14*, 313–327. [CrossRef]
32. Yuan, C.Y.; Simon, R.; Mady, N.; Dornfeld, D. Embedded temporal difference in life cycle assessment: Case study on VW Golf A4 CAR. In Proceedings of the IEEE International Symposium on Sustainable System & Technology, Phoenix, AZ, USA, 18–20 May 2009. [CrossRef]
33. Zhai, Q.; Ciardo, K.; Yuan, C.Y. Temporal discounting for life cycle assessment: Perspectives and mechanisms. In Proceedings of the 17th CIRP International Conference on Life Cycle Engineering, Hefei, China, 19–21 May 2010. [CrossRef]
34. Kendall, A.; Chang, B.; Sharpe, B. Accounting for time-dependent effects in biofuel life cycle greenhouse gas emissions calculations. *Environ. Sci. Technol.* **2009**, *43*, 7142–7147. [CrossRef] [PubMed]

35. Reap, J.; Roman, F.; Duncan, S.; Bras, B. A survey of unresolved problems in life cycle assessment. Part 1: Goal & scope and inventory analysis. *Int. J. Life Cycle Assess.* **2008**, *13*, 290–300. [CrossRef]
36. Huppes, G. Methods for life cycle inventory of a product. *J. Clean. Prod.* **2005**, *13*, 687–697. [CrossRef]
37. Owens, J.W. Life-Cycle assessment in relation to risk assessment: An evolving perspective. *Risk Anal.* **1997**, *17*, 359–365. [CrossRef]
38. Chen, S.; Sun, Z.; Li, S.; Liu, Y.; Shi, X. Research and application status of dynamic life cycle assessment. *China Environ. Sci.* **2018**, *38*, 4764–4771. [CrossRef]
39. Müller, A.; Wörner, P. Impact of dynamic CO2 emission factors for the public electricity supply on the life-Cycle assessment of energy efficient residential buildings. *IOP Conf. Ser. Earth Environ. Sci.* **2019**, *323*, 012036. [CrossRef]
40. Su, S.; Li, X.; Zhu, Y. Dynamic assessment elements and their prospective solutions in dynamic life cycle assessment of buildings. *Build. Environ.* **2019**, 248–259. [CrossRef]
41. Batouli, M.; Mostafavi, A. Service and performance adjusted life cycle assessment: A methodology for dynamic assessment of environmental impacts in infrastructure systems. *Sustain. Resilient Infrastruct.* **2017**, *2*, 117–135. [CrossRef]
42. Negishi, K.; Tiruta-Barna, L.; Schiopu, N.; Lebert, A.; Chevalier, J. An operational methodology for applying dynamic Life Cycle Assessment to buildings. *Build. Environ.* **2018**, 611–621. [CrossRef]
43. Hu, M. Dynamic life cycle assessment integrating value choice and temporal factors-A case study of an elementary school. *Energy Build.* **2018**, 1087–1096. [CrossRef]
44. Keiron, P.; Roberts David, A.; Turner, J.; Anne, M.; Stringfellow Bello, I.; Powrie, W.; Watson, G. SWIMS: A dynamic life cycle-Based optimisation and decision support tool for solid waste management. *J. Clean. Prod.* **2018**, 547–563. [CrossRef]
45. Bixler, T.S.; Houle, J.; Ballestero, T.P.; Mo, W. A dynamic life cycle assessment of green infrastructures. *Sci. Total Environ.* **2019**, 1146–1154. [CrossRef] [PubMed]
46. Cardellini, G.; Mutel, C.L.; Vial, E.; Muys, B. Temporalis, a generic method and tool for dynamic Life Cycle Assessment. *Sci. Total Environ.* **2018**, 585–595. [CrossRef] [PubMed]
47. Onat, N.C.; Kucukvar, M.; Tatari, O. Uncertainty-Embedded dynamic life cycle sustainability assessment framework: An ex-Ante perspective on the impacts of alternative vehicle options. *Energy* **2016**, 715–728. [CrossRef]
48. Onat, N.C.; Kucukvar, M.; Tatari, O. Integration of system dynamics approach toward deepening and broadening the life cycle sustainability assessment framework: A case for electric vehicles. *Int. J. Life Cycle Assess.* **2016**, *21*, 1009–1034. [CrossRef]
49. Louis, J.; Pongracz, E. Life cycle impact assessment of home energy management systems (HEMS) using dynamic emissions factors for electricity in Finland. *Environ. Impact Assess. Rev.* **2017**, 109–116. [CrossRef]
50. Beloin-Saint-Pierre, D.; Levasseur, A.; Margni, M.; Blanc, I. Implementing a dynamic life cycle assessment methodology with a case study on domestic hot water production. *J. Ind. Ecol.* **2016**, *21*, 1128–1138. [CrossRef]
51. Zhang, B.; Chen, B. Dynamic Hybrid Life Cycle Assessment of CO2 Emissions of a Typical Biogas Project. *Energy Procedia* **2016**, 396–401. [CrossRef]
52. Pehnt, M. Dynamic life cycle assessment (LCA) of renewable energy technologies. *Renew. Energy* **2006**, *31*, 55–71. [CrossRef]

International Journal of
*Environmental Research and Public Health*

MDPI

*Article*

# Matrix of Architectural Solutions for the Conflict between Transport Infrastructures, Landscape and Urban Habitat along the Mediterranean Coastline: The Case of the Maresme Region in Barcelona, Spain

**Anna Martínez, Xavier Martín * and Jordi Gordon**

IAR Group, School of Architecture La Salle, Ramon Llull University, 08022 Barcelona, Spain; a.martinez@salle.url.edu (A.M.); jordi.gordon@salle.url.edu (J.G.)
* Correspondence: xavier.martin@salle.url.edu

**Abstract:** Maresme is a littoral region of Barcelona (Spain) in which the railway and an important road run along the coastline with a high landscape impact. Over time, several facilities connected to these transport infrastructures have appeared, such as industries, malls, marinas or train stations. These activities profit from the easy connection but create a barrier between the inhabitants and the sea. This research follows three aspects identified in a large variety of locations along the Mediterranean coast: longitudinal mobility, transversal accessibility and landscape discontinuities. The first territorial analysis defines a series of urban problematics classified by category. Then, the most representative case studies are developed by means of urban and architectural projects. The comparative analysis of these proposals provides a catalogue of design strategies which can be combined as criteria for solving multiple conflicts detected in the region. The result of this project is a methodology based on a matrix of general guidelines to ease the solving of local conflicts in a homogeneous way for the whole territory. The final aim is to re-establish order and continuity in the Mediterranean littoral skyline, fostering sustainable mobility and recovering public space for inhabitants.

**Keywords:** urban regeneration; littoral landscape; Mediterranean architecture; sustainable mobility; transport infrastructure; greenway

check for updates

**Citation:** Martínez, A.; Martín, X.; Gordon, J. Matrix of Architectural Solutions for the Conflict between Transport Infrastructures, Landscape and Urban Habitat along the Mediterranean Coastline: The Case of the Maresme Region in Barcelona, Spain. *Int. J. Environ. Res. Public Health* **2021**, *18*, 9750. https://doi.org/10.3390/ijerph18189750

Academic Editors: M. Dolores Esteban, José-Santos López-Gutiérrez, Vicente Negro and José Marcos Ortega

Received: 30 July 2021
Accepted: 13 September 2021
Published: 16 September 2021

**Publisher's Note:** MDPI stays neutral with regard to jurisdictional claims in published maps and institutional affiliations.

## 1. Introduction

This research aims to provide architectural guidelines to solve mobility conflicts between transport infrastructures, landscapes and urban habitats, which are common in the seafront of the Mediterranean littoral [1]. Traditionally, free space between the sea and mountains has been a place for exchanges between civilizations and where the most intense activities have taken place [2]. For this reason, over time, it has been occupied by rural and fishing villages [3] as well as by industries and transport infrastructures [4].

Furthermore, the tourism boom has produced deep transformations in the economy and the way of living for these settlements [5], which have consequently increased accessibility conflicts around the sea due to urban sprawl and gentrification [6]. Recently, the maritime façade has once again become a busy place full of interactions between locals and foreigners, such as on cruise ships or leisure boats or through migrants [7]. In addition, civil society demands the use of the coast as an open public space, free and without physical or visual interruptions [8]. Finally, climate change drives us to rethink the way we occupy coastal territories and move through them in terms of space, time and energy consumption [9].

The present research is in the field of architecture, focusing on three particular action approaches: territory, building and constructive or climatic systems [10]. For this reason, common tools of architectural design are used to analyze, register, classify, synthesize, propose and show the methods. However, its purpose is not solving a professional architectural project but developing a research methodology in relation to an architectonic design process [11]. The practice of the project is systematized and settled from an abstract dimension, following architectural design methods based on synthesis and cause-effect relations [12].

Within this scope, the Maresme coastal region (Barcelona, Spain) is presented as a representative case study because of its conflicts between landscape and mobility (Section 2). This region consists of unique geography located between the mountains and the sea in a narrow piece of land occupied by a sequence of compact urban settlements, tourism sprawl, industrial parks and agricultural fields. In addition, due to the proximity of Barcelona, two large and intensive transport infrastructures have emerged along the seashore: the railway and the National II road (N-II). Due to their linearity, access to the beach and maritime landscape from coastal villages is almost blocked, affecting citizens and tourists. Therefore, the final purpose of this work is to foster more sustainable and healthier mobility systems (e.g., cycling, walking and swimming) [13] not only at a local level but also in relation to the capital city of Barcelona [14].

This research can be understood as an opportunity to solve these problems in a coherent way with the littoral landscape not only in one specific location, but also in other similar places all along the Mediterranean coastline following the methods presented [15–21]. Furthermore, in these littoral areas, there is a common conflict between the management of the different implicated administrations [22] (e.g., municipalities, provinces and government) and also between different infrastructures (e.g., railways, roads and ports). Hence, this research plans to establish unified criteria for the whole littoral that are able to solve most of the detected problematics. As a main result, it provides a tool to facilitate the application of solutions by different stakeholders with similar principles: the matrix of standardized design solutions.

This article shows the cartographic analysis (Section 3.1.), the development of the case study of the Maresme region (Section 4.1.) and the set of architectural guidelines defined as a matrix and the schemes of the standard solutions (Section 5). The visual and universal characteristics of these proposals generate a common base of understanding which aims to promote public debate about the aforementioned conflicts. Furthermore, this is the first step in the process to re-establish the order and continuity of the seafront [23], enhancing healthy mobility and recovering the accessibility of public spaces to citizens [24].

## 2. Hypothesis, Objectives and Case Study: The Maritime and Urban Landscapes of the Maresme Region

This research is focused on answering the following questions. First, how can architectural designs contribute to generating more sustainable mobility and reducing motor transport intensity, such as by train or car? Secondly, which public space design patterns are useful to provide free, open and high-quality landscape accessibility to inhabitants? Finally, is it possible to generate a system of solutions with unified criteria to add order and dignity to the maritime facade based on solving issues in different locations with similar casuistry at a local scale instead of a unique larger project?

The aim of this work is to offer regeneration strategies for mobility and landscape conflicts, understood as guidelines which can be used for different administrations at different scales and with different deadlines but with unified criteria [25]. The following list notes the most representative objectives that set the scope of this research:

- Freedom of movement: the rationalization of mobility and accessibility, ensuring the continuity of pedestrian routes and road paths; pacification of circulation systems; promotion of sustainable and healthy mobility (cycles and pedestrians) and introduction of new types of transport to diversify and unload its intensity [26].
- Sight continuity: the consideration of long-distance views between urban and coastal landscapes; preservation of the continuity of the horizon from inland and definition of the maritime façade and its appearance from the sea [27].

- Compatibility of uses: strengthening the heterogenous character of the littoral; promotion of economic diversity considering existing uses and setting their physical and temporal combination with the new ones and transformation of obsolete buildings and areas with potential for urban regeneration [28].
- Character of place: preservation of historic and natural sites, namely those important for the memory of the place; identification of traditional characteristics related to urban and agricultural structures and the introduction of these activities to the new dynamics of the territory [29].
- Mediterranean culture: the reinforcement of traditional sustainability criteria used by local cultures and following their ways of living and the adoption of the Mediterranean passive systems and materials, including vegetation, to face climate conditions in the design of buildings and public spaces [30].

The research methodology is applied to the case study of Maresme. This region extends along the northern littoral of Barcelona in a narrow territory between the littoral mountain range and the Mediterranean Sea [31]. Its coast length is about 50 km, and the width between the sea and the mountains varies between 5 and 15 km [32] (Figure 1).

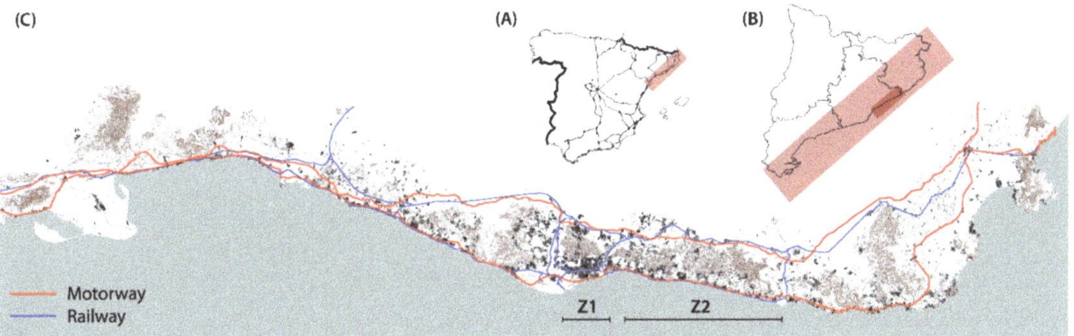


**Figure 1.** Index map of the Catalan coast and location of the Maresme region (Z2) in relation to Barcelona (Z1) (Source: authors' own). (**A**) Location of the Catalan coast in Spain. (**B**) Location of the Maresme region on the Catalan coast. (**C**) Index map detail: topography, hydrography, urban settlements and transport infrastructures.

Traditionally, in this region, transport infrastructures have been placed parallel and close to the coastline, following its uniform topographic level [33]. The first railway built in Spain has connected Barcelona and Mataró (capital of Maresme) since 1848 [34], and the N-II road connects Madrid and France through Barcelona [35]. Both infrastructures have undergone different transformations (e.g., extensions, bridges or roundabouts), but the route along the seafront has been maintained. Coastal villages are located all along the road and connected inland following streams and riverbeds, which are the most representative landscape elements in Maresme [36].

Maresme is a territory with a deep agricultural and industrial tradition [37], although nowadays it is oriented to the third sector, including services and tourism [38]. From the mid-20th century, the expansion of leisure activities fostered the occupation of this territory by means of secondary residences, which finally developed into a stable population [39]. Villages have extended inland and over the hillsides in an urban sprawl configuration. This situation implies an important increase in daily traffic between Maresme's villages and the city of Barcelona, both by train and by car [40].

In relation to the sea, its coastline is a continuous facade that has been compacted by the systematic construction of industrial parks and public facilities in both margins of the urban settlements. Despite this repeated pattern on the littoral front, the region is divided into three landscape units which define their particular values in relation to identity, geography and activities: Baix Maresme, Alt Maresme and Baixa Tordera [41]. The geographic range of this research starts in Montgat and ends in Malgrat de Mar, where the railway goes inland. A series of similar landscape elements link 10 municipalities located along the seafront (Figure 2). It is a region practically built in all its length, and rural spaces between the urban settlements are mainly occupied by crops and campsites [42].

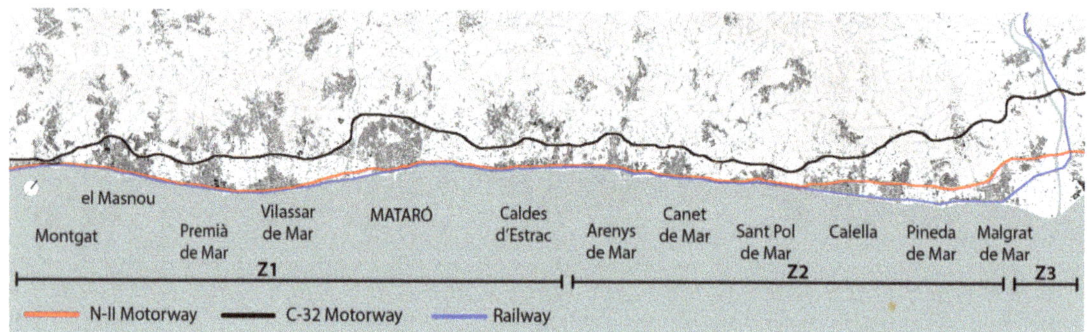


**Figure 2.** Transport infrastructures follow the coastline of Maresme and relate three landscape units: Baix Maresme (Z1), Alt Maresme (Z2) and Baixa Tordera (Z3) (Source: authors' own).

At the edge of this fringe, the N-II road and the railway run longitudinally and in continuity with the coastline. In addition, some cycling lanes and pathways follow this route, but they are fragmented and have several topographic and physical obstacles [43]. This grouping of infrastructures has an average width of about 40 m, but in some specific spots it can be 180 m, with significant unevenness between the interior side and the seaside. Thus, transversal accessibility from the inhabited area toward the sea is irregular due to the multiple types of obstacles and usually narrow public spaces of low comfort quality (Figure 3A). All along this region, due to its particular orography, the presence of riverbanks and streams which flow into the beach is also common, sometimes with significant floods [44]. These hydrographic elements successively cross underneath the infrastructures and are commonly used by citizens to reach the sea [45].

**Figure 3.** Fieldwork views of the Maresme littoral (Source: authors' own). (**A**) Canalized stream mouth below the railway. (**B**) Factory located between the railway and beach. (**C**) Beach bar and facilities close to the railway.

Finally, the seafront has been traditionally occupied by industries due to the easy connectivity by both land and sea (Figure 3B). However, nowadays, very few are still working, and they have instead been replaced by modern industrial parks in the urban peripheries. These historic factory buildings are part of the culture of the site, and even though some of them have been demolished, the most representative ones are being refurbished or are due to be repurposed for new public uses (e.g., the Anis del Mono factory in the nearest coast of Badalona or the flour factory Ylla-Aliberch in Mataró) [46].

Recently, huge commercial malls have occupied areas of this coastal fringe, mainly in the boundaries of the villages. These spots also comprise transport nodes, roundabouts and parking lodgments to provide access to public services and allow inhabitants to access the beach. Both typologies use an important amount of land, despite being located in a privileged landscape facing the sea. Furthermore, the successive presence of nautical clubs and marinas creates restrictions on the freedom of movement and continuous circulation of inhabitants in relation to the beach (Figure 3C).

Finally, regarding climate change, in the Catalan coast, there is a common meteorological phenomenon which produces strong storms from the sea toward the land (llevantades, related to the levant wind). This rough weather drags sand and destroys beaches and seafronts, flooding constructions located along the coast. The latest notorious example is Storm Gloria in January 2020 [47].

## 3. Methodology

### 3.1. Mediterranean Strategies Methodology

This research project followed the methodology established in the research line "Mediterranean Strategies" of the Integrated Architectural Research (IAR) Group, applied by the authors in previous similar studies [48]. This was developed in an inductive process which sought to obtain new ways of acting from the specific study and resolve some representative cases, which were chosen after a deep analysis of the site. The aim was to set a series of guidelines that could be applied to all the identified cases along the territory and also in other similar regions or locations. This investigation focuses on the architectural design, based on the graphic registry as a tool for representation and idea development [49].

Thus, the research methodology was based on the stages noted in Table 1.

**Table 1.** Research process based on IAR's "Mediterranean Strategies" methodology.

| Methodology | Analytical Stage | Proposal Stage |
|---|---|---|
| **Analysis** | (1) Analysis of the territory, history, culture, economy and traditions | (2) Identification and classification of typological problematics |
| **Regeneration projects** | (3) Selection of representative case studies and context analysis | (4) Proposal of transformation, specific models and solutions and typologies |
| **Intervention strategies** | (5) Four types of solutions: site, project, construction and climate | (6) Definition of action guidelines for unitarian territorial regeneration |

#### 3.1.1. Analysis and Classification of Problematics

The first step in the research was the cartographic analysis of the territory (Figure 4). For this purpose, current situation maps were drawn based on fieldwork and on digitalization of existing cartographies [50,51]. Visits to the site were recurrent in order to validate the online information. Drawings and schemes were developed following a common graphic criterion, based on the selection of collected information and in relation to the expected use of these mapping documents (e.g., project requirements, analytical phases, scales and formats) [52].

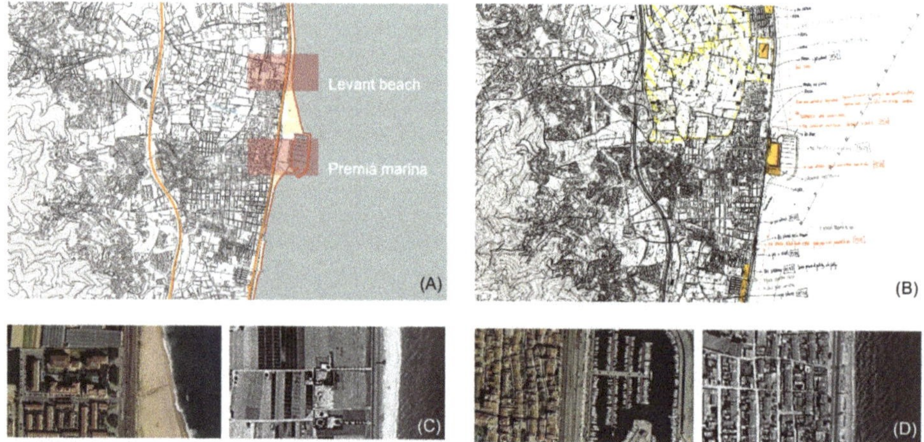


**Figure 4.** Cartographic analysis based on fieldwork and archive mining (Source: authors' own). (**A**) Digital mapping of the Premià de Mar (Sea Premià) area. (**B**) Fieldwork draft map of Premià de Mar. (**C**,**D**) Sample of diachronic analysis by comparison between the current aerial view and the view from an American flight (1956) [51].

At the same time, other intangible elements of the territory were analyzed in order to determine the causes of its current degradation: the history, economy, traditions and culture of the place. The combination of these two approaches on the site eased the detection of its different problematics, which were identified and classified by typologies. This phase ended with the selection of representative study cases [53].

#### 3.1.2. Regeneration Projects

Based on the classified problematics, some study cases were selected as a representative sample of the typological conflicts to be solved. Working on a specific study case instead of a whole territory provided better defined and detailed solutions [54]. In addition, developing specific proposals from urban and architectural processes provided solutions that were more related to the context. Despite the use of architectural design tools, the proposed solutions have an abstract character, since they are based on universal elements [55]. They are presented with visual drawings and models to be comprehensible for all stakeholders.

The proposals of these particular projects are intended to be developed from the strict conditions of the place and paying attention to urban parameters and current constructive systems. With these criteria, new uses and facilities can be proposed with no alteration of their presence in the urban landscape while fostering sustainable solutions in relation to the Mediterranean weather conditions [56].

#### 3.1.3. Intervention Strategies

Finally, the different solutions proposed for each case study are compared and synthesized as action guidelines. These strategies recognize the site integration criteria, the project definition, constructive execution and climate protection. These actions can be applied to all different locations within the territory in a unified sense but are flexible enough to be adapted to each site [57]. The final result of the research is not an architectonic project but a series of basic guidelines to be developed in each of the locations by different professionals and administrations. The methodological process allows for defining these universal criteria and aims to establish strategies to reinforce the quality of the littoral landscape and foster its public use for future generations [58].

### 3.2. Research Lines' Definitions

The first territorial analysis of the Maresme case study highlighted certain problematics that are repeated all along the coast. The study was organized into three research lines related to three different approaches to the territory: longitudinal continuity, transversal permeability and urban landscape. Each line was defined by specific objectives in relation to the identified problematics, mobility and the geographic context. The multiple combinations of these three approaches and their solutions will create an adaptative system to solve all the different conflicts detected by means of unified criteria.

#### 3.2.1. Longitudinal Continuity

Infrastructures and facilities are usually connected following a route parallel to the coastline (Figure 5). However, there are relevant differences between the main roads of fast mobility (car, train or bus) and secondary pathways of slow and light mobility (pedestrians or bicycles). There is also a distinction between long-distance trips, mainly related to Barcelona, and the short-distance ones between villages within Maresme.

**Figure 5.** Longitudinal continuity studies the effects of the infrastructures and the civic ways which connect different urban settlements all along the coast (Source: authors' own).

Railways run along a uniform topographic level, usually delimited by fences on both sides. Depending on the area, this infrastructure consists of one or two lines in both directions. Every village has a station or a stop, and sometimes even two of them. Most of the platforms have been lengthened to support longer trains, as the frequency of services and volume of users are very high during rush hours [59]. Platforms are usually accessed by underground corridors. Nowadays, ticket validation is automated, so most of the traditional buildings have been transformed or closed to the public.

Despite being situated at the seafront, it is normal that in the short term, there is not a clear plan to get rid of this type of railway infrastructure [60]. In the city of Mataró, there is an urban project to modify its route inland, but it is currently unaffordable in terms of costs and political management [61]. In any case, there is a plan to transform this railway into a tram that is slower and less aggressive in relation to the landscape [62]. Therefore, the case study projects take the following consideration into account: keeping the same route but with a lower traffic intensity.

The N-II road has never had its route modified within the Maresme region. It connects all villages in the area and extends toward Barcelona to the south and toward Girona and France to the north. This road has two-way traffic, with three to four lanes in urban areas and up to six lanes in some inter-urban zones. Inland connections and village access are solved with roundabouts and crossroads. Here, there is a plan for directing part of the traffic supported by the N-II road to other existing highways inland, such as the Comarcal 32 (C-32) motorway, which is faster and wider [63]. This way, in the future, this road will be converted to a civic pathway, adequate for local or regional traffic. This future condition is also taken into account in the case study proposals.

Both infrastructures (railway and N-II road) are continuous elements with a strong presence, sometimes even located at different topographic levels. On the contrary, pedestrian and cycle pathways along the seaside are discontinuous and fragmented but with great potential. These pathways are recognized as part of the history of the region, being very attractive in terms of landscape and ability to foster sustainable and healthy mobility [64]. These routes are commonly located in the narrow space between the railway and the sea. Obstacles might be produced by natural elements (e.g., streams or topography), by buildings or facilities (e.g., stations, marinas), by infrastructure (e.g., underground passages) and also by different urban treatments in the boundaries between municipalities.

This research line recognizes the necessity of giving continuity to these relatively small and healthy longitudinal itineraries, which are mostly used by the inhabitants in short trips or even related to leisure and tourism activities. However, these could also support longer distances and higher volumes if they became easy to use, comfortable and secure. The design and measures of these external lanes have to be developed considering the storms coming from the sea and the water level rising [65].

The specific objectives of this research line consist of pacifying and diversifying the different types of traffic and ensuring the coherence of their routes parallel to the coastline. For this purpose, two parallel and continuous ways can be disposed of in relation to the central railway: a civic pathway on the inner side and a landscape-friendly greenway next to the sea on the outer one.

### 3.2.2. Transversal Permeability

Maresme's hydrographic river basin is singular due to the fast overflow of rainwater and conflicts generated by the inefficiency of the stream mouths, which need to cross below coastline infrastructures toward the sea [66]. This problem is similar for the inhabitants and tourists' mobility from villages facing the beach. They also have to cross the coastline infrastructures, which are sometimes highly uneven, implying an unacceptable lack of safety (Figure 6).

**Figure 6.** Transversal permeability studies the connectivity between the inland area and the beach, following the streams which cross infrastructures through underground passages (Source: authors' own).

This transversal connectivity for water, vehicles and pedestrians is solved in many different ways all along the coastline but usually with no order and without any comfort quality. Some streams have already been canalized, but most of them still flow into the beach, as do some sewer pipes and local overflow channels. Furthermore, access for vehicles to the coast is made possible through underground passages connected to the urban road system. There are few of these in relation to pedestrian ones, and the conditions for construction (e.g., measures, slope, curves and lane width) usually require a large amount of land to be occupied with no other use. Furthermore, these vehicle passages do not take into account compatibility with pedestrian circulations.

Finally, the majority of pedestrian access routes to the beach are underground passages. These can be public or private and are regularly distributed along the coast. Most of them are former sewer pipes, so they do not meet the minimal measure requirements nor do they provide comfort or safety. Sometimes, railway platforms are also used to cross the infrastructure, and there are very few examples of elevated footbridges.

The main objective of this second research line is to ensure transversal accessibility to the coast from inland urban settlements, rationalizing their location and enhancing the quality of underground passages. The plan is to set a regular pattern of passages along the coast in both urban and inter-urban areas equally and especially in the crossroads between longitudinal infrastructures and transversal axes (e.g., streets and pathways). It is also important to provide spatial and comfort quality for these passages, fostering their use as a public space and an agent of dynamic urban activity between both sides of the railway. It will be necessary to include design solutions to keep mobility safe even when floods happen. Solutions for extreme flooding and natural disasters are not part of the aims of this project.

#### 3.2.3. Urban Landscape

The case studies identified in this research line refer to isolated or extensive constructions of different sizes and also to urban voids which produce important discontinuities in the seafront (Figure 7). These are equally located on both sides of the infrastructures in urban and inter-urban areas. Some examples of these constructions are disused buildings, those undergoing a transformation process (e.g., factories and railway stations), enclosures of private complexes (e.g., nautical clubs, marinas and campsites) and parking lodgments related to the beach, commercial areas and stations.

**Figure 7.** Urban landscape studies obsolete enclosures and buildings with multiple typologies, which become discontinuities in a general view of the littoral landscape (Source: authors' own).

Regeneration of obsolete buildings as well as the definition of links between urban voids and natural free spaces provide better landscape conditions [67]. Due to their capacity to support new uses if needed, these areas of conflict might become nodes of activity and provide a great opportunity to organize longitudinal and transversal circulations [68].

This research line aims to recover the quality and continuity of the seafront by means of balancing the presence of infrastructure, buildings and open spaces. Projects will enhance the value of the littoral landscape, considering all the elements which are part of this [69] (i.e., not only the physical ones, but also visual, cultural, urban and historical references, always in relation to the place).

Table 2 shows a classification of the identified problematics along the Maresme region, organized according to these three research lines. This is the first step to recognize the needs of the territory and to set a complete image to ease the identification of the most representative case studies to be developed (Table 2).

**Table 2.** Identification of typological conflicts classified by research lines (Source: authors' own).

| Longitudinal Continuity | Transversal Permeability | Urban Landscape |
| --- | --- | --- |
|  |  |  |
| L1_Parking lodgments | T1_Beach stream mouths | P1_Railway stations |
|  |  |  |
| L2_Inadequate pavement | T2_Vehicle underground passage | P2_Abandoned factories |
|  |  |  |
| L3_Seafront not urbanized | T3_Pedestrian underground ways | P3_Parking lodgments |
|  |  |  |
| L4_Streams | T4_Elevated footbridges | P4_Temporary beach buildings |
|  |  |  |
| L5_Railway stations | T5_Railroad crossing | P5_Nautical clubs and facilities |
|  |  |  |
| L6_Broken pathways | T6_Station underground passages | P6_Footbridges |

**Table 2.** *Cont.*

| Longitudinal Continuity | Transversal Permeability | Urban Landscape |
|---|---|---|
| L7_Vehicle underground passage | T7_ Pedestrian private passages | P7_Roundabouts |
| L8_End of pathway | T8_Natural streams | P8_Marinas |
| L9_Topographic change | T9_Elevated vehicle passages | P9_Factories and commercial |
| L10_Seafront facilities | T10_Natural footbridges | P10_Abandoned urban voids |
| | | P11_Campsites |

### 3.3. Setting of Action Variables

Once the problem typologies were set, a series of variables could be defined to clarify the landscape conditions of the site. Based on a general overview, these variables were related to the transversal section of the coastline and the landscape elements which were part of it: topography, hydrography, constructions or activities. A combination of these variables and problematics would establish the selection of a varied and valid sample of case studies, which could be developed by means of an urban and architectural project.

Four variables were defined in relation to the coastline conditions (Table 3):

- Variable 1: the typology of occupations in the inner part of the infrastructures in relation to urban settlements and rural areas, considering activities and constructions;
- Variable 2: uses and landscape typologies in the outer part of the infrastructures in relation to the coastline and activities related with the sea, leisure and landscape;
- Variable 3: the topographic difference between both sides of the infrastructures, which is variable along the coastline and defines the conditions for the connection between villages and the sea;
- Variable 4: the distance from the sea and the first line occupied on the inner side of the infrastructures, which is variable along the coastline and defines the relation with the urban habitat from villages toward the sea.

**Table 3.** Classification of the action variables and landscape elements (Source: authors' own).

| Variable 1 | | Variable 2 | |
|---|---|---|---|
|  V1A_Large urban consolidated | 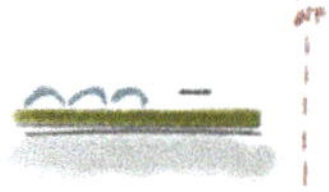 V1F_Crops |  V2A_Marina |  V2F_Beach |
|  V1B_Small urban consolidated |  V1G_Transport |  V2B_Fish harbor |  V2G_Parking/Rocks |
|  V1C_Isolated urbanconsolidated | 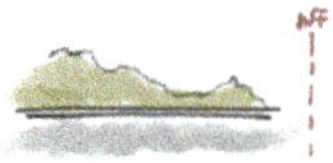 V1H_Leisure nature |  V2C_Seafront/Beach | 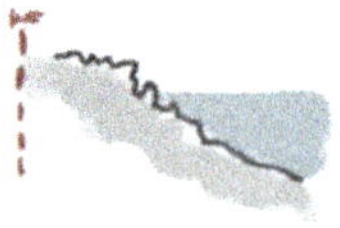 V2H_Rocks |
|  V1D_Urban campsite | |  V2D_Circulations/Parking/Beach | 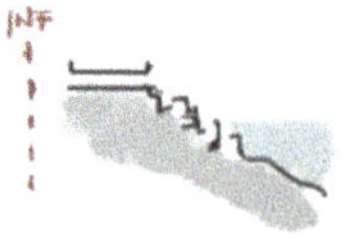 V2I_Seafront/Rocks |
| 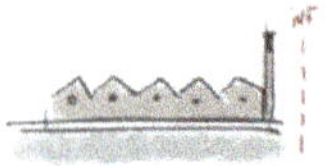 V1E_Industry | | 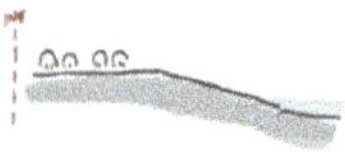 V2E_Parking/Beach |  V2J_Seafront/Rock/Beach |
| **Variable 3** | Topographic difference between both sides | | |
| **Variable 4** | Distance between inhabited urban front and the beach | | |

## 4. Results

### 4.1. Urban Regeneration Projects

From the territorial analysis and fieldwork focusing on the three research lines, some representative case studies were selected as a sample to provide solutions to the identified typological problematics and action variables. Each one of these locations was developed by means of an urban regeneration project. A comparison of the solutions provided would allow for establishing a series of general guidelines to be applied to the whole region. Working with particular locations eases the development of real projects, following a methodology from the most general to the most particular and reverting again to a larger scale with the standard strategies.

The proposed solutions have an abstract character to them, as they are based on universal relations between architectonic elements, such as pedestrian walkways, squares, underground passages, courtyards, stairs or ramps. The selection of case studies was based on a combination of typological problematics and action variables, both of which were related to landscape elements, the historical evolution of each site, the potential for urban regeneration and access to required information (Table 4). To develop this phase of the research, four case studies were selected, and the projects are detailed in the following subsections (Figure 8). The descriptions of each case study include some architectural works presented as project references related to materiality, climatic systems or structure.

**Table 4.** Case study selection with identification of problematics, research lines and variables.

| Case Study | Problematics [1] | Variables |
|---|---|---|
| Downtown of Premià de Mar | L3<br>T1 + T3<br>P4 + P10 | V1H + V2C + V3 + V4 |
| Stop on the beach of Vilassar-Cabrera | L1 + L3 + L7<br>T1 + T2 + T3 + T6<br>P3 + P4 + P5 | V1A + V2D + V3 + V4 |
| Railway station of Mataró | L1 + L5 + L10<br>T3 + T5 + T6<br>P1 + P10 | V1B + V2A + V3 + V4 |
| Stream mouth in Llavaneres | L3 + L4 + L7 + L10<br>T8 + T3 + T2 + T1<br>P4 + P5 + P10 | V1H + V2C + V3 + V4 |

[1] Longitudinal continuity (L), transversal permeability (T) and urban landscape (P).

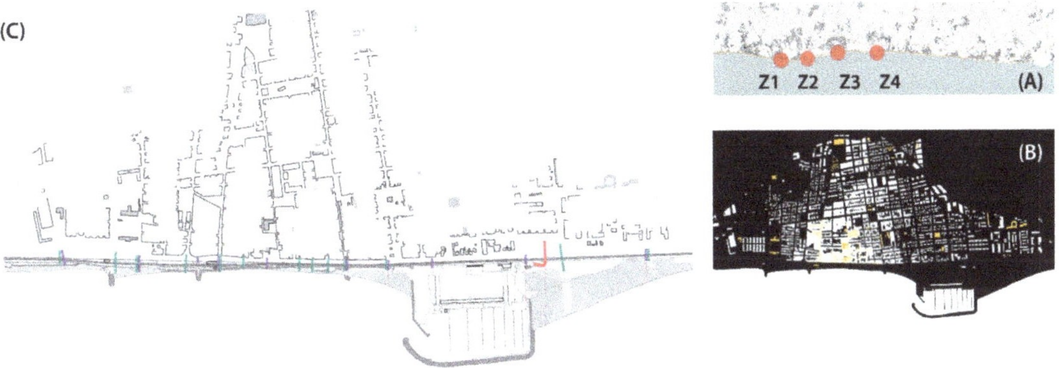


**Figure 8.** Graphics of the first approaches from urban regeneration projects (Source: authors' own). (**A**) Location of case studies in the Maresme region. (**B**) Figure and ground diagram of Premià de Mar, with urban structure, public spaces and buildings related to the coastline. (**C**) Transversal street and underground passage catalogue of Premià de Mar.

#### 4.1.1. Downtown of Premià de Mar

This case study was located in the downtown area of Premià de Mar and an example of a linear solution (greenway and civic way) and of public urban spaces connected to crossroads. This village was selected due to its average size in the region, which is accessible for pedestrians and has a compact, historic downtown that sets the maritime façade.

Urban analysis detected the existence of multiple underground passages, most of them being former sewer pipes, which connected streets, squares and facilities to the sea. To give a solution to this aspect, the project focused on a generic section for both a civic way and a greenway, each one located on a different side of the railway. Two public squares appear at the crossroads: one with a more urban character at the end of the downtown street and the other with a more landscaping-based character in the peripheries [70].

The idea of using underground passages to foster accessibility toward the sea was based on several aspects, including sustainable responsibility by reusing the existing passages and improving their design quality and also through landscape integration by aiming to preserve clean views over the zero level and the horizon, which defines the seafront. In addition, these passages need to meet basic functional, comfort and security requirements, considering the relevant risk of flooding.

In the littoral square, the project included some facilities related to the beach (e.g., showers, benches and bike rentals). There was also an opportunity to consider the construction of lightweight docks to protect the beaches from sea storms and also provide a new public transport system of boats similar to Venetian vapporetos to diminish the traffic of land infrastructures [71]. Constructive and architectonic criteria applied to solve sections of longitudinal pathways and public squares need to be replicable in the whole littoral [72]. These need to be flexible systems that can be adapted to the diverse conditions of each place and made of industrialized and easily maintainable materials (Figure 9).

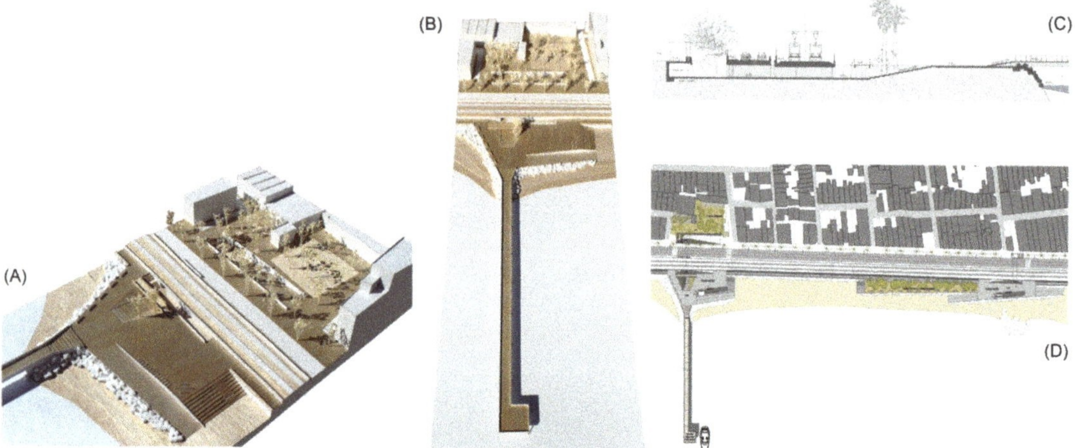


**Figure 9.** Urban and architectural project for the case study of downtown Premià de Mar (Source: authors' own). (**A**,**B**) Model pictures of the dock and public squares. (**C**) Transversal section through an underground passage. (**D**) General floorplan with the urban and landscape squares, a lightweight dock and a seaside greenway.

4.1.2. Stop on the Beach of Vilassar-Cabrera

The case study of a stop on the beach of Vilassar-Cabrera is an example of parking lodgments along the seafront related to stations and other facilities. This is a common typology of railway stations in the Maresme region. It is located on the outer part of the railway, so the solution is easily repeatable in other stations. Urban voids and uses of parking and public services were considered in this project [73].

Urban analysis showed that this is an inter-urban station with a high influx of users that need to access it by car. On the sand dunes, there is a parking lodgment which can only be reached through an underground passage. Citizens, tourists and patrons of restaurants and other leisure activities also park in this area. The seafront has a discontinuous skyline, with several buildings located in different alignments and a lack of urban and architectural ambience. Furthermore, the pedestrian underground passages are too narrow and distributed irregularly.

The seafront can be solved in a continuous way all along the outer part of the railway. This axis is oriented toward pedestrians, bicycles and service vehicles. All diverse beach establishments are designed with a system of prefabricated constructions to unify the maritime façade [74]. These lightweight constructions are distributed in contact with the seafront to help organize parking lodgments, commercial activities and the platforms of the stop.

In this case, a construction system solved both facades and roofs at the same level as the railway lines, with attractive open views toward the landscape. These were lightweight and prefabricated constructions made of timber structures and lattices. Due to the usual overlapping with the maritime public domain boundaries, these buildings needed to be temporary and easy to dismantle [75]. Finally, the underground access paths to the platforms were widened and illuminated through patios, following the same criteria of the previous case study [76]. The underground passage for vehicles was also transformed with a more urban treatment related to the adjacent public square, making it adequate for pedestrians (Figure 10).

**Figure 10.** Urban and architectural project for the case study of a stop on the beach of Vilassar-Cabrera (Source: authors' own). (**A**) General elevation of the maritime façade. (**B**) Transversal section through the underground passage. (**C**) Model picture of the prefabricated constructions. (**D**) General floorplan with different elements used to add order to the maritime seafront: prefabricated buildings, greenway and urban public spaces.

#### 4.1.3. Railway Station of Mataró

The case study of the railway station of Mataró is an important example, as it is located in the largest town in the Maresme region. This focused on the regeneration of an obsolete historic building to enhance connectivity and to act as a creator of civic activity in the city [77].

Urban analysis detected that this is a traditional, disused building; however, it is located in a strategic position: in the fringe between an urban settlement and the sea. Access to platforms and to the seaside is produced by underground passages with no spatial or comfort qualities. Most of them are former sewer pipes.

The project proposes to widen the existing underground passages and to connect all of them with the basement of the railway station [78]. With this solution, all of these elements can be unified, creating a huge underground lobby with courtyards and vegetation. As such, all the underground connections (e.g., the nautical club, beach, platforms and station) are supported by one unique space that is bigger and better controlled in terms of maintenance and security. This space is not an underground passage anymore but a building with natural light, ventilation and views of the outside [79].

Regarding the existing building of the railway station, it can be refurbished and recovered as a public facility for the city. The ground floor opens to the civic way, and the roof is also accessible to the public, with attractive views over the sea. The intermediary levels can be transformed for public or private use for the third sector (i.e., services, associations, coworkers and start-ups).

This project is thought to be a representative case of obsolete urban buildings. Their integral architectonic regeneration, with the addition of new public uses, transforms these spots into new centralities and nodes of urban activity. In addition, this typology also solves the longitudinal continuity and transversal permeability issues in a unique and clear type of intervention (Figure 11).

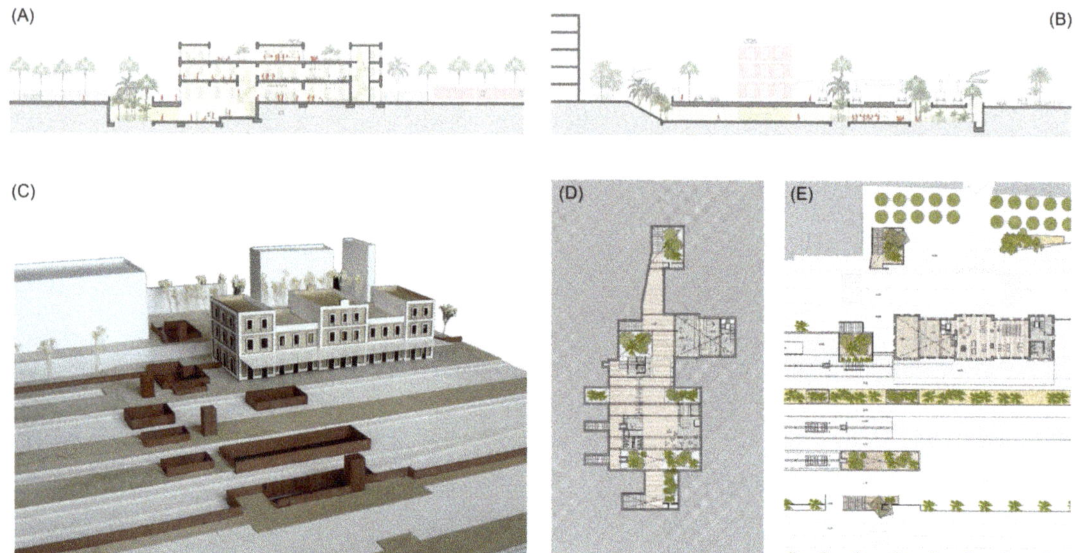


**Figure 11.** Urban and architectural project for the case study of the railway station of Mataró (Source: authors' own). (**A**) Section of the historic building and lobby. (**B**) Transversal section of the underground passage. (**C**) Model picture of the station. (**D**,**E**) Basement and ground floor plans with green courtyards and an underground structure following the railway platforms.

#### 4.1.4. Stream Mouth in Llavaneres

The case study of the stream mouth in Llavaneres is an example of crossing between a longitudinal greenway and the ending of an important hydrographic element used as a public space while it is not flooded. The objective was to generate a set of pathways to enhance the transversal mobility inland but in a healthier and more sustainable way [80]. At the same time, as this stream is a public space, eventual floods and the effects on existing constructions were considered (e.g., bridges, restaurants, stations, parking lodgments and facilities).

Urban and landscape analysis detected that this type of stream usually had little water, so it could be regularly used as a public space [81]. However, there are sometimes floods that fill the stream and convert it into a river. The mouth of the Llavaneres stream is a paradigmatic case with its width (about 40 m) and the two bridges that provide continuity of the railway and the N-II road, which solve several topographic unevenness issues. Inside this hydrographic element, there are some spontaneous parking lodgments, and on the beach, there are several disorganized facilities and establishments.

This project aimed to treat all these crossing spaces with an urban character, solving the connection between the longitudinal greenway and the transversal streams all along the coast [82]. Public urban spaces created around the stream mouth and close to the beach can be designed as public squares with strong materials that can support eventual floods [83]. The stream, due to its dimensions and importance, is projected as a green corridor extending to the coastline that is designed using nature-based systems [84].

All the parking lodgments and facilities related to the beach were organized by means of lightweight constructions, using the same prefabricated wooden system in the previous case studies. Due to topographic unevenness, this project included a footbridge over the railway to ease accessibility to platforms and to the beach. Additionally, the public square was equipped with facilities related to sports and healthy activities (e.g., bike rentals, showers and urban furniture) and to renewable energy production (e.g., solar panels and wind turbines) (Figure 12).

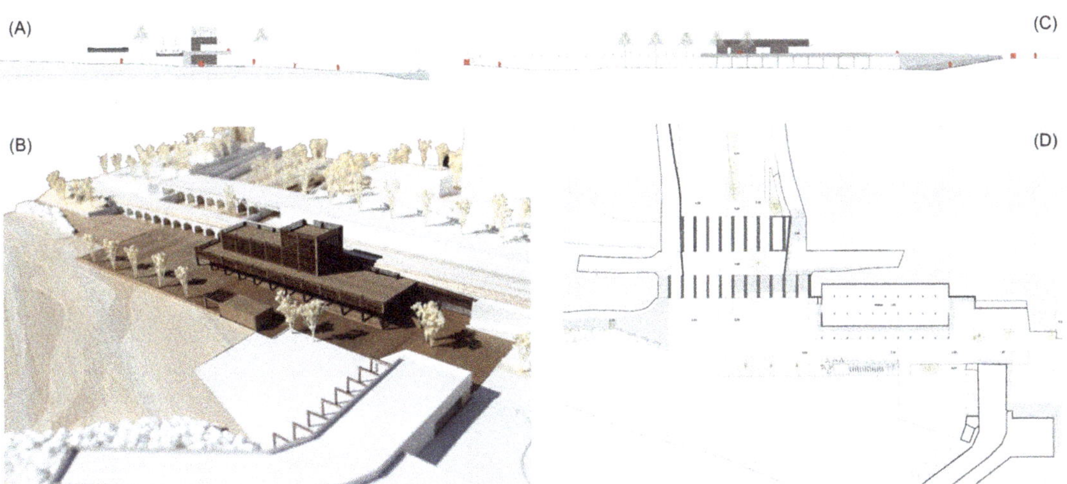


**Figure 12.** Urban and architectural project for the case study of the stream mouth in Llavaneres (Source: authors' own). (**A**) Transversal section of the underground passages. (**B**) Model picture of prefabricated constructions. (**C**) General elevation of the maritime façade. (**D**) General floorplan with green treatment of the stream, floodable squares and light-weight constructions for local services and the railway station.

### 4.2. Intervention Strategies

Specific resolution of the four urban and architectural projects, understood as typological case studies, would make it possible to visualize diverse situations in a common graphic register (e.g., drawings, 3D models and physical models). Comparison of these projects establishes a compound of action guidelines. These are intervention strategies which can be applied to similar locations along the Maresme littoral or along other coastal regions following the same research methods. These strategies are classified following the three research lines which structure the whole methodology.

#### 4.2.1. Longitudinal Continuity Strategies

These solutions are focused on transforming the N-II road in a civic way and ensuring continuity of the seaside greenway:

- Reduce the road width, with one lane in each direction and an auxiliary third one;
- Widen and urbanize the urban sidewalks to ease the introduction of large courtyards for better access to underground passages;
- Widen the sidewalk between the N-II road and the railway by getting rid of the concrete protections (New Jerseys), and small courtyards and staircases can provide light and access to underground passages;
- Ensure continuity for pedestrians and bicycles by means of a two-way greenway along the seaside;
- Connect the seafront with the inner side of the infrastructures, with a regular distribution of underground passages at a local scale and green corridors through the main streams at a regional scale;
- Transform railway protections at the seaside (i.e., in contact with the longitudinal greenway) using stone gabions and vegetation;

- Foster longitudinal light mobility with a regular distribution of facilities related to bicycles, beach services and renewable energy generation;
- Promote alternative public transport systems such as bike rentals or ferries between Maresme villages and Barcelona;
- Construction of lightweight docks to protect beaches from sea storms, provide service to ferries and show a new sight of the coast as a leisure and tourism landmark.

#### 4.2.2. Transversal Permeability Strategies

These solutions are focused on fostering accessibility to the sea from the urban settlements located further inland:

- Transform existing underground passages instead of building new ones;
- Use the most suitable existing passages in relation to flood prevention (i.e., those that are not directly facing a street toward the sea);
- Provide universal accessibility to all underground passages by means of staircases, ramps and elevators;
- Prevent the direct entrance of rainwater by means of massive handrails and topographical modification of access routes to underground passages;
- Improve the ambience of these passages to make them more secure and comfortable by widening them and adding natural light and ventilation, and as the N-II road is reduced, the passages' lengths will decrease equally;
- Enlarge access paths to underground passages, with two on both sides and one in the middle;
- Use vertical elements in the access points, such as elevators or wind turbines, to create a recognizable landmark repeated regularly along the coast;
- Add regular footbridges related to light constructions to solve high unevenness;
- Offer new uses for the basements of public buildings related to underground passages, such as adding courtyards and vegetation (e.g., station lobbies and access to platforms);
- Modify underground passages for vehicles to move from motorway aesthetics toward a more urban one, with pavement and furniture such as those in public spaces.

#### 4.2.3. Urban Landscape Strategies

These solutions are focused on providing a unified image to the littoral landscape that is easy to understand from both sides (inland and sea):

- Add new activities to obsolete buildings, parking lodgments and huge commercial complexes in the seafront to highlight them as catalysts for urban regeneration;
- Recognize historic railway stations as part of the collective memory of each place and connect their basements to underground passages, creating activity nodes;
- Open the ground floors of public buildings toward adjacent urban spaces to enhance recognition and active use of these facilities;
- Organize facilities, parking lodgments, stations, commercial areas and beach services by means of lightweight prefabricated constructions that are easy to dismantle;
- Rethink the urban character of nautical clubs and marinas to enhance the continuity of seaside greenways as landmark elements;
- Transform the boundaries of private enclosures (such as campsites, sports fields and nautical clubs) into wide edges with vegetation related to the littoral landscape.

## 5. Discussion

Once the architectural strategies are defined and classified in each of the three research lines, the last step of the methodology consists of defining a matrix of standardized design solutions. This document is a diagnosis tool to be used during the brainstorming stage of urban and architectural projects developed to solve the problematics detected along Maresme's littoral. Each one of these proposed solutions was related to a category regarding its effects on mobility and landscape: 1_Longitudinal, 2_Transversal, 3_Crossing and 4_Building.

This first stage of research concluded with a definition of eight standard architectural solutions organized by pairs in each category. As a catalogue, these strategies could be applied along the territory with the aim to solve the multiple combinations between typological problematics and action variables, both of which were identified at the beginning of this research.

The matrix of standardized design solutions is graphically defined in Figure 13 as a scheme which shows the two infrastructure sides (urban above and maritime below) and different types of transversal passages. The conceptual schemes of the eight standardized solutions are drawn in red. In addition, each one of the architectonic strategies is represented with a generic virtual model (3D views, floorplans and sections), which provides the recommended dimensions and materials to be applied in forthcoming architectural projects promoted by local administrations (Figure 13).

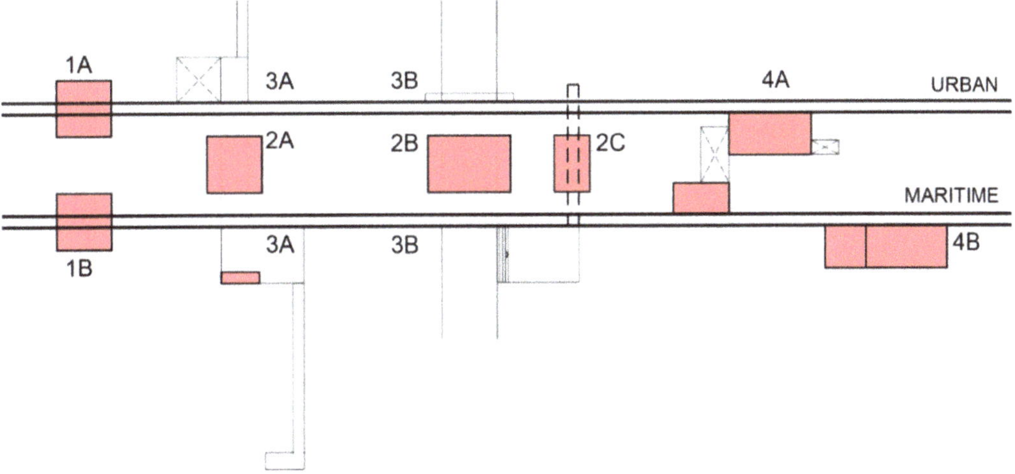


**Figure 13.** Matrix of standardized design solutions (Source: authors' own).

1A_ Longitudinal/Urban represents the typological solution for the urban civic way. The urban sidewalk is widened with pavement and trees. The N-II road is reduced, with two lanes for circulation and a third auxiliary one for bus stops, parking and stop and go. Protections with the railway are kept. This is the basic typological section which is extensible to the whole littoral (Figure 14A).

1B_Longitudinal/Maritime represents the typological solution for the greenway along the seaside. It consists of a two-way pathway for both pedestrians and bicycles. When the topography is reduced or too steep, the solution provides concrete prefabricated panels which can be fixed, cantilevered or solved as a bench with a handrail. In those sections with more terrain, the solution includes vegetation, urban furniture and facilities. This is also an extensible solution for the whole littoral (Figure 14B).

2A_Transversal/Underground/Non-floodable is the definition of a typology of passage understood as the transformation of an existing one. It contains three access courtyards: two in the extremes and one in the center. Passages are accessible by means of elevators, ramps and staircases located following a longitudinal circulation. Canopies for water protection and wind turbines are located close to access points to be recognized as landmarks. Flood prevention is taken into account by their location far from the main streams and also by including solid handrails and podiums. Interior wall treatment of the tunnels is proposed with light-colored ceramics that are easy to maintain (Figure 15A).

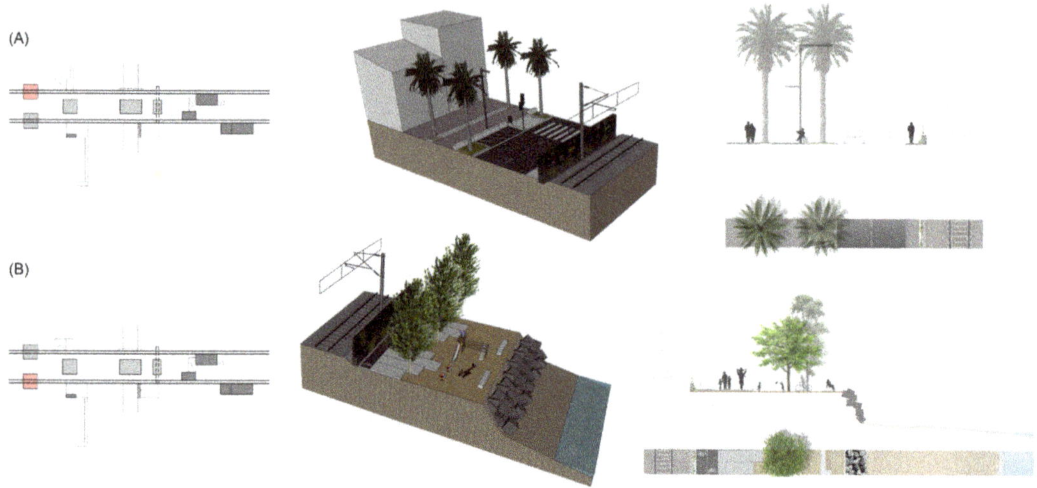


**Figure 14.** Standardized solutions for longitudinal typologies (Source: authors' own). (**A**) Urban side and civic way. (**B**) Maritime side and greenway.

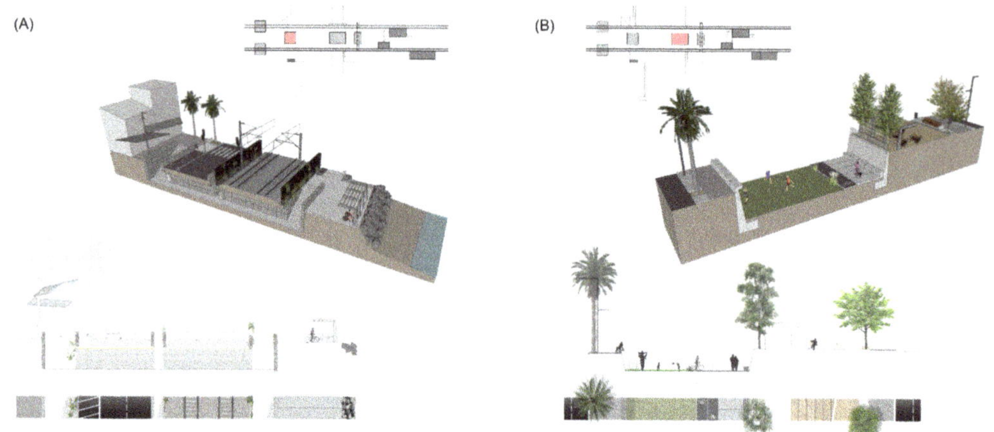


**Figure 15.** Standardized solutions for transversal typologies (Source: authors' own). (**A**) Underground public space, non-floodable. (**B**) Underground public space, floodable.

2B_Transversal/Underground/Floodable is a solution that focuses on the design of streams with eventual water flow but of a high intensity. These are planned to be green connections for pedestrians and bicycles toward the interior of the region. The riverbank edges are designed with concrete walls and vegetal treatment or also with steps for leisure activities. The access points are based on ramps and staircases linked to sidewalks at the street level, where prefabricated constructions can be distributed with restaurants and other facilities protected from water. The riverbank is treated with autochthonous Mediterranean vegetation, as are the upper sidewalks (Figure 15B).

3A_Crossing/Non-floodable/Urban and Maritime is a typology that solves the crossing of longitudinal and transversal axes, which can be combined along the littoral. The solution provides two types of squares in relation to their location: urban or maritime. These public spaces are wide and have clear geometries. Materiality is defined with soft and hard pavements, vegetation, urban furniture and facilities for bikes, ferries and beach services. Access to underground passages is solved with large courtyards which can be covered with canopies. Urban squares, usually located downtown, are connected to the docks (Figure 16A).

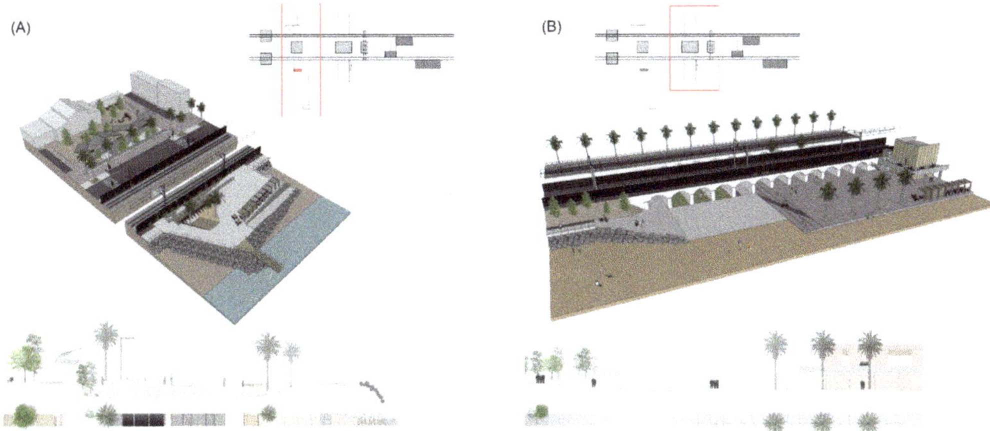


**Figure 16.** Standardized solutions for crossing typologies (Source: authors' own). (**A**) Non-floodable crossing, both urban and maritime. (**B**) Floodable crossing, both urban and maritime.

3B_Crossing/Floodable/Urban and Maritime is a typology that solves the crossing of longitudinal greenways and transversal streams. These are nodes for pedestrians and bicycles but also access points for service vehicles (e.g., for maintenance and restaurant supplies at the beach level). The solution considers the real possibility of having this urban space flooded. For this reason, the topography is adapted with paved platforms with concrete prefabricated panels and vegetation along the edges. Wooden modular constructions can be distributed to add order to existing commercial activities and parking lodgments (Figure 16B).

4A_Building/Urban is a typology focused on the transformation of existing buildings, with the addition of new uses and the creation of a connection between longitudinal and transversal mobilities. Buildings are developed in the basement, sharing spaces with existing underground passages for access to railway platforms or to the beach. These spaces are understood as lobbies with courtyards, skylights and vegetation. The rest of the building is refurbished to support public uses. The ground floor can be opened toward the civic way and the city. This solution is mainly for obsolete railway stations, but it can be applied to other buildings, such as factories or facilities (Figure 17A).

4B_Building/Maritime is a typology focused on defining a modular system to build dismountable constructions in order to unify the urban landscape along the longitudinal greenway. This system can organize all different kinds of uses and activities on the coastline, including restaurants, commercial businesses, nautical clubs, beach services, railway stops and parking lodgments. In areas of high unevenness, isolated elements can be hidden with canopies that add new covered areas to host outdoor activities. These constructions are prefabricated, being made of timber structures and lattices, fabrics and vegetation (Figure 17B).

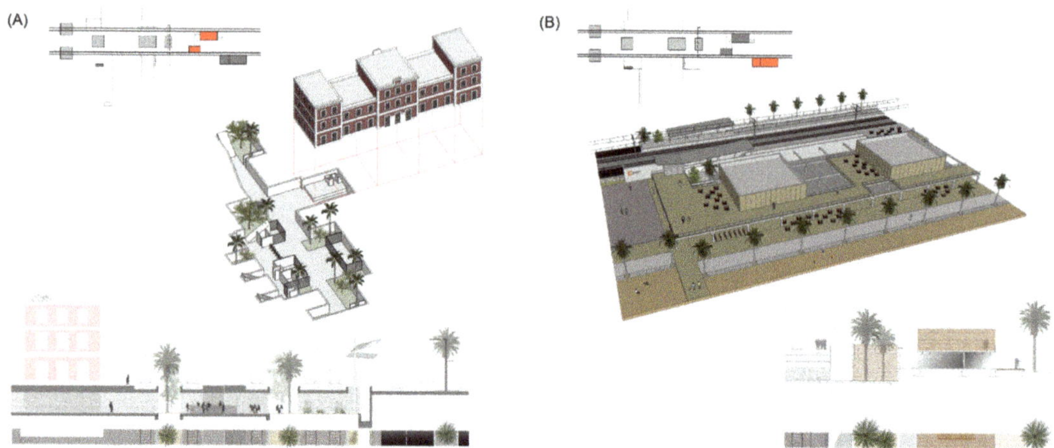


**Figure 17.** Standardized solutions for building typologies (Source: authors' own). (**A**) Urban side, civic way. (**B**) Maritime side, greenway.

## 6. Conclusions

This research provides a method to obtain standard solutions in order to solve coastline mobility conflicts produced by linear infrastructures. This process occurs in three dimensions: (1) ensuring longitudinal continuity by means of a greenway all along the seaside, (2) reinforcing transversal permeability between the inland areas and the seaside by means of a sequence of connections beneath the linear infrastructures and (3) unifying the qualities of the littoral urban landscape by organizing its maritime façade, its public spaces and its buildings and supporting social activities.

Application of these guidelines will result in a pleasant commuting route for inhabitants and tourists to be used not only for daily local trips or toward Barcelona, but also during weekends and holidays. In fact, many administrations in the county are focused on these ideas to regenerate the coastline in a more civic way. For example, the Mataró municipality is currently developing a proposal for a new underground station [85]. Furthermore, the Catalan government has promoted free circulation through inland highways, which has great potential for unloading the traffic on the N-II road in order to convert it into a civic way [86].

A relevant aspect of this research was to introduce the architectonic project practice into research methods. The four representative proposals ease the identification and further application of generic urban design elements, such as civic roads, greenways, urban and maritime squares, underground passages or modular structures. Due to their standard nature, these solutions can be combined to solve conflicts in all different locations along the studied territory. The matrix of standardized design solutions is a tool to identify the starting point for each specific urban project under a homogeneous criterion for the whole region. Consequently, the most suitable solutions for each site need to be adapted in relation to the contextual conditions by means of a professional architectural project.

In conclusion, this research provides a method to define the basic system of solutions and also the process to combine them so as to be easily replicable. Considering the Maresme region as a case study, the urban or architectural design that will follow the research will regenerate the coastal public spaces, converting them into urban references all located at the crossings between transversal and longitudinal ways. In a wider scope, the intention of this research was to assist administrations with solving local problematics but with common criteria. Therefore, there is no need for a large-scale urban project, which is usually expensive and complex. On the contrary, the intervention strategies proposed are economically sustainable and affordable for a wide range of administrations and stakeholders, and they are also flexible enough to be developed through varied deadlines set by different municipalities.

This research provides a common base of discussion defined by comprehensible tools to explain and develop urban projects. With the idea of solving the issues of a territory as a whole by means of small interventions, all the solutions provided are coherent and organized in a common design tool. This overview establishes general design bases to organize the whole maritime façade of the Maresme region in relation to its particular conflicts while taking into account its landscape values. However, while the standard solutions identified might be coherent for Maresme, they may not be suitable for other Mediterranean territories. The value of this research is in its capacity to be repeated in other regions by following the process and methods to set a customized matrix for each location which is closely related to its specific conflicts and its landscape conditions.

In order to answer the three aforementioned research questions (see Section 2), this study established a specific methodology. First, the littoral fringe was analyzed in a homogeneous way by registering and classifying the problematics both virtually and with repeated fieldwork visits. In addition, intangible aspects were taken into account to understand the current problematics, such as the historic, social or cultural characteristics of the region. Second, in-depth analysis of the specific site was required prior to selection and proposal of the case study projects to achieve universal solutions. In this phase, basic design tool decisions were important to finally come to adequate but generic solutions, such as work scale, materiality or modeling parameters. Finally, the last phase involved synthesis of the solutions by reducing the design proposals to a set of essential elements, dimensions and materials.

Throughout this process, the use of graphic tools is fundamental, and decisions are taken in terms of the research methods from the analysis (e.g., digital cartographies, schematic mapping and handmade sketches) to the design (e.g., models, site plans and 3D images) and, finally, the matrix of solutions (e.g., standard sections, cuts and 3D sketches). By means of graphic aspects, this research establishes tools to register, compare, decide, propose and show a proposal through an intelligible code for both citizens and administrations.

The next step in this research aimed to register all the problematics and variables identified in a general map of the region by means of Geographical Information Systems (GISs) [87]. This forthcoming document will include georeferenced locations related to the characteristics of each place. Once the locations were properly identified in the map, the matrix of standardized design solutions would add new data regarding which intervention strategies could be applied in each place. This process would set the basis for further architectonic projects promoted by different municipalities (Table 5). To provide the maximum potential to this second phase of the research, this GIS mapping tool should be designed with a clear layout and be linked with official cartographies in order to foster its use by administrations at different scales and also by public and private stakeholders.

**Table 5.** Standardized solutions applied to case studies.

| Case Studies | Standardized Solutions |
|---|---|
| Downtown of Premià de Mar | 1A + 2A + 3A |
| Stop on the beach of Vilassar-Cabrera | 1A + 2A + 4B |
| Railway station of Mataró | 1A + 2A + 4A |
| Stream mouth in Llavaneres | 1A + 2B + 3B + 3C |

**Author Contributions:** Conceptualization, A.M., X.M. and J.G.; methodology, A.M., X.M. and J.G.; formal analysis, A.M., X.M. and J.G.; visualization, A.M., X.M. and J.G.; writing—original draft preparation, A.M., X.M. and J.G.; writing—review and editing, A.M., X.M. and J.G. All authors have read and agreed to the published version of the manuscript.

**Funding:** This research project was funded by Direcció General d'Infraestructures de Mobilitat Terrestre del Departament de Territori i Sostenibilitat (Generalitat de Catalunya), grant numbers PTOP-2015-562 and 2015-URL-Proj-047. The APC was funded by Secretaria d'Universitats i Recerca del Departament d'Empresa i Coneixement (Generalitat de Catalunya), grant number 2017 SGR 1327.

**Institutional Review Board Statement:** Not applicable.

**Informed Consent Statement:** Not applicable.

**Data Availability Statement:** Not applicable.

**Acknowledgments:** The authors would like to express their gratitude to the stakeholders that participated in this study and contributed with their valuable insights, and especially to the collaborators of the IAR research group: Robert Terradas, Amador Ferrer, Joan Briz, Núria Martí, Anna Peguero, Xavier Ramis and Jordi Mas.

**Conflicts of Interest:** The authors declare no conflict of interest. The funders had no role in the design of the study; in the collection, analyses, or interpretation of data; in the writing of the manuscript, or in the decision to publish the results.

## References

1. Ferrer, A.; Martínez, A.; Rentería, I.; Terradas, R. *Estrategias Mediterráneas*; Claret: Barcelona, Spain, 2015; pp. 12–13.
2. Braudel, F. *El Mediterráneo*; Espasa Calpe: Madrid, Spain, 1987.
3. Subbotin, O. Architectural and planning principles of organization and reconstruction of coastal areas. *Mater. Sci. Forum* **2018**, *931*, 750–753. [CrossRef]
4. Martínez, A.; Rueda, C.; Gordon, J.; Ospina, J. Estratègies per a la Regeneració Sostenible D'assentaments Turístics a la Costa Mediterrània. Available online: http://hdl.handle.net/2099.2/3543 (accessed on 7 June 2021).
5. Cinelli, I.; Anfuso, G.; Privitera, S.; Pranzini, E. An Overview on Railway Impacts on Coastal Environment and Beach Tourism in Sicily (Italy). *Sustainability* **2021**, *13*, 7068. [CrossRef]
6. Guardiola, F. L'obsessió per Reconciliar la Ciutat Amb el Mar. Available online: https://www.raco.cat/index.php/Espais/article/download/91807/169983 (accessed on 7 June 2021).
7. Martínez, A.; Vives, L. Stratégies Méditerranéenes pour la régéneration architectonique du bord de la mer. *Classeur. Mare Nostrum* **2017**, *2*, 341.
8. Gehl Studio San Francisco. *Planing by Doing: How Small, Citizen-Powered Projects Inform Large Planning Decisions*; Gehl Studio: San Francisco, CA, USA, 2016; p. 2.
9. Woodroffe, C.; Muray-Wallace, C. Sea-level rise and coastal change: The past as a guide to the future. *Quat. Sci. Rev.* **2012**, *54*, 4–11. [CrossRef]
10. Renteria, I. Intervención en los edificios. In *Estrategias Mediterráneas*; Claret: Barcelona, Spain, 2015; pp. 102–105.
11. Gascón, C.; Rovira, T. *Pautas de Investigación*; Ed. UPC: Barcelona, Spain, 2007.
12. Ehrenzweig, A. *El orden Oculto del Arte. Por el Proceso Creativo*; Labor: Barcelona, Spain, 1973.
13. Heinrichs, D.; Jarass, J. Designing healthy mobility in cities: How urban planning can promote walking and cycling. *Bundesgesundheitsbl* **2020**, *63*, 945–952. [CrossRef]
14. Clopés, J. Efectos de la dinámica litoral inducidos por la actividad humana en áreas urbanizadas: El caso del Maresme (Barcelona). *Territoris* **2008**, *7*, 57.
15. Martí, P.; Mayor, C.; Melgarejo, A. Waterfront landscapes in Spanish cities: Regeneration and urban transformations. *WIT Trans. Built. Environ.* **2018**, *179*, 45–56.
16. Versaci, A.; Cardaci, A. Waterfront di Messina. Maregrosso, cuore della città, tra evoluzione e involuzione. *Agribus. Paesaggio Ambiente* **2018**, *21*, 28–35.
17. Parrinello, G.; Bécot, R. Regional Planning and the Environmental Impact of Coastal Tourism: The Mission Racine for the Redevelopment of Languedoc-Roussillon's Littoral. *Humanities* **2019**, *8*, 13. [CrossRef]
18. González, M. El Serrallo de Tarragona: De barrio marinero a enclave temático para turistas y visitantes. *Arx. d'Etnogr. Catalunya* **2011**, *11*, 115–130.
19. Gluvačević, D. The power of cultural heritage in tourism-example of the city of Zadar (Croatia). *Int. J. Sci. Manag. Tour.* **2016**, *2*, 3–24.
20. Balaguer, E. Desarrollo Urbanístico y Calidad de Vida en Copenhague: El Caso de Orestad. Master's Thesis, Universitat Politècnica de Catalunya, Barcelona, Spain, 28 June 2012.
21. Rosenberg, E. Inventing the seashore: The Tel Aviv-Jaffa promenade. In *Contemporary Urban Landscapes of the Middle East*; Routledge: London, UK, 2016; pp. 75–97.
22. Roca, E.; Villares, M.; Oroval, L.; Gabarró, A. Public perception and social network analysis for coastal risk management in Maresme Sud (Barcelona, Catalonia). *J. Coast. Conserv.* **2015**, *19*, 693–706. [CrossRef]
23. Solá-Morales, M. *Las Formas del Crecimiento Urbano*; Edicions UPC: Barcelona, Spain, 1996.

24. Busquets, J.; Domingo, M.; Eizaguirre, X.; Moro, A. *Les formes Urbanes del Litoral Català*; Diputació de Barcelona: Barcelona, Spain, 2003.
25. Silva, P. Tactical urbanism: Towards an evolutionary cities' approach? *Environ. Plan. B Plan. Des.* **2016**, *43*, 1040–1051. [CrossRef]
26. Gehl, J.; Gemzoe, L. *Nuevos Espacios Urbanos*; Gustavo Gili: Barcelona, Spain, 2002; pp. 10–21.
27. Jacobs, A. Great Streets. *ACCESS Magaz.* **1993**, *1*, 23–27.
28. Jacobs, J. *Muerte y Vida de las Grandes Ciudades*; Capitán Swing: Madrid, Spain, 2020.
29. Hernández, J. La resonancia del lugar. Arquitectura contemporánea y contexto. In *Arquitectura y Ciudad. La Tradición Moderna Entre la Continuidad y la Ruptura*; Arte y Estética: Madrid, Spain, 2007; pp. 12–39.
30. Goldfinger, M. *Arquitectura Popular Mediterránea*; Gustavo Gili: Barcelona, Spain, 1993; pp. 8–26.
31. Cisteró, X.; Camarós, J. Les rierades al Maresme. *L'Atzavara* **2014**, *23*, 61–79.
32. AAVV. *Gran Geografía Comarcal de Catalunya. El Vallès i el Maresme*; Enciclopèdia Catalana: Barcelona, Spain, 1982; pp. 348–353.
33. Terribas, A.; Lavin, J. El Desdoblamiento Ferroviario en el Maresme Norte. Available online: http://hdl.handle.net/2099.1/26513 (accessed on 23 July 2021).
34. Botey, C.; De Castellarnau, X. La cultura del tren: El ferrocarril Barcelona–Mataró. Trobada d'Entitats de Recerca Local i Comarcal del Maresme. *Sessió d'Estudis Mataronins* **2019**, *13*, 217–232.
35. Alejandre, V. *La N-II y sus Precedentes Camineros. Itinerarios Históricos y vías de Comunicación Entre Madrid/Toledo y Zaragoza (de la Antigüedad al siglo XX)*; Centro de Estudios Bilbilitanos-Institución Fernando el Católico: Calatayud, Spain, 2017; pp. 261–350.
36. Serra, J.; Bautista, R.; Maia, L.; Montori, C. Sistemas de protección de costas (i): Regeneración de playas. El ejemplo del Maresme (1987–1998). *Geogaceta* **1998**, *25*, 187–190.
37. Biel, M. Mataró: Del Tèxtil a Ciutat del Coneixement. Available online: https://ddd.uab.cat/pub/tfg/2014/125853/TFG_Mireia_Biel_1269002.pdf (accessed on 17 August 2021).
38. Gras, F. Anàlisi de L'impacte Socioeconòmic del Turisme: El cas dels dos Maresmes. Bachelor's Thesis, Facultat de Geografia i Història, Universitat de Barcelona, Barcelona, Spain, 2020. Available online: http://hdl.handle.net/2445/149598 (accessed on 12 June 2021).
39. Parcerisas, L.; Marull, J.; Pino, J.; Tello, E.; Coll, F.; Basnou, C. Land use changes, landscape ecology and their socioeconomic driving forces in the Spanish Mediterranean coast (El Maresme County, 1850–2005). *Environ. Sci. Policy* **2012**, *23*, 120–132. [CrossRef]
40. The Catalan Government. Pacte per la Mobilitat Sostenible al Maresme. Public Agreement. Available online: https://web.gencat.cat/ca/actualitat/detall/Pacte-per-la-Mobilitat-Sostenible-al-Maresme (accessed on 15 July 2021).
41. VVAA. *Catàlegs de Paisatge de Catalunya*; Observatori del Paisatge: Barcelona, Spain, 2019.
42. The Catalan Government. Pla d'Espais d'Interès Natural (PEIN), Regulatory Decret 328/1992. Available online: http://sig.gencat.cat/visors/enaturals.html (accessed on 12 May 2021).
43. Sala i Martí, P.; Bretcha, G.; Puigbert, L. *La Carretera en el Paisatge*; Observatori del Paisatge: Olot, Spain, 2020; Available online: http://www.catpaisatge.net/fitxers/publicacions/carreteres/carreteres.pdf (accessed on 25 July 2021).
44. Díaz, J.; Maldonado, A. Transgressive sand bodies on the Maresme continental shelf, western Mediterranean Sea. *Mar. Geol.* **1990**, *91*, 53–72. [CrossRef]
45. Ballesteros, C.; Jiménez, J.A.; Viavattene, C. A multi-component flood risk assessment in the Maresme coast (NW Mediterranean). *Nat. Hazards* **2018**, *90*, 265–292. [CrossRef]
46. Duran, A.M.; Guerra, J.G.; Martín, X.; Piquer, A.P. The Case of "Anis del Mono" Factory. Regenerating Coastal Heritage in Catalonia. *Athens J. Mediterr. Stud.* **2016**, *2*, 3. [CrossRef]
47. Romero, G.; Nadal, J.; Ribas, A. L'impacte del Temporal Glòria al Litoral de l'Alt-Maresme i al Delta de la Tordera. Diagnosis, Aprenentatges i Propostes D'actuació; Academic Study. Master's Thesis, Universitat Autònoma de Barcelona, Barcelona, Spain, 2020. Màster Universitari en Polítiques i Planificació per a les Ciutats, L'ambient i el Paisatge. Available online: https://ddd.uab.cat/record/236113 (accessed on 10 June 2021).
48. Mas, V.; Temes, R.; Serrano, B.; Jiménez, C.; Azulay, M. *Research and Development Project, ERAM Estrategias para la Regeneración Sostenible de Asentamientos Turísticos en la Costa Mediterránea*; Universitat Politècnica de València: Valencia, Spain, 2015; Available online: https://riunet.upv.es/handle/10251/47918 (accessed on 2 September 2021).
49. Van Veelen, P. Adaptive planning for resilient coastal waterfronts: Linking flood risk reduction with urban development in Rotterdam and New York City. In *A+BE | Architecture and the Built Environment*; TU Delft: Delft, The Netherlands, 2016. [CrossRef]
50. Gordon, J.; Martin, X.; Peguero, A. El trabajo de campo. In *strumento esencial en el proyecto de investigación. In Estrategias Mediterráneas*; Claret: Barcelona, Spain, 2015; pp. 100–101.
51. Institut Cartogràfic i Geològic de Catalunya (ICGC). Available online: https://www.icgc.cat/ (accessed on 28 August 2021).
52. Ferrer, A. El dibujo del territorio como forma de conocimiento. In *Estrategias Mediterráneas*; Claret: Barcelona, Spain, 2015; pp. 80–83.
53. Martí, C. *Las Variaciones de la Identidad. Ensayo Sobre el Tipo en Arquitectura*; Fundación Arquia: Barcelona, Spain, 2014.
54. Martínez, A. Restablecer el carácter del lugar. In *Estrategias Mediterráneas*; Claret: Barcelona, Spain, 2015; pp. 92–95.
55. Alexander, C. *A Pattern Language: Towns, Buildings, Constructions*; Oxford University Press: Oxford, UK, 1977.
56. Fang, C. *Waterfront Landscapes*; Design Media Publishing Limited: London, UK, 2011.

57. Marshall, W.; Duvall, A.; Main, D. Large-scale tactical urbanism: The Denver bike share system. *J. Urban. Int. Res. Placemaking Urban Sustain.* **2016**, *9*, 135–147. [CrossRef]
58. Martínez, A.; de Rentería, I.; Martín, X.; Gordon, J. Estrategias de regeneración del paisaje urbano en el litoral del Maresme (Barcelona). *ISUF-H Ciudad. Formas Urbanas Perspect. Transversales* **2018**, *7*, 97.
59. Martínez-Sala, E. Oportunitats pel pla Territorial del Maresme Davant d'una Reforma Infraestructural Tren-Autopista-N-II. Final Thesis, ETS d'Enginyeria de Camins, Canals i Ports de Barcelona. 2010. Available online: http://hdl.handle.net/2099.1/12609 (accessed on 5 July 2021).
60. Dawson, D.; Shaw, J.; Gehrels, W.R. Sea-level rise impacts on transport infrastructure: The notorious case of the coastal railway line at Dawlish, England. *J. Transp. Geogr.* **2016**, *51*, 97–109. [CrossRef]
61. The Catalan Government. La Línia Orbital Ferroviària: Vertebració del Territori Metropolità, Infraestructures Ferroviàries de Catalunya. Available online: https://ifercat.gencat.cat/index.php?contenido=22 (accessed on 28 July 2021).
62. El fin del Peaje de la C-32 Transforma el Maresme, Published on 3 October 2020, La Vanguardia. Available online: https://www.lavanguardia.com/local/maresme/20201003/483806389958/fin-peaje-c-32-transforma-maresme.html (accessed on 26 August 2021).
63. Garcia-López, M.A. Urban spatial structure, suburbanization and transportation in Barcelona. *J. Urban Econ.* **2012**, *72*, 176–190. [CrossRef]
64. Banister, D. The sustainable mobility paradigm. *Transp. Policy* **2008**, *15*, 73–80. [CrossRef]
65. The Catalan Government. *Manual for the Design of Cyclepaths in Catalonia*; Departament de Política Territorial i Obres Públiques: Barcelona, Spain, 2008.
66. Walton, A. Economic Analysis on the Impacts of Climate Change in the Maresme Coast. Bachelor's Thesis, Universitat Politècnica de Catalunya, Barcelona, Spain, 2016. Available online: http://hdl.handle.net/2117/103223 (accessed on 23 June 2021).
67. Rowe, C.; Koetter, F. *Ciudad Collage*; Gustavo Gili: Barcelona, Spain, 1981; pp. 147–177.
68. Balletto, G.; Ladu, M.; Milesi, A.; Borruso, G. A Methodological Approach on Disused Public Properties in the 15-Minute City Perspective. *Sustainability* **2021**, *13*, 593. [CrossRef]
69. Sauer, I.; Roca, E.; Villares, M. Beach users' perceptions of coastal regeneration projects as an adaptation strategy in the western Mediterranean. *J. Hosp. Tour. Res.* **2019**. [CrossRef]
70. Martinez, C.; Pemjean, R. La línea del agua. El monumento a la Partisana de Carlo Scarpa en Venecia. *Rita* **2019**, *11*, 36–44. [CrossRef]
71. Crane, S. Mediterranean Dialogues. Le Corbusier, Fernand Pouillon and Roland Simounet. In *Modern Architecture and the Mediterranean, Vernacular Dialogues and Contested Identities*; Lejeune, J., Sabatino, M., Eds.; Routledge: New York, NY, USA, 2009; pp. 95–99.
72. Muntañola, J. Mind, Land & Society: A New Architecture for a Better Enviroment. Available online: https://upcommons.upc.edu/handle/2099.3/36773 (accessed on 30 May 2021).
73. Gastaldi, F.; Camerin, F. El proceso de remodelación del waterfront de Génova y los proyectos de Renzo Piano desde los años 80 hasta el Blue Print. *ACE–Archit. City Environ.* **2017**, *11*, 33–64. Available online: http://hdl.handle.net/2117/101730 (accessed on 16 July 2021).
74. Austin, G. Case study and sustainability assessment of Bo01, Malmö, Sweden. *J. Green Build.* **2013**, *8*, 34–50. [CrossRef]
75. The Catalan Government. Pla Director Urbanístic del Sistema Costaner (PDUSC), Regulatory DOGC 6722. Available online: https://portaldogc.gencat.cat/utilsEADOP/PDF/6722/1374519.pdf (accessed on 12 May 2021).
76. Olgivay, V. *Arquitectura y Clima. Manual de Diseño Bioclimático Para Arquitectos y Urbanistas*; Gustavo Gili: Barcelona, Spain, 2010.
77. Rodriguez, G.; Brebbia, C. *Coastal Cities and Their Sustainable Future*; WIT Press: Southampton, UK, 2015; pp. 125–136.
78. Urzelai, J.; Arbonés, E. Integración paisajística del TRAM a su paso por Serra Grossa. *Arquitectura* **2009**, *356*, 21–27. Available online: https://dialnet.unirioja.es/servlet/articulo?codigo=3006836 (accessed on 6 July 2021).
79. Loubes, J.P. *Arquitectura Subterránea. Aproximación a un Hábitat Natural*; Gustavo Gili: Barcelona, Spain, 1985.
80. Hoyos, F. El Modelo Barcelona de Espacio Público y Diseño Urbano. Paseo Sant Joan (1979–2011). Master's Thesis, Universitat de Barcelona, Barcelona, Spain, 2012. [CrossRef]
81. Batlle, E. El Jardí de la metròpoli al Llobregat = The Metropolis garden on the Llobregat river. *Paisea* **2013**, *1*, 68–73. Available online: http://hdl.handle.net/2117/20960 (accessed on 22 June 2021).
82. Cornu, R. Les Portefaix et la transformation du port de Marseille. *Ann. Midi* **1974**, *86*, 181–201. [CrossRef]
83. Frantzeskaki, N. Seven lessons for planning nature-based solutions in cities. *Environ. Sci. Policy* **2019**, *93*, 101–111. [CrossRef]
84. Raymond, C.; Frantzeskaki, N.; Kabisch, N.; Berry, P.; Breil, M.; Nita, M.; Calfapietra, C. A framework for assessing and implementing the co-benefits of nature-based solutions in urban areas. *Environ. Sci. Policy* **2017**, *77*, 15–24. [CrossRef]
85. Bote Confia que la Nova Estació de Mataró Será Una Realitat cap al 2026, Published on 21 January 2021, Capgròs. Available online: https://www.capgros.com/actualitat/mataro/bote-confia-nova-estacio-mataro-sera-realitat-cap-2026_801773_102.html (accessed on 19 March 2021).

86. La N-II se Pacifica con el fin de Peajes en la C-32, Published on 28 August 2021, El Periódico. Available online: https://www.elperiodico.com/es/economia/20210828/pacificacion-carretera-n-ii-gratuidad-autopistas-catalanas-c-32-12020468 (accessed on 1 September 2021).
87. Valente, R.; Stamatopoulos, L.; Donadito, C. Environmental design criteria through Geoindicators for two Mediterranean coastlands. *CSE–City Saf. Energy* **2014**, *2*, 63–76. [CrossRef]

International Journal of
*Environmental Research and Public Health*

MDPI

*Article*

# Socio-Economic Context and Community Resilience among the People Involved in Fish Drying Practices in the South-East Coast of Bangladesh

**Sabrina Jannat Mitu [1,†], Petra Schneider [2], Md. Shahidul Islam [1], Masud Alam [3], Mohammad Mojibul Hoque Mozumder [4], Mohammad Mosarof Hossain [1] and Md. Mostafa Shamsuzzaman [1,*,†]**

1 Department of Coastal and Marine Fisheries, Sylhet Agricultural University, Sylhet 3100, Bangladesh; mitusabrina42@gmail.com (S.J.M.); islamms2011@yahoo.com (M.S.I.); mosarofsau@gmail.com (M.M.H.)
2 Department for Water, Environment, Civil Engineering and Safety, University of Applied Sciences, Magdeburg-Stendal, Breitscheidstraße 2, D-39114 Magdeburg, Germany; petra.schneider@h2.de
3 Department of Agricultural Statistics, Sylhet Agricultural University, Sylhet 3100, Bangladesh; malam.stat@sau.ac.bd
4 Fisheries and Environmental Management Group, Helsinki Institute of Sustainability Science (HELSUS), Faculty of Biological and Environmental Sciences, University of Helsinki, 00014 Helsinki, Finland; mohammad.mozumder@helsinki.fi
* Correspondence: shamsuzzamanmm.cmf@sau.ac.bd
† These authors contributed equally to this work.

**Citation:** Mitu, S.J.; Schneider, P.; Islam, M.S.; Alam, M.; Mozumder, M.M.H.; Hossain, M.M.; Shamsuzzaman, M.M. Socio-Economic Context and Community Resilience among the People Involved in Fish Drying Practices in the South-East Coast of Bangladesh. *Int. J. Environ. Res. Public Health* **2021**, *18*, 6242. https://doi.org/10.3390/ijerph18126242

Academic Editors: M. Dolores Esteban, José-Santos López-Gutiérrez, Vicente Negro and José Marcos Ortega

Received: 6 April 2021
Accepted: 7 June 2021
Published: 9 June 2021

**Publisher's Note:** MDPI stays neutral with regard to jurisdictional claims in published maps and institutional affiliations.



**Abstract:** The south-east coast, specifically the Cox's Bazar region, of Bangladesh has achieved a tremendous impetus for producing a large volume of dried fish by involving thousands of marginalized coastal people. This study aimed to assess the socio-economic profile, livelihood strategies, and resilience of the communities engaged in fish drying on the south-east coast using a mixed-methods approach and an Analytic Hierarchy Process (AHP). The study's findings revealed that communities involved in drying were socio-economically undeveloped due to their lower literacy, unstable incomes, and labor-intensive occupations. Apart from notable child labor employed in fish drying in Nazirertek, female workers had relatively higher participation than males. Nevertheless, the female workers had less control over their daily wages and reported working at USD 3.54–5.89 per day, which was relatively lower than male workers who received USD 4.15–8.31 per day. Through fish drying activities, very few workers, producers, and traders were found to be self-reliant. In contrast, the livelihoods of the workers were not as secure as the processors and traders. In addition to suffering from various shocks and constraints, dried fish processors and workers, dried fish traders, off-season income, an abundance of fish species, fish drying facilities, trader's association, and social interrelationship played a significant role in maintaining community resilience. The study recommends appropriate interventions to alternative income diversification options, strong collaboration between communities, local authorities, and government for sustainable livelihoods and better community resilience.

**Keywords:** coastal fisheries; dry fish; livelihood; vulnerability; AHP

## 1. Introduction

Dried fish accounts for the 4th most significant share of fish consumed in Bangladesh [1] and is much relished by the country's people for its flavor, texture, and taste [2]. It is an accessible and low-cost food source and can contribute a large percentage of protein and significant micronutrients to the diet of poor people [3]. Fish drying is the most extensive fish processing activity in Bangladesh's coastal region that contributes significantly to livelihoods and nutrition, especially for poor and marginalized communities in coastal and inland areas [4]. These activities are of great importance to Bangladesh, as more than 17 million people, including 1.4 million women, depend on fish farming, processing, and handling [5]. After harvesting, more than one-third of the landings are used for drying all year round [6,7]; therefore, these drying practices have provided solvency to thousands of coastal populations.

The processing and trade of dried fish are becoming a promising and profitable industry, offering the processors, traders, and other stakeholders opportunities to make much money in the fisheries sector. As a result, dried fish has demand both on the national and international markets. In contrast, the export of dried fishes has increased from 517 metric tons (value 94 million USD) in the fiscal year 2001–2002 to 3144 metric tons (value 5.01 million USD) in 2018–2019 [8]. However, in the context of global food security and livelihoods of small-scale fishers in developing countries, it is dispiriting that the importance of dried fish and the people, directly and indirectly, involved in drying are poorly understood and rarely recognized [1]. Therefore, the study of the socio-economic condition and community's resilience is essential. Moreover, it explains the actual situation of the population in a particular region and allows seeing how individuals or families fit into society through economic and social measures. Finally, such studies help to take appropriate initiatives for the proper management of communities.

Fish drying activities on the south-east coast vary considerably according to the weather conditions. Nevertheless, the negative impact of climate change is a severe concern at regional, national, and global levels, affecting most reclaims' sustainability perspectives, including the aquatic environments, ecosystems, and the dependent societies [9–12]. The economies and livelihoods of communities dealing with dried fish's processing and trade are affected by climate variability. In contrast, the communities are vulnerable to extreme weather conditions (tidal storms, heavy rains, cloudy weather) [13]. These situations raise the need to address sustainable livelihoods and community resilience among those engaged in fish drying. Hence, the purpose of this study is to assess the socio-economic conditions and the key indicators of strength for the communities involved in fish drying practices on the south-east coast of Bangladesh. Many studies are available on the socio-economic conditions of fishers in Bangladesh [14–18]. However, scanty research work has been done on the fish drying communities except for their socio-economic conditions, such as labor well-being's in dried fish value chains [1], the efficiency of dry fish marketing [6], and quality analysis of dry fish [19]. The study's findings will help the selected communities, different organizations, and government bodies to formulate policies for improving the socio-economic conditions of fish drying communities.

## 2. Materials and Methods

### 2.1. *Profile of the Study Sites*

The study was carried out at two fish drying areas along the south-east coast: Nazirertek (under Ward no. 1 of Cox's Bazar Sadar) and Chitapara (under Ward no. 2 of Cox's Bazar Sadar) located in the Cox's Bazar district of Bangladesh (Figure 1). The main criterion for selecting these two areas was the community's reliance on fish drying.

Nazirertek (one of the most extensive fish drying yards in Bangladesh) has been built on approximately 200 acres of land at the Bhakkhali River's mouth, Cox's Bazar. Processing of dried fish begins in mid-August, and if the weather remains good, the process continues until mid-April/May of the following year (also called peak season). During the peak season, 20,000 workers (most of them women) work in different *Shutki mahals* (fish drying yards), and about $45.34-54.43 \times 10^5$ metric tons of dried fish are produced in the *Shutki Palli* (fish drying village) at a market price of around USD 24 million [20]. Traditional solar energy methods are used for large-scale fish species drying, including Chhuri (*Trichiurus haumela*), Laitta (*Harpadon nehereus*), Faishya (*Setipinna phasa*), Poa (*Argyrosomus regius*), and Surma (*Scomberomorus guttatus*). Fish drying is mainly done in *Khola* (Bengali name of fish processing facility), where raw fish is spread on bamboo mats on the floor or placed on bamboo scaffolding or shelves for drying.

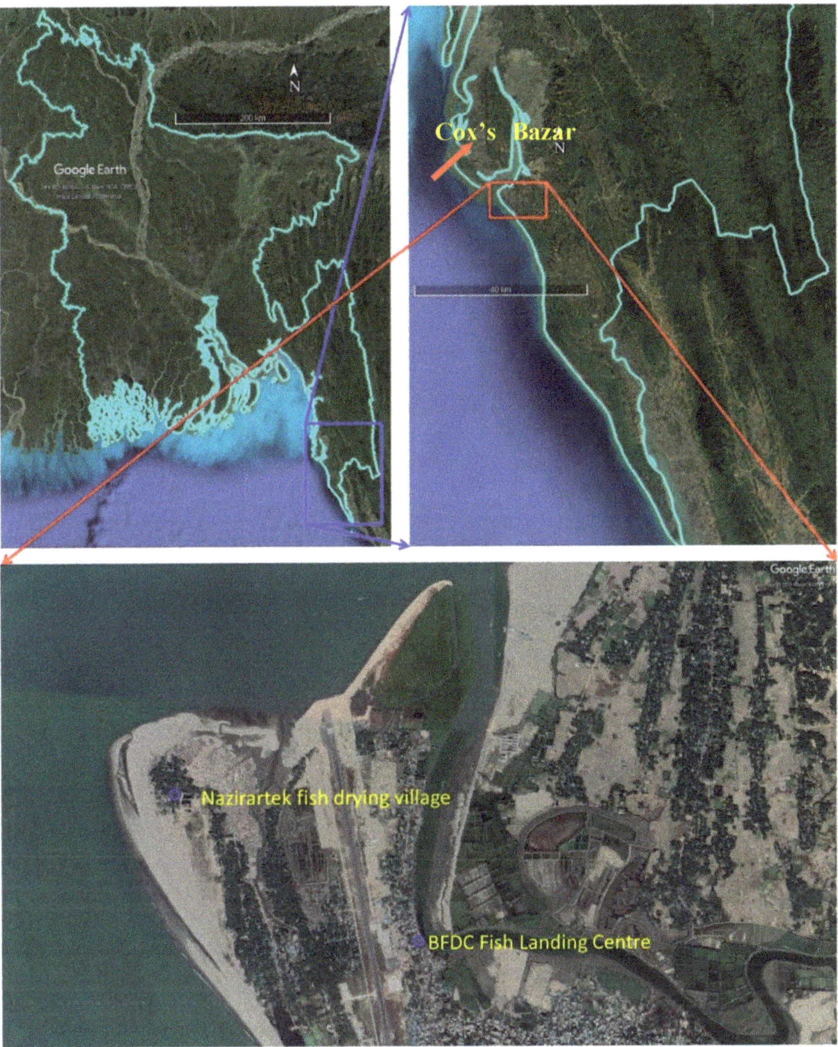


**Figure 1.** Location of the study area in Cox's Bazar district of Bangladesh (Google Maps).

On the other hand, Chitapara is in Cox's Bazar town near the Bangladesh Fisheries Development Corporation (BFDC) fishery ghat. In this area, rooftop fish drying practices are carried out on a small scale using bamboo shelves through solar energy. As a result, dried and salted-dried fish and high-value byproducts (fin, swim bladder) are also produced, which are in significant demand from neighboring countries and the tribal communities.

### 2.2. Empirical Data Collection Methods

A total of 250 dried fish processors, workers, traders, and fishers involved in fish drying activities directly or indirectly from Nazirertek ($N$ = 215) and Chitapara ($N$ = 35) were randomly selected to perform the study from September to December 2019. A mixed approach was applied, including individual interviews ($N$ = 250), interviews with key informants (15), and ten focus group discussions (FGD) with checklists (Table 1).

**Table 1.** An overview of empirical data collection methods.

| Tools | Participants | Sample Size | Research Issues/Objectives |
|---|---|---|---|
| Individual interview (II) | Dried fish traders, processors, workers, fishers involved in fish drying practices | $N$ = 250 Male-153 Female-97 | Socio-demographic factors-age, religion, gender, marital status, family type, household size, education, occupation, housing condition, income, savings, credit access, etc. |
| Key informant interview (KII) | Members of dried fish trader's association, knowledgeable persons in the communities, Fisheries Scientific Officers, NGO workers. | 15 | Knowledge and experience persons often play a vital role in the community. Cross-checked interviews validate the collected data. |
| Focus group discussion (FGD) | Elderly and young male and female workers, community leaders, widow, experienced fishermen | 10 (5–8 participants) | Semi-structured data gathering method that takes advantage of group dynamics and allows respondents to discuss critical issues |
| PRA tools | Male and female respondents of the community | 20 | Daily activity chart, seasonal variations |

After developing a semi-structured questionnaire, a face-to-face survey was conducted with $N$ = 250 respondents to gather qualitative and quantitative information. This study examined several socio-economic indicators hypothesized to reflect the livelihood activities, economic conditions, and food security of people involved in fish drying activities. Through the interviews, a range of qualitative information was obtained by asking communities about livelihood diversity, underlying constraints and vulnerabilities, and mechanisms for coping with the financial crisis and seasonal fluctuations. In addition, direct observations and interviews with key informants, processors, traders, and fish drying workers helped to gather information about the processing and trade of dried fish, the use of preservatives, hygienic and sanitation conditions that constitute a significant concern for public health. Therefore, to analyze these qualitative data, a content analysis method that interprets and encodes different transmitted materials (e.g., documents, articles, books, audios, interviews, and images) through classification, tabulation, and evaluation was employed [21]. Later, all qualitative and quantitative data were entered into Microsoft Excel Spreadsheet 2013 and then analyzed in IBM SPSS version 22, such as descriptive statistics and chi-square test.

### *2.3. Secondary Data Collection*

The secondary data were collected from relevant published books, scholarly articles, relevant literature, and newspapers [20] through an online search, e.g., Google scholar.

### *2.4. Sustainable Livelihood Framework*

To understand fish drying communities' resilience based on dependency upon the available assets, this study applied the "Sustainable livelihood approach" (SLA). SLAs are a way of understanding the needs of the poor and identifying the significant constraints and positive strength for their resilience. Based on the sustainable livelihood framework, a fishery-based livelihood embraces several components: (a) livelihood assets (owned or accessed by people, i.e., human, financial, physical, natural, and social capital), (b) vulnerability context (risk factors surrounding livelihoods); (c) transforming structures and processes (the structures associated with a formal organization, e.g., government, NGOs, laws and rights, social relations and participation) (d) livelihood strategies (the range and combination of activities people undertake or do to achieve livelihood goals such as productive activities, investment strategies); and (e) livelihood outcomes (achievement or output of the people's livelihood strategies) [22]. Through open-ended interviews and FGD, much information was gathered about access to different types of capital, livelihood strategies and decision-making processes, local institutions, and their ability and willingness to respond to changing vulnerability contexts. Themes were identified and classified into manageable categories of different variables: physical capital, financial capital, social capital, strength, threats, and outcomes.

### *2.5. Community Resilience Assessment*

In terms of community resilience, the term 'socio-economic stability' denotes how the community can maintain their livelihoods and desired living standards without outside support, following undesirable shocks [21]. As the fish drying communities in Cox's Bazar region are vulnerable to seasonality and extreme weather conditions, it is essential to know how the communities manage resilience in the face of change.

The procedure for assessing the resilience of fish drying communities is presented in Figure 2. A multi-criteria decision model-the adaptive analytical hierarchy process (AHP) was used for resilience assessment of the communities involved in fish drying (Figure 2). The model structure for assessing communities' resilience was based on a three-level hierarchical structure that breaks down all criteria into sub-models. To cluster a hierarchy, it was first decided which criteria to group, based on the similarity of these criteria in terms of the functions they perform or the features they share [23].

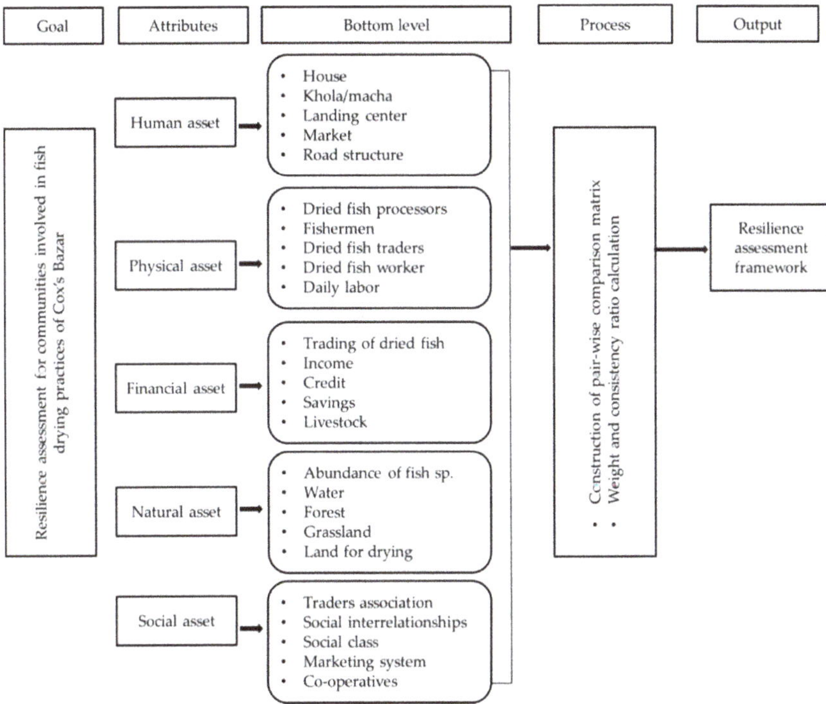


**Figure 2.** The adapted analytical hierarchy process (AHP) for resilience assessment of communities involved in fish drying practices in the Cox's Bazar region of Bangladesh.

The top or first hierarchy level represents the goal of the multi-criteria decision-making analysis process. In contrast, the intermediate or second hierarchy level lists the respective evaluation criteria compared pair-wise to assess their relative weight. Each of these clusters was considered a sub-model. Finally, the bottom level of the hierarchy contains criteria for evaluation. All of these criteria (sub-attributes) were identified to create a pair-wise comparison matrix that assesses the relative importance of the different measures to evaluate the resilience of communities involved in fish-drying practices.

Weight and Score

The development of weight was based on a pair-wise comparison matrix. A comparison of the relative importance of the two criteria was involved in determining resilience for specified objectives (Table 2). To use this procedure, the weights needed to sum up to 1. Ratings are systematically scored on a 17-point continuous scale from 1/9 (least significant) to 9/9 (most important) [23]. In this research, scores were assigned in rank order according to the number of factors involved in evaluating resilience assessment without repetition. Consistency ratios (CR) of 0.0030 to 0.0145 for the table were well within the balance of less than or equal to the 0.10 recommended by Saaty [23], signifying a small probability that the weights were developed by chance [24].

**Table 2.** The relative importance of two criteria [23].

| 1/9 | 1/8 | 1/7 | 1/6 | 1/5 | 1/4 | 1/3 | 1/2 | 1 | 2 | 3 | 4 | 5 | 6 | 7 | 8 | 9 |
|---|---|---|---|---|---|---|---|---|---|---|---|---|---|---|---|---|
| Extremely | | Very strongly | | Strongly | | Moderately | | Equally | Moderately | | Strongly | | Very strongly | | Extremely | |
| Less important | | | | | | | | | More important | | | | | | | |

The present study identified 25 essential criteria of livelihood assets to assess communities' resilience, such as human assets (dried fish processors, fishermen, dried fish traders, fish drying workers, and fish drying worker), financial assets (trading of dried fish, daily income, credit, savings, and livestock rearing), natural assets (abundance of dried fish species, enough land for drying, water, forest, and grassland), social assets (co-operatives, dried fish traders' association, social interrelationship, social class, and marketing system), and physical assets (house, *khola/macha*, landing center, market, and road structure) (Figure 2). Weights were given following the effectiveness of the criteria. The weight for each factor was determined by pair-wise comparisons in the context of a decision-making process known as the analytical hierarchy process [23,25], which was also suggested by other authors [21,26–28]. The assessment of resilience at each level of the factor was determined from survey findings and professional judgment.

### 2.6. Statistical Analysis

Multinomial logistic regression analysis, Principal Component Analysis (PCA), and community resilience analysis were performed through SPSS Version 22 (IBM) and MS Excel Spreadsheet 2013.

#### 2.6.1. Likelihood Ratio Test

Likelihood ratio tests between various categorical variables (has two or more categories, also known as qualitative variables) indicate a significant association between the variable and the socio-economic state of the community. In this study, descriptive statistics (cross-tabulation analysis) were applied to perform likelihood ratio tests between various categorical variables through SPSS software. The categorical demographic variables such as age, gender, religion, marital status and educational status, occupational profile, family type, training facility, access to resources, food security, food type, and savings scheme were chosen to find significant associations.

2.6.2. Binary Logistic Regression Analysis

A binary logistic regression model was used to examine how socio-economic and demographic variables affect fish drying activities' food safety. In the study, respondents were considered to have food securities if they were able to intake enough food three times a day. Otherwise, respondents were considered food insecure, if they could not manage enough food for their families in one day or eat twice a day. As the dependent variable (having food security or not) is dichotomous, the model was used to identify the factors affecting the odds ratio of the food status of the communities. The odds ratio refers to people's probability of having food insecurity (Pi) to predict people's food insecurity (1 − Pi). The dependent variable used in the study is the dummy variable that takes the value of one for having (food security); 0 otherwise (food insecure).

The logistic model of the relationship between the respondent's food security status variable and its explanatory variables is specified as Equation (1):

$$\text{In}[Pi/(1 - Pi)] = \beta_0 + \beta_1 X_{1i} + \beta_2 X_{2i} + \ldots\ \ldots \quad (1)$$

where subscript i denotes the i-th observation in the sample, P is the probability of the outcome, $\beta_0$ is the intercept, and $\beta_1$, $\beta_2$ ... are the coefficients associated with each explanatory variable, $X_1$ (age), $X_2$ (Gender), $X_3$ (occupation)......

2.6.3. Multinomial Logistic Regression Analysis

A multinomial logistic regression model is generally used when there is a categorical dependent variable where the dependent variable is nominal and has more than two categories. It uses one category as a referenced category (any one of them) and compares other categories with a reference category by taking log odds. This study's dependent variable was the respondents' socio-economic status (SES), and the independent variables were the socio-demographic variables. The respondents were ranked into poor, middle, and rich classes to identify the key determinant of the livelihood strategy according to their income, land ownership, and utility services. Empirically, the MLR in this study can be expressed as Equation (2):

$$\begin{aligned} \text{Log} = \frac{\text{Prob}(Y_i = j)}{\text{Prob}(Y_i = j')} &= \alpha + \beta_1(\text{age}) + \beta_2(\text{marital}) + \beta_3(\text{ocupation}) + \beta_4(\text{house}) \\ &+ \beta_5(\text{owner}) + \beta_6(\text{drinking}) + \beta_7(\text{income}) + \beta_8(\text{credit}) \ldots\ldots\ldots \end{aligned} \quad (2)$$

J is the identified cluster, poor class, and middle class, and j′ is the reference cluster, Rich class.

## 3. Results

### *3.1. Socio-Demographic Profile of the Communities*

The main occupations of all respondents in the selected community were fish drying, fishing, and fish trading, and they spent the busiest time in about nine months engaging in drying. During the survey, various age groups of dry fish producers, workers, and traders varying from 5 to 60 years were found to be involved in fish drying where most of them (38% at Nazirertek and 49% at Chitapara) belonged to the age group of 30–40 years. The communities' religious status revealed that the fish drying business was mainly dominated by Muslims (94%, Nazirertek, and 86%, Chitapara). Simultaneously, the minorities were the Hindu communities who had been seen to be involved in fishing and other businesses (Table 3).

**Table 3.** Summary of the communities' socio-demographic profile involved in fish drying activities in the study areas ($N$ = 250).

| Characteristics | Categories | Nazirertek ($n$ = 215) | Chitapara ($n$ = 35) |
|---|---|---|---|
| | | Frequency (%) | Frequency (%) |
| | Household profile | | |
| Age (years) | 5–20 | 23 (11) | - |
| | 20–30 | 24 (11) | 7 (20) |
| | 30–40 | 82 (38) | 17 (49) |
| | 40–50 | 73 (34) | 8 (22) |
| | 50–60 | 13 (6) | 3 (9) |
| Religion status | Hindu | 13 (6) | 5 (14) |
| | Muslim | 202 (94) | 30 (86) |
| Marital status | Married | 123 (57) | 28 (80) |
| | Unmarried | 36 (17) | 2 (6) |
| | Divorced | 11 (5) | 2 (6) |
| | Widowed | 45 (21) | 3 (8) |
| Occupational profile | Dry fish processors/owner | 64 (30) | 12 (35) |
| | Fish drying worker | 123 (57) | 13 (37) |
| | Fishermen | 10 (5) | 5 (14) |
| | Dried fish traders | 18 (8) | 5 (14) |
| Level of education | Illiterate | 78 (36) | 10 (29) |
| | Can sign only | 62 (29) | 13 (37) |
| | Primary | 51 (24) | 11 (31) |
| | Secondary | 24 (11) | 1 (3) |
| Family type | Joint | 88 (41) | 13 (37) |
| | Nuclear | 127 (59) | 22 (63) |
| Family size | A small family (2 to 4) | 62 (29) | 10 (29) |
| | Medium family (5 to 7) | 125 (58) | 14 (40) |
| | Large family (8 to 10) | 22 (10) | 7 (20) |
| | Very large family (above 10) | 6 (3) | 4 (11) |
| Number of children | 1 to 2 | 101 (47) | 10 (29) |
| | 3 to 4 | 108 (50) | 23 (65) |
| | 5 to 6 | 6 (3) | 2 (6) |
| Children going to school/not | School going children | 123 (57) | 31 (89) |
| | Non-going children | 92 (43) | 4 (11) |
| Earning member of the family | One | 92 (43) | 10 (29) |
| | Two | 58 (27) | 23 (65) |
| | Three | 65 (30) | 2 (6) |
| Residential status | Migrant | 171 (79) | 12 (35) |
| | Non-migrant | 44 (21) | 23 (65) |
| Having an alternative occupation | Yes | 24 (11) | 9 (26) |
| | No | 191 (89) | 26 (74) |
| | Housing and basic facilities | | |
| Housing structure | Buildings | 11 (5) | 12 (34) |
| | Semi pacca | 45 (21) | 3 (9) |
| | Tin & wood | 90 (42) | 18 (51) |
| | Straw roof & bamboo | 69 (32) | 2 (6) |
| Sanitary facilities | Pacca | 11 (5) | 4 (12) |
| | Open/kacha | 19 (9) | 4 (11) |
| | Pit latrine | 105 (49) | 8 (23) |
| | Semi pacca/pacca | 80 (37) | 19 (54) |
| Drinking water facility | Govt. tube well | 136 (63) | 14 (40) |
| | Own tube well | 79 (37) | 21 (60) |
| Electricity facilities | Yes | 194 (90) | 32 (91) |
| | No | 21 (10) | 3 (9) |
| Having social securities (Insurance) | Yes | 17 (8) | 6 (17) |
| | No | 198 (92) | 29 (83) |

**Table 3.** *Cont.*

| Characteristics | Categories | Nazirertek ($n$ = 215) | Chitapara ($n$ = 35) |
|---|---|---|---|
| | | Frequency (%) | Frequency (%) |
| | Ownership of house and land | | |
| Ownership of the house | Owner | 90 (42) | 25 (72) |
| | Rented | 108 (50) | 7 (20) |
| | Leased | 17 (8) | 3 (8) |
| Agricultural land ownership | Less than 5 decimal | 37 (17) | 4 (11) |
| | No land | 178 (83) | 31 (89) |
| | Access to common property resources | | |
| Have access to other resources | Yes | 77 (36) | 16 (46) |
| | No | 138 (64) | 19 (54) |
| Credit access | Self-sufficient | 95 (44) | 18 (51) |
| | Borrowed from NGOs/Bank | 24 (11) | 5 (15) |
| | Borrowed from co-operatives | 77 (36) | 7 (20) |
| | Borrowed from Neighbors | 19 (9) | 5 (14) |
| Participation in training programs | Yes | 65 (30) | 9 (26) |
| | No | 150 (70) | 26 (74) |

Data obtained from the survey showed that most of Nazirertek (36%) communities had no formal education, whereas in Chitapara, 37% of respondents could only write their names. However, 29% of Nazirertek respondents could only sign, while the other 24% and 11% had primary and secondary education, respectively. In Chitapara, these percentages were 37%, 31% and 3%, respectively. Housing conditions of a community indicate the level of well-being or economic status of the people. In Nazirertek, 42% of respondents lived in houses made of wood with a tin shed, and the other 32% lived in straw roof houses, whereas in Chitapara, 34% lived in buildings, and 51% lived in houses made of tin and wood (Table 3).

Migration is defined as the movement of people from one place to another within the state's boundary to take up employment or establish residence. The survey showed that 79% of respondents from Nazirertek were migrants, while in Chitapara, the number of migrants was lower (35%) than Nazirertek (Table 3). At Nazirertek, it was found that 44% of respondents did not need any financial help or did not take any loan, whereas 36% of respondents borrowed money from co-operatives, 11% from NGOs or banks, and 9% borrowed from their neighbors. In Chitapara, these percentages were 51%, 15%, 20% and 14%.

### *3.2. Gender Perspectives on the Livelihoods of Fish Drying Communities*

In Nazirertek, most of the fish drying workers (16% of male and 41% of female) were employed by the dry fish producers (30%) and traders (8%), while women (especially widows, divorcee) made up the majority of the dried fish workers. As a result, most male workers (29%) received USD 4.12–5.89 day$^{-1}$, and only a few (6%) received USD 5.89–8.25 day$^{-1}$ by working from 10–12 hrs. On the other hand, more than half of the women (51%) received USD 3.54–4.12 day$^{-1}$, and only 5% received USD 4.12–5.89 day$^{-1}$ (Figure 3).

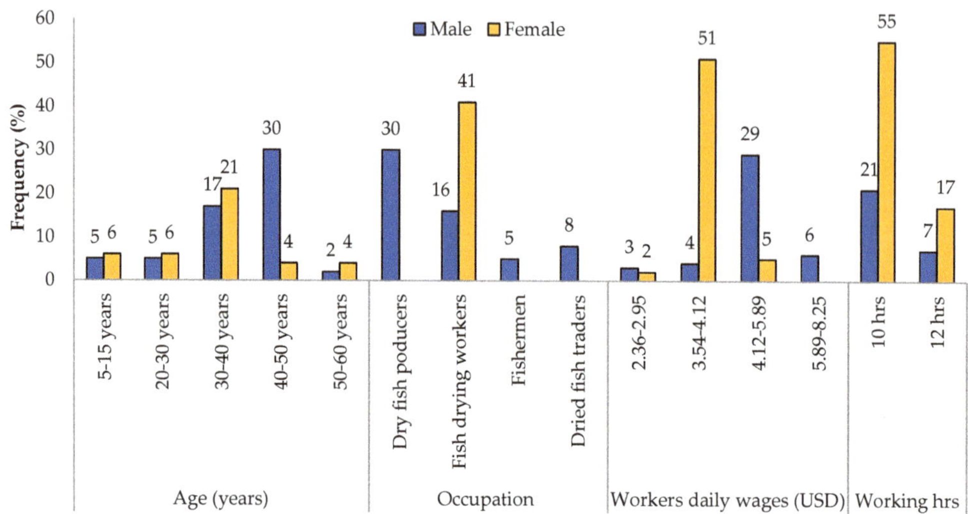


**Figure 3.** Gender participation profile of the fish drying community living in Nazirertek.

Most of the child workers in Nazirertek belonged to the age group 5–15 years (5% male, 6% female), while male children earned USD 2.02–2.52 day$^{-1}$ (3%) to USD 3.54–4.12 day$^{-1}$ (4%) and female earned (2%) USD 2.36–2.95 day$^{-1}$ (Figure 3). Figure 4 shows that no children were engaged in drying fish in Chitapara. In contrast, male workers received maximum wages (62%) ranging from USD 2.35–3.53 day$^{-1}$, and the minimum wages were received by female workers (38%) ranging from USD 1.76–2.35 day$^{-1}$.

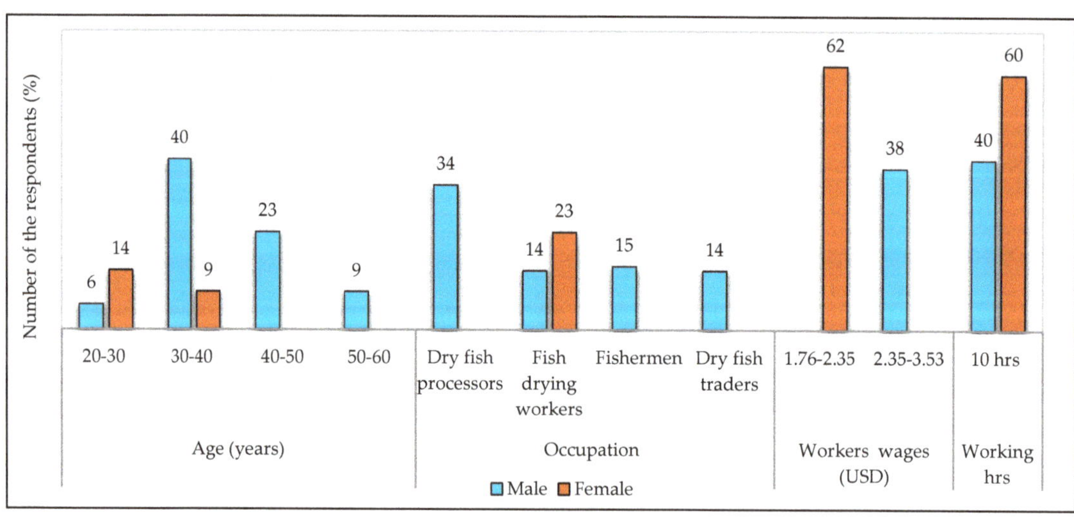


**Figure 4.** Gender participation profile of the fish drying community living in Chitapara.

### 3.3. Reasons behind Employing Children in Drying

Respondents were asked why children were employed, in which 57% of the dried fish processors and owners of Nazirertek reported engaging children at the request of their parents; 39% thought the area was suitable for child labor, and 4% said some parents force their child to work (Figure 5). No children in Chitapara have been found to participate in fish-drying activities, which may be due to their parent's self-reliance in sustaining their livelihoods.

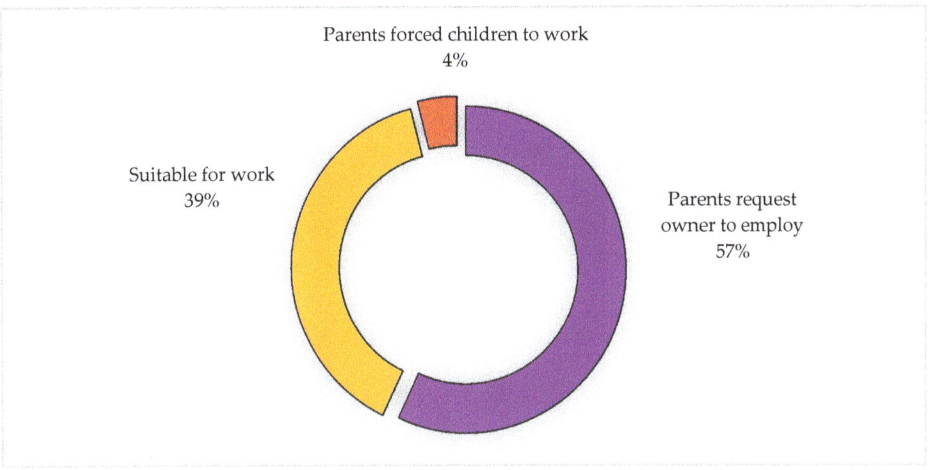


**Figure 5.** Reasons behind employing children for fish drying activities in the study areas.

### 3.4. Health Issues and Treatment Facilities

About 29% of respondents at Nazirertek reported suffering from back pain/rheumatism, whereas at Chitapara, 26% of respondents reported suffering from swelling of the eyes. In addition, swelling of the eyes, skin disease, asthma, diarrhea or fever, anemia, and night blindness were also reported to be suffered by respondents (Figure 6). Moreover, the health facilities for the communities were deplorable; and 48% of respondents from Nazirertek relied on dispensaries for treatment and 49% of the respondents of Chitapara received treatment from Govt. hospital (Figure 6).

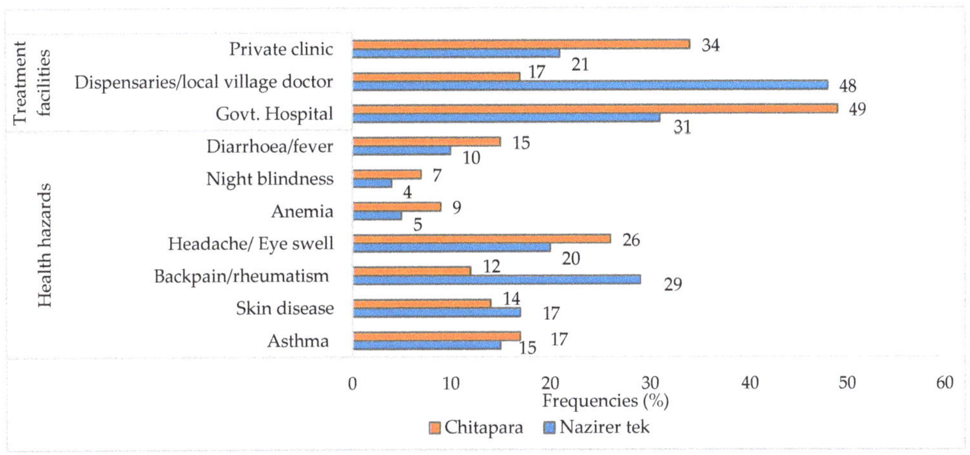


**Figure 6.** Occupational hazards and treatment facilities of the communities involved in fish drying practices.

### 3.5. Income during the Drying Season (Peak Season)

In the fish drying season, 30% of dried fish producers in Nazirertek earned USD 2359–5898 year$^{-1}$, whereas the highest income was earned by dried fish traders (4%), varying from USD 5898–11,794 year$^{-1}$. Among the fish drying workers, 21% earned USD 825–943 year$^{-1}$, whereas fishermen's annual income ranged from USD 1061–1187 year$^{-1}$ (Figure 7). In Chitapara, fish drying communities run the business on a small scale, so their income was not very high compared to Naziertek, dried fish traders, or producers.

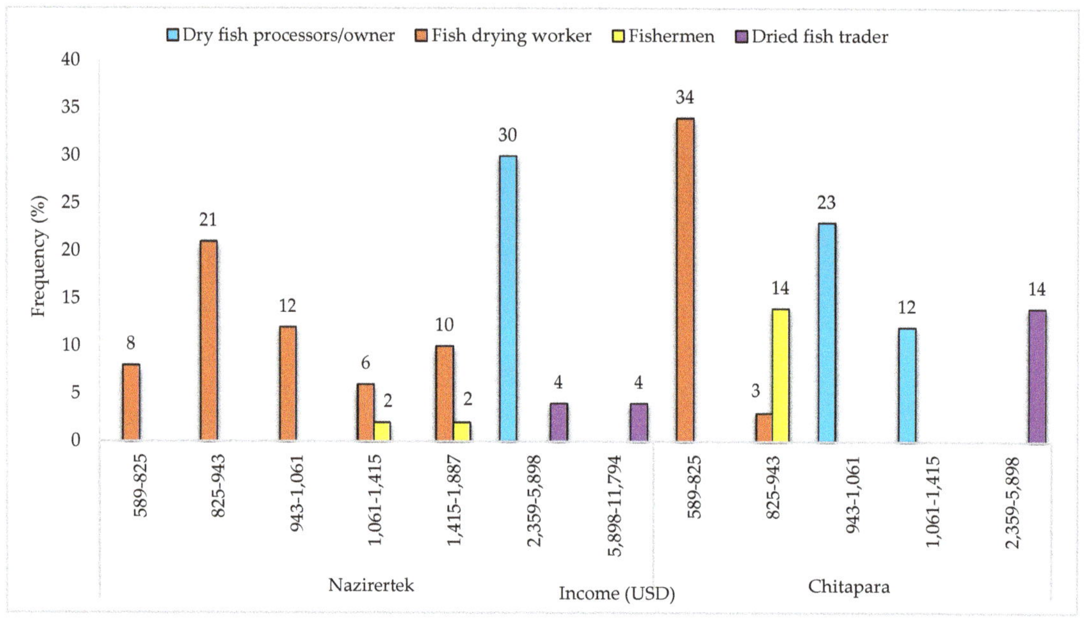


**Figure 7.** The annual income of the communities according to their occupations.

### 3.6. Food Consumption and Security Status

The intake of a poor-quality diet was relatively higher in Nazirertek than in Chitapara (Table 4). The study revealed that about 95% and 86% of the respondents, both from Nazirertek and Chitapara, were able to take three meals daily, whereas the poor workers had to skip a meal a day during the off-season. The situation of food security between the two villages was relatively higher in Chitapara than in Nazirertek.

Binary logistic regression analysis gave the most suitable model with seven variables (age, gender, alternative occupation, household size, income, sole earning member of the family, and house structure) that significantly impact communities' food security. As shown in Table 5, the age of the respondents has a positive coefficient (0.744) and a significant effect ($p < 0.05$ level) in ensuring the food security status of the communities. This indicates that very old respondents are more likely to have food security, which may be due to their work experience accumulated with age. In addition, the results showed that having an alternative occupation was positive and significant at the ($p < 0.05$) level. This specifies that the higher the alternative income-earning activities, the higher is the probability that the community would be food secure. Hence, a unit increase in alternative income levels will increase communities' likelihood of being food secure by 9.528.

**Table 4.** Nutrition and food consumption ratio of the respondents of the studied areas.

| Types | Variables | Nazirertek ($n$ = 215) | Chitapara ($n$ = 35) | Remarks |
|---|---|---|---|---|
| | | Frequency (%) | Frequency (%) | |
| Meal (times/day) | Three times/day | 205 (95) | 30 (86) | During the off-season, most of the workers had to skip one meal in a day and took two meals/day |
| | Four times/day | 10 (5) | 5 (14) | |
| Having Food security | No | 114 (53) | 9 (26) | |
| | Yes | 101 (47) | 26 (74) | |
| Variation of food intake daily | Nutritious diet (Rice, fish/meat, vegetables/pulses/egg/milk) | 85 (40) | 19 (54) | During the off-season, most of the workers had to take low quality fish and vegetables six days/week and rarely ate meat/milk/egg |
| | Poor quality diet (Rice, low-quality fish, pulses, vegetables, meat once/twice a month) | 130 (60) | 16 (29) | |

**Table 5.** Determinants of the factors affecting the food security of people participating in fish drying activities.

| | Socio-Demographic Factors | Co-Efficient B | S.E. | Wald | Sig. | Exp (B) |
|---|---|---|---|---|---|---|
| Step 1 | Age of the respondents | 0.744 | 0.321 | 5.365 | 0.021 | 2.103 |
| | Gender participation profile | −1.705 | 0.921 | 3.426 | 0.064 | 0.182 |
| | Occupation | −0.648 | 0.730 | 0.787 | 0.375 | 0.523 |
| | Alternative occupation | 2.254 | 0.953 | 5.592 | 0.018 | 9.528 |
| | Education | −0.361 | 0.451 | 0.639 | 0.424 | 0.697 |
| | Household size | 0.793 | 0.509 | 2.422 | 0.120 | 2.210 |
| | Ownership of the house | −0.603 | 0.735 | 0.674 | 0.412 | 0.547 |
| | Having Livestock | 0.458 | 0.631 | 0.527 | 0.468 | 1.581 |
| | Income during peak season | 0.879 | 0.363 | 5.856 | 0.016 | 2.408 |
| | Credit access | 0.048 | 0.261 | 0.033 | 0.855 | 1.049 |
| | Sole earning member of the family | −1.724 | 0.799 | 4.652 | 0.031 | 0.178 |
| | Social status | 1.007 | 1.076 | 0.876 | 0.349 | 2.737 |
| | Marital status | 0.126 | 0.309 | 0.166 | 0.684 | 1.134 |
| | Housing structure | −0.614 | 0.293 | 4.407 | 0.036 | 0.541 |
| | Constant | −3.568 | 4.108 | 0.754 | 0.385 | 0.028 |

Significant at, 5%, 1%.

The likelihood that a household will have food security depends on the source of household income. The result showed that the annual income of communities has a significant (found significant at the $p < 0.05$ level) effect on safeguarding food security in the communities with a positive coefficient (0.879). A unit increase in income level increases respondents' probability of becoming food secure up to 2.408. Having a single earning family member has a negative significant association (significant at the $p < 0.05$ level) to ensuring food security for the dried fish communities. The results showed that a decrease in the family's earning members would increase food insecurity. The housing structure had a negative coefficient (−0.614), which had little effect on food security, but it was found to be significant at the ($p < 0.05$) level. Table 5 showed that people living in tin, wood, thatched-roof, and bamboo houses are less likely to have food security than those living in buildings and semi-pacca houses. This may be that most people, especially workers living in bamboo and straw-roofed homes, must pay rent during the drying season due to most of their income being spent buying food items and on rental houses.

### 3.7. Public Health Concern

During the study period, the rate of using pesticides to dry fish was relatively low (2% at Nazirertek and 1% at Chitapara); on the other hand, pre-processing of raw fish during drying was unhygienic (Figure 8). For drying, low-value fish are first brought from the fish landing center to the drying facility (*Khola*). The fish are then washed with water as needed and dumped on bamboo mats. Fish are then mixed with salt before being sorted, which is when the fish's quality significantly deteriorates. After sorting, the fish are washed with water in a bamboo basket or plastic bucket or drums, and spread on a bamboo mat or shelf to dry. Unfortunately, the bamboo mats and baskets used are often dirty and are not washed after one drying cycle and before drying the next batch. Apart from public health concerns, such conditions promote the attack of blowflies that infest fish during drying, especially in the rainy season when rain makes drying difficult.

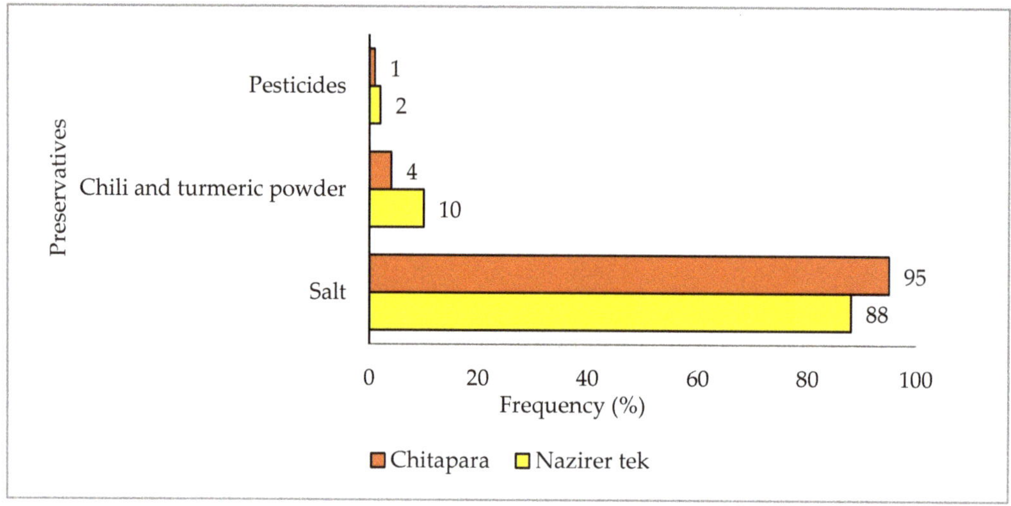


**Figure 8.** Frequency of using preservatives during fish drying and processing.

### 3.8. Livelihood Constraints and Vulnerability Context

As the fish drying communities on the south-east coast are highly vulnerable to seasonality and extreme weather conditions, these people face many socio-economic constraints to sustaining their livelihoods, such as capital crisis, lack of social securities, and poor institutional support for borrowing money (Figure 9).

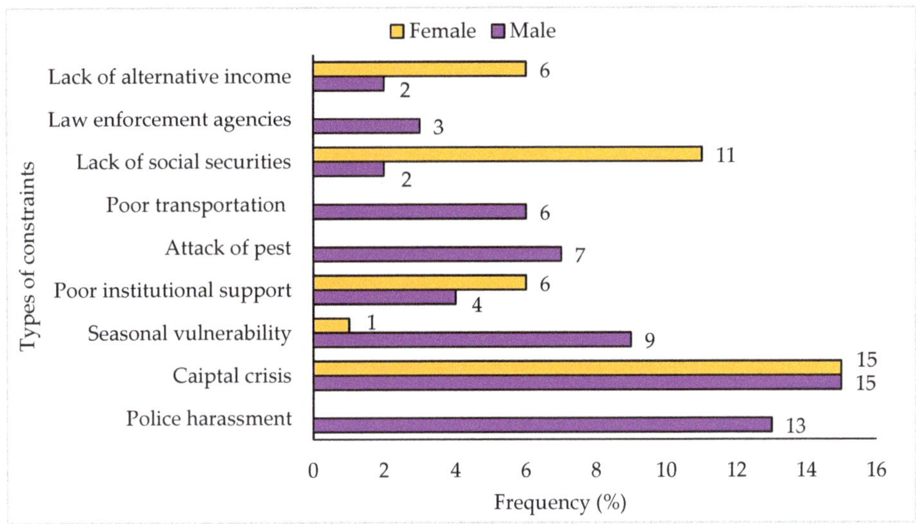


**Figure 9.** Significant constraints to diversified livelihoods faced by male and female respondents of the community.

### 3.9. Coping Mechanisms in the Off-Season

Fish drying practices were mainly carried out from August to April (called peak season), and during these nine months, the traders, processors, and workers were found to pass time by drying fish. However, during the monsoon season (also known as the off-season), there was no fish drying activity. In the off-season, the situation became unbearable where both men and women pursued different initiatives to support their livelihoods. Figure 10 illustrates that about 4% of men and 18% of women stay at home and do agronomical work. In comparison, 24% of the male respondents prepared *Macha* (Bengali name of fish drying facility) for drying fish, and 15% of the female respondents spent time dedicating themselves to raising poultry, sewing clothes, and doing handicrafts.

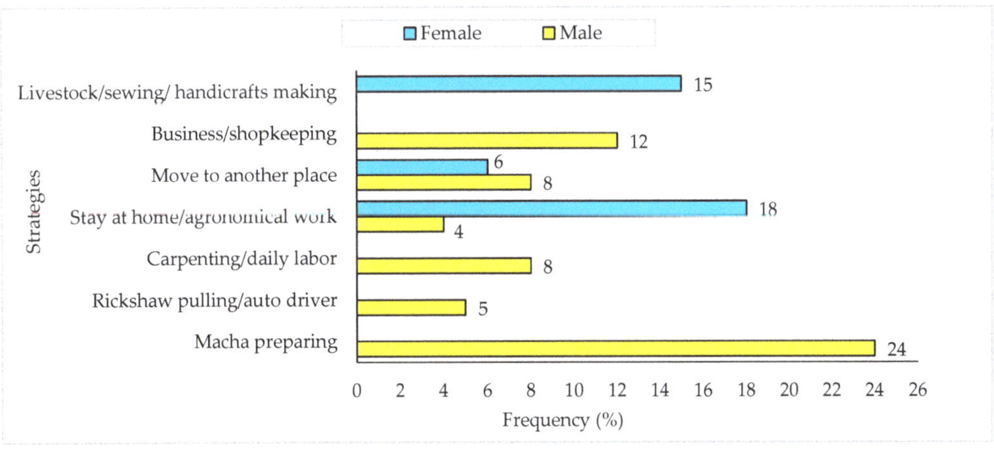


**Figure 10.** Coping mechanisms evolved by the communities during the off-season.

### 3.10. Assessment of Resilience to Communities' Livelihoods Vulnerabilities

This study applied the 'Sustainable Livelihood Framework' to understand the resilience of communities based on the level of dependence on existing assets (Figure 11). In assessing resilience, the identified human assets were the dried fish processors, fishermen, dried fish traders, fish drying workers, and daily labor, managing the fish drying communities with their experiences and professional knowledge. The indicators of physical assets were houses, *Macha* for fish drying, fish landing center, markets, and road structures, and the loss of these physical assets leads to complete suspension of fish drying activities.

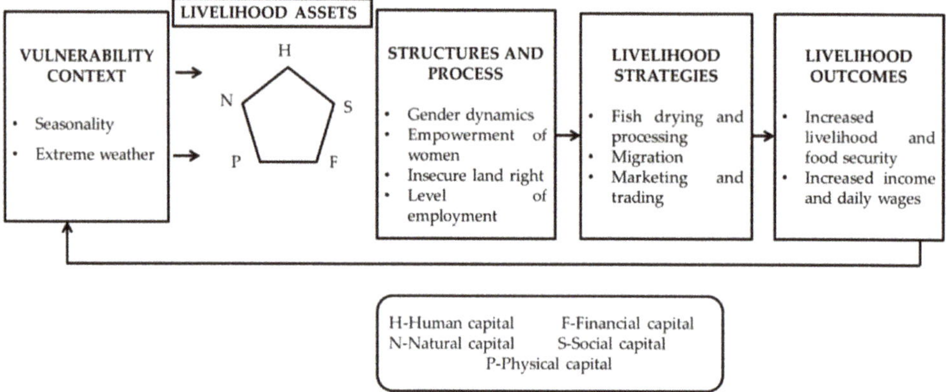


**Figure 11.** Sustainable livelihood framework of the communities.

The availability of dried fish species and enough land for drying were the communities' main livelihood options. The other natural resources that played a vital role in the communities' resilience were water, forest, and grassland, whereas grassland protects against land erosion. Production and trading of dried fish, daily income, and food expenditure were identified as critical financial assets that play an indispensable role in community resilience. However, the role of credit, illegal tax, and livestock cannot be ignored. The communities' social assets, including co-operatives, dried fish traders' association, social interrelationship, and social class, play an essential role in maintaining economic growth and human well-being (Table 6).

The effectiveness of sub-attributes in each asset is summarized in Figure 12. The results showed that the criteria of dried fish processors, dried fish workers, house, *khola/macha* for drying, selling of dried fish, income during the off-season, and the abundance of fish species, enough land for fish drying, trader's association, and social interrelationship were relatively high. Additionally, those criteria are effective by 20–40% compared to other livelihood assets that indicate the most increased role in resilience assessment, whereas fishers, dried fish traders, market, road structure, savings, livestock, water, marketing network, and co-operatives were found with 10–19% relative effectiveness, indicating a moderate role in building resiliency. The sub-attributes of daily labor, landing center, credit access, forest, grassland, and social class had less than 10% relative effectiveness, indicating the least significant resilience assessment.

**Table 6.** A pair-wise comparison matrix for assessing the relative importance of different criteria for resilience assessment of communities involved in fish drying practices to livelihood vulnerabilities in the study areas (numbers show the row factor rating close to the column factor).

| Assets/Capitals | Stakeholders Involved in Fish Drying Practices | | | | | Weight |
|---|---|---|---|---|---|---|
| Human capital | Dried fish processors | Fishermen | Dried fish traders | Dried fish laborer | Daily labor | |
| Dried fish processors | 1 | 1/4 | 1/3 | 1/2 | 1/5 | 0.065 |
| Fishermen | 4 | 1 | 2 | 1/2 | 1/3 | 0.201 |
| Dried fish traders | 3 | 1/2 | 1 | 1/3 | 1/2 | 0.190 |
| Dried fish worker | 2 | 1/3 | 3 | 1 | 1/5 | 0.126 |
| Daily labor | 5 | 2 | 2 | 3 | 1 | 0.417 |
| | | Consistency ratio (C.R): 0.0083 | | | | |
| Physical capital | House | Khola/macha for drying | Landing center | Market structure | Road structure | |
| House | 1 | 1/2 | 1/4 | 1/3 | 1/5 | 0.066 |
| Khola/macha for drying | 2 | 1 | 2/3 | 1/4 | 1/2 | 0.132 |
| Landing center | 4 | 2 | 1 | 3 | 2 | 0.386 |
| Market structure | 3 | 2 | 1/6 | 1 | 1/2 | 0.162 |
| Road structure | 5 | 3 | 2/3 | 1/2 | 1 | 0.253 |
| | | Consistency ratio (C.R): 0.0145 | | | | |
| Financial capital | Dried fish sold | Income during offseason | Credits | Illegal tax | Livestock | |
| Selling of dried fish | 1 | 1/2 | 1/3 | 1/4 | 1/5 | 0.073 |
| Income during offseason | 2 | 1 | 1/2 | 1/3 | 1/2 | 0.128 |
| Credit access | 3 | 2 | 1 | 3/2 | 2 | 0.327 |
| Illegal tax | 4 | 2 | 1/3 | 1 | 2 | 0.260 |
| Livestock | 5 | 2 | 1/2 | 1/3 | 1 | 0.212 |
| | | Consistency ratio (C.R): 0.0055 | | | | |
| Natural capital | Fish abundance | Water | Forest | Grassland | Enough land for fish drying | |
| The abundance of fish species | 1 | 1/2 | 1/4 | 1/6 | 1/3 | 0.067 |
| Water | 2 | 1 | 2/3 | 2 | 3 | 0.202 |
| Forest | 4 | 2 | 1 | 1 | 2 | 0.301 |
| Grassland | 6 | 2 | 1/2 | 1 | 4 | 0.327 |
| Enough land for fish drying | 3 | 1/4 | 1/5 | 1/3 | 1 | 0.103 |
| | | Consistency ratio (C.R): 0.0140 | | | | |
| Social capital | Traders association | Social interrelation | Social class | Marketing system | Co-operatives | |
| Traders association | 1 | 1/2 | 1/4 | 1/3 | 1/5 | 0.069 |
| Social interrelationship | 2 | 1 | 2/3 | 1/2 | 1/2 | 0.146 |
| Social class | 4 | 2 | 1 | 3 | 2 | 0.358 |
| Marketing system | 3 | 2 | 1/4 | 1 | 1 | 0.202 |
| Co-operatives | 5 | 2 | 2/3 | 1/3 | 1 | 0.226 |
| | | Consistency ratio (C.R) 0.0089 | | | | |
| Overall | Human | Physical | Financial | Natural | Social | |
| Human | 1 | 1/2 | 1 | 2 | 1/3 | 0.136 |
| Physical | 2 | 1 | 3 | 5 | 1/2 | 0.301 |
| Financial | 1 | 1/3 | 1 | 2 | 1/2 | 0.140 |
| Natural | 1/2 | 1/4 | 1/3 | 1 | 1/4 | 0.070 |
| Social | 3 | 1 1/2 | 2 | 4 | 1 | 0.353 |
| | | Consistency ratio (C.R) 0.0030 | | | | |

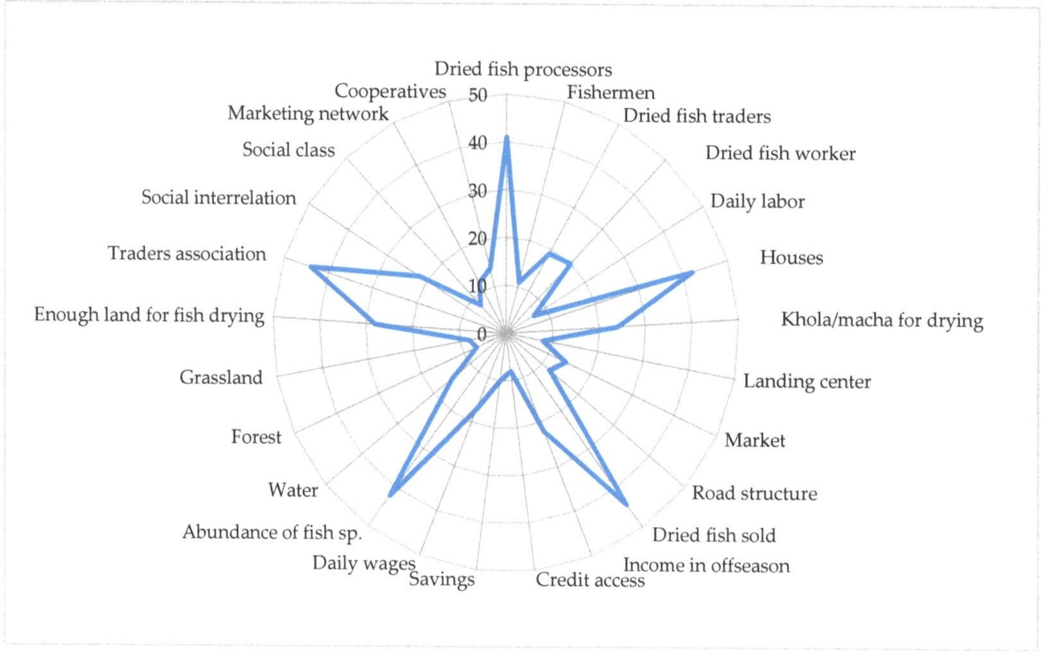


**Figure 12.** The relative importance (%) of livelihood assets in determining the community's resilience.

### 3.11. Likelihood Ratio Test between Different Pairs of the Categorical Variable

Likelihood-ratio tests between categorical variables showed that all other variables, except religion and alternative occupation, had a significant association (1%, 5%) with the community's socio-economic status (Table 7).

**Table 7.** Likelihood ratio test between different categorical variables.

| Categories | Value | df | Significant |
|---|---|---|---|
| Age of the respondents | 120.738 | 8 | 0.000 |
| Gender | 156.561 | 2 | 0.000 |
| Religion status | 0.509 | 2 | 0.775 |
| Marital status | 128.494 | 6 | 0.000 |
| Occupational profile | 326.315 | 6 | 0.000 |
| Alternative occupation | 4.796 | 2 | 0.051 |
| Family type | 13.901 | 2 | 0.001 |
| Residential status | 13.650 | 2 | 0.001 |
| Level of education | 81.302 | 6 | 0.000 |
| Ownership of houses | 239.006 | 4 | 0.000 |
| Electricity facilities | 12.828 | 2 | 0.002 |
| Drinking water facilities | 213.039 | 2 | 0.000 |
| Agricultural land | 6.103 | 2 | 0.047 |
| Having Livestock | 11.943 | 2 | 0.003 |
| Subsistence Production | 11.337 | 2 | 0.003 |
| Annual Income (USD) | 333.197 | 12 | 0.000 |
| Training facilities | 14.700 | 2 | 0.001 |

### 3.12. Multinomial Logistic Regression Analysis According to Socio-Economic Status

This analysis was accustomed to determining whether demographic variables influence poor, middle, and rich class communities' livelihood strategies. Table 8 represents the parameter estimate for the final mode while the "Rich class" category had been taken in the reference group. The odds ratio coefficients of the model's demographic variables were calculated, and then the estimated coefficients of the two classes were compared with the reference category. The results from the logistic analysis indicated that out of 15 hypothesized variables, three socio-demographic variables (drinking water facilities, income in the peak season, and credit access) were found to have a significant influence on the livelihood strategies of the "Poor class" respondents and seven variables (education, sole earning member of the family, having children, drinking water and treatment facilities, income in the peak season and credit access) had a significant influence on the livelihood strategies of the "Middle class" communities at 1%, 5%, significance level. When compared with the other demographic variables, the income of the poor class respondents had an Odds Ratio (OR) = 0.007 (95% CI 0.000 to 0.300), $p = 0.010$ and credit access to services had OR = 20.389 (95% CI 2.613 to 159.060), $p = 0.004$; drinking water facilities had an Odds Ratio (OR) = 0.003 (95% CI 0.002 to 0.267), $p = 0.011$. From the "middle class" categories, the results showed that the educational status of the respondents had an Odds Ratio (OR) = 2.424 (95% CI 0.982 to 5.985), $p = 0.055$; sole earning member of the family had Odds Ratio (OR) = 1.240 (95% CI 0.943 to 1.631), $p = 0.124$; number of children had an Odds Ratio (OR) = 8.155 (95% CI 1.426 to 46.651); income of the respondents had Odds Ratio (OR) = 0.399 (95% CI 0.171 to 0.930), $p = 0.033$; drinking water facilities had an Odds Ratio (OR) = 110 (95% CI 0.020 to 0.592), $p = 0.010$; treatment facilities of the respondents had an Odds Ratio (OR) = 0.155 (95% CI 0.063 to 0378), $p = 0.142$; and credit access to services had an OR = 4.277 (95% CI 1.658 to 11.034), $p = 0.003$.

**Table 8.** Multinomial logistic regression analysis according to the economic status of the studied areas.

| Parameter Estimates | | | | | | | |
|---|---|---|---|---|---|---|---|
| Socio-Economic Status | | B Coefficient | Std. Error | Sig. | Exp (B) Odd Ratio | 95% Confidence Interval for Exp (B) | |
| | | | | | | Lower Bound | Upper Bound |
| Poor class | Intercept | −4.281 | 11.076 | 0.699 | | | |
| | Age of the respondents | 2.232 | 1.766 | 0.206 | 9.315 | 0.292 | 296.953 |
| | Marital status | 3.491 | 2.452 | 0.154 | 32.811 | 0.269 | 4006.702 |
| | Occupation of the respondents | 0.661 | 2.073 | 0.750 | 1.938 | 0.033 | 112.612 |
| | Level of education | −0.044 | 0.941 | 0.962 | 0.956 | 0.151 | 6.045 |
| | Sole earning member of the family | −0.707 | 0.612 | 0.248 | 0.493 | 0.149 | 1.635 |
| | Housing structure | −0.094 | 1.125 | 0.933 | 0.910 | 0.100 | 8.257 |
| | Ownership of the house | −2.147 | 2.894 | 0.458 | 0.117 | 0.000 | 33.973 |
| | Number of children | 3.148 | 2.227 | 0.157 | 23.297 | 0.296 | 1832.378 |
| | School going children | 3.366 | 1.809 | 0.063 | 28.964 | 0.835 | 1004.358 |
| | Drinking water facilities | −5.728 | 2.250 | 0.011 | 0.003 | 0.002 | 0.267 |
| | Sanitary facilities | 1.793 | 1.220 | 0.142 | 6.007 | 0.549 | 65.690 |
| | Types of disease | 0.609 | .982 | 0.535 | 1.838 | 0.268 | 12.609 |
| | Treatment facilities | 0.897 | 2.012 | 0.656 | 2.453 | 0.048 | 126.644 |
| | Income in peak season | −4.948 | 1.910 | 0.010 | 0.007 | 0.000 | 0.300 |
| | Credit access | 3.015 | 1.048 | 0.004 | 20.389 | 2.613 | 159.060 |
| Middle class | Intercept | −0.894 | 4.715 | 0.850 | | | |
| | Age of the respondents | 1.097 | 0.759 | 0.148 | 2.995 | 0.677 | 13.245 |
| | Marital status | 2.555 | 2.330 | 0.273 | 12.870 | 0.134 | 1237.758 |
| | Occupation of the respondents | 0.103 | 0.329 | 0.753 | 1.109 | 0.582 | 2.112 |
| | Level of education | 0.885 | 0.461 | 0.055 | 2.424 | 0.982 | 5.985 |
| | Sole earning member of the family | 0.320 | 0.157 | 0.042 | 1.377 | 1.011 | 1.874 |
| | Housing structure | 0.066 | 0.252 | 0.792 | 1.069 | 0.653 | 1.750 |
| | Ownership of the house | −2.469 | 1.677 | 0.141 | 0.085 | 0.003 | 2.267 |
| | Number of children | 2.099 | 0.890 | 0.018 | 8.155 | 1.426 | 46.651 |
| | School going children | 0.963 | 0.641 | 0.133 | 2.620 | 0.746 | 9.202 |
| | Drinking water facilities | −2.207 | 0.859 | 0.010 | 0.110 | 0.020 | 0.592 |
| | Sanitary facilities | 0.431 | 0.410 | 0.293 | 1.539 | 0.689 | 3.437 |
| | Types of disease | −0.164 | 0.296 | 0.579 | 0.849 | 0.476 | 1.515 |
| | Treatment facilities | −1.866 | 0.456 | 0.000 | 0.155 | 0.063 | 0.378 |
| | Income in peak season | −0.919 | 0.432 | 0.033 | 0.399 | 0.171 | 0.930 |
| | Credit access | 1.453 | 0.484 | 0.003 | 4.277 | 1.658 | 11.034 |

## 4. Discussion

The availability of marine fish, and the drying, processing, and trade of these fish have brought solvency to many poor coastal populations and increased socio-economic well-being. However, they also exposed them to many constraints that drive the need to improve community resilience. The communities engaged in fish drying activities on the south-east coast are socio-economically backward due to their labor-intensive occupation, lower literacy level, unstable income, dependency on seasonal drying, lack of access to resources, and alternative income. Due to the relatively low education rate in the dry fish community, people have always been lagging in improving sustainable livelihoods, income diversification, modern technology adaptation, and socio-economic welfare. Among the two villages, the illiteracy rate was relatively higher (36%) in Nazirertek than in Chitapara (29%). At the same time, more than one-third of the respondents in Chitapara could only write their names (Table 3). This result coincides with the study conducted by [29], which expressed that 25% of dried fish producers in Barisal and 40% of dried fish producers in the Kuakata region were uneducated. While people once had the myth that more families could earn more money [30] due to poverty and daily expenses, most community members in the present study prefer to have a nuclear family rather than a joint family. Household profiles of fish-drying communities on the southeastern coast revealed that most people had to manage a large family of 5–7 people dominated by adult members, exceeding the national average household size (4.5 people per household). This result is consistent with the study of [31], reporting that half of the fishermen in the Noakhali area belong to 5–6 families.

The participation of women among workers was higher than men (Figures 4 and 5). Many widows and divorcee women had become self-reliant by working in the fish drying yards. Fish drying workers were employed seasonally by producers and traders, while some worked as permanent workers and casual workers. Men managed virtually all the dried fish processing activities, but most of the work was done by female workers, such as tying pairs of dried fish and grading, sorting, and rotating dried fish. More or less similar results were reported by Almaden [16], where the author noted that women were mainly engaged in fish processing and men were involved in catching fish. During the study, females were asked if they were more discriminated against than men. In that case, a mixed reaction was observed among them, and some replied they were subjected to discrimination, while others said that men were doing heavier work than women and get paid more than women. A study conducted by Roy et al. [18] reported that gender discrimination was widespread among Indian Sundarban.

The housing pattern, ownership of land, and houses is an important indicator to assess the socio-economic well-being of a community. It was observed that half of the respondents (50%) of Nazirertek dwell in rented houses, where 72% of traders and processors in Chitapara had their own houses to live in (Table 3). The causal workers of Nazirertek stated that during the peak season, they lived in poorly constructed rental houses near the drying sites built or leased out by the owners/traders. They also revealed that they found it challenging to share a single room with their 5–6 family members and pay USD 7.07–8.26/month, whereas this scenario was less in Chitapara. On the contrary, it specified that 78% of fry and fingerling traders had their own houses, and 22% lived in rental houses [32].

Access to sanitary, clean, and safe drinking water is regarded as an essential fundamental in society. Despite the local authorities trying to secure basic facilities, the community's basic facilities were less than satisfactory. Most of the study areas' communities have electricity facilities, but the other facilities (drinking, sanitary) enjoyed by the Nazirertek communities were not as good as Chitapara. Asif et al. [32] pointed out that the sanitation facility of traders in the Jessore region was good. However, not all fishermen on Nijhum Dip Island have access to electricity facilities and instead use solar power [33]. Therefore, these results do not correlate with the current findings.

Migration is an essential determinant of livelihood strategies. The study revealed that most producers or traders were displaced by natural disasters from Kutubdia, Chakaria,

Moheshkhali, Myanmar (Rohingyas) (Table 3) and lived in Nazirertek permanently for 15 to 32 years. During peak season, seasonal fish-drying workers also move to Nazirertek, searching for work with their families and living in houses built by processors and traders near the drying area. Therefore, seasonal fluctuations are not the only factor driving communities to migrate. Further, different migration factors such as lack of employment opportunities and social insecurity in the local community force them to move elsewhere. Similar observation stated that fishers and their families on the south-central coast pushed highly to nearby places due to socio-economic vulnerabilities, insecurities, and unemployment [30].

Most of the communities had been observed to work long hours in the hot sun. As a result, they suffered from various health issues, including headache/swelling of the eyes, back pain, rheumatism, and dark skin due to the seasonal changes and exposure to sunlight (Figure 6). There was one Upazila Health Complex in the study area, but the medical facilities the community enjoyed were unsatisfactory as they had little capacity to pay for medical care. According to the study, more than one-third of respondents, as dry fish producers, traders, and fishers, had access to a fishery office (Table 3), institutional organization, sea, market, and firewood. In contrast, most of the fish drying workers had no access to an institutional organization or microcredit access due to their social status in the communities. Singh et al. [34] reported that about 82.50% of Coastal Odisha women had access to the market, and less than one-third of males and females had access to institutional credit. Thus, these findings are not relevant to the present study.

Among the people involved in drying, dried fish producers and traders had better livelihoods than workers due to their high annual incomes, occupation, ownership of their assets, and seasonal investments in fish drying. However, most of the workers were landless and poor and were exploited by producers and traders while working under the supervision of producers and traders. Belton et al. [1] also made a similar observation, who noted that workers from very different social origins were employed in fish drying under various production relations mixtures. The authors also concluded that this had a significant impact on workers' lives but often led to the exploitation of subgroup workers, which adversely affected social welfare.

Communities engaged in fish drying activities were found to work hard throughout the day to manage their food and livelihoods. However, most of them had difficulty meeting the necessities of life. As the income of dry fish producers or traders was high (Figure 7), it was easier for them to meet their livelihoods and basic requirements. On the contrary, much of the income of poor fishermen and workers was spent managing food, treatment, and education. They were found to suffer from food shortages, and to withstand this situation they were compelled to reduce their meal frequency to two meals per day and try to consume less-expensive food items (Table 4). Reducing meal frequency and fish consumption reflects their low-income level and lack of alternative livelihood opportunities during the off-season. Similar findings were also reported by Rana et al., Mondal et al. [17,35].

The age of the respondents, having an alternative occupation, income, sole earning member of the family, and house structure affects the community's food security in many ways (Table 5). Income is an essential component of economic access to food at both the communities and individual levels. From Table 5, it can be concluded that the higher the annual income of the participants, the more likely the participant has food security. In the off-season, most people were out of work because drying and processing were their main occupations; nevertheless, they had to rely on other alternative professions to maintain their livelihoods and manage their food. Hence, earning family members also affect the food security status of the communities. Simultaneously, the more earners they have in the family, the more money they can achieve and the more they can manage their families. Thus, the community will have more food security. A similar observation was made by Omotayo and Aremu [36], showing that age, gender, and household size significantly impact food security status among rural households in the North West State of South

Africa. Moreover, this study's findings also correspond to Maharjan and Joshi [37], who concluded that providing income-generating opportunities to economically active age people can significantly reduce food insecurity.

Fish drying activities in the study areas were highly seasonal, and most of the respondents, especially the fish drying workers, were employed on a seasonal basis. This pattern and level of employment harmed community livelihoods, including capital crisis and social insecurities in the workplace (Figure 10). According to Marimuthu [38], the inland fishermen were found to face many problems such as employment patterns, lack of transportation facility, no safety, or high risk, which was more or less similar to the present study. Furthermore, most of the communities largely depended on fish drying activities. However, during the off-season, the communities' situation became so unbearable that both men and women had to pursue different initiatives to support their livelihoods (Figure 10). To cope with the financial crisis, they had to borrow money from various institutions and relatives for livelihood maintenance or other purposes, including marriages, festivals, medical, and different basic needs [39,40].

Community resilience embraces the idea of how biophysical and socio-economic systems can respond to any changes, unpleasant shocks, and seasonality [28]. As solar energy is the fundamental element to drying fish, harsh weather environments such as windstorms, heavy rain, cloudy days, and destructive current can be a significant barrier to the communities' resilience. The resilience study revealed that the fish drying communities were less resilient to livelihood vulnerabilities than other communities because they relied on natural resources, solar energy, and migration (Figure 12). The dry fish processors play a crucial role in maintaining resilience, where the workers work as 'food for work, cash for work.' In the case of community resilience, the experienced and skilled dry fish producers and workers, production and trading of dried fish, enough land for fish drying, availability of dried fish species, traders' association, and social interrelationship among the communities has been found to play a crucial role in enhancing community resilience (Table 6). Hossain et al. [24] also made a similar observation on the resilience assessment of fishing communities at Nijhum Dwip Island.

As far as quality control and public health are concerned, it has been observed that most of the processors, traders, and workers throughout the study period use salt, chili, and turmeric powder as preservatives rather than pesticides to dry and process fish. Although the extent of pesticide application has been reduced, only a tiny part of the community has reported spraying pesticides (*Sobicron, Nogos*) to the dried fish to reduce pest attacks on heavy rains and cloudy days. The residual effects of these pesticides can be very detrimental to human health. So far, dried fish cannot reach a broader range of wealthy consumers, especially health-conscious consumers.

When conducting the study, the main limitations were dialect issues and bad weather conditions, which made communication with the respondents very difficult. Therefore, the most fundamental challenge for the dried fish sector will be ensuring more sustainable fisheries management. Other challenges in this sector include the insecurity of ownership faced by many drying processors or traders due to their vulnerability to climate change. Fish drying activities in Nazirertek are carried out on *khas* (government-owned land), which can be evacuated due to economic activities and may affect communities' working and living conditions largely dependent on drying. On the other hand, dried fish for preparing fish meals in aquaculture feed can play a significant role in the fish feed industry.

A more direct approach should focus on diversification of livelihoods and access to basic amenities, including health care, education, clean drinking water, hygiene improvement, and dietary supplements to sustain the livelihoods of communities. Moreover, the local community leader should organize alternative employment opportunities or training facilities for workers, especially in the off-season, to diversify their livelihoods. A safe place should be provided for the dried fish producers and traders to expand their fish drying activities, such as a private zone. Governments with local authorities need to take the necessary steps to increase dried fish exports in domestic and international markets. In addition, processors and workers who follow traditional fish drying techniques should be trained for improved sun drying, hygiene, and public health.

## 5. Conclusions

Across the country, the poor and marginalized coastal people are involved in drying fish and work hard to meet the growing demands of dried fish with their skill and flawless efforts. Nevertheless, their livelihood patterns are less diversified and rarely recognized. Most widows, divorcees, and unmarried female workers were found to be self-reliant on the southeastern coast by working in fish drying yards. Moreover, most workers involved in dry fishing experienced livelihood insecurity as they are landless, poor, and unskilled with a limited work environment and exploited by processors and traders. To boost the community resilience of the dry fish workers, it is vital to expand alternative occupation and social protection, providing community empowerment for making resource-use decisions and institutional, organizational, and government support. Therefore, this study's findings will contribute as a base knowledge for government and local management authorities for new, practical, and equitable management for communities.

**Author Contributions:** S.J.M.: Writing, original draft preparation, P.S.: Data curation, Funding acquisition, and editing, M.S.I.: Visualization, M.A.: Data analyzing, M.M.H.M.: Reviewing, editing, M.M.H.: Visualization and M.M.S.: Conceptualization, methodology, writing and reviewing. All authors have read and agreed to the published version of the manuscript.

**Funding:** This research received no external funding.

**Institutional Review Board Statement:** Not applicable.

**Informed Consent Statement:** Written informed consent has been obtained from the participants to publish this paper.

**Acknowledgments:** The authors would like to extend their gratitude to the Ministry of Science and Technology, Government of Bangladesh for National Science and Technology (NST) Fellowship which enabled the authors to carry out the research. The authors are also thankful to the traders, processors, workers for their cordial help during field works.

**Conflicts of Interest:** The authors declare no conflict of interest.

## References

1. Belton, B.; Hossain, M.A.; Thilsted, S.H. Labour, identity and well-being in Bangladesh's dried fish value chains. In *Social WellBeing and the Values of Small-Scale Fisheries*; Springer: Cham, Switzerland, 2018; pp. 217–241.
2. Paul, P.C.; Reza, M.S.; Islam, M.N.; Kamal, M. A review on dried fish processing and marketing in the coastal region of Bangladesh. *Res. Agric. Livest. Fish.* **2018**, *5*, 381–390. [CrossRef]
3. Gupta, T.; Chandrachud, P.; Muralidharan, M.; Namboothri, N.; Johnson, D. The Dried Fish Industry of Malvan. In *Dried Fish Report Final*; 2020; pp. 1–12. Available online: https://www.dakshin.org/wp-content/uploads/2020/08/Malvan-Dried-Fish-Report-Final.pdf (accessed on 10 February 2021).
4. Hossain, M.A.; Haque, M.E.; Alam, M.S. Livelihood and food security status of fisher's community in the northern districts of Bangladesh. *Int. J. Nat. Soc. Sci.* **2015**, *2*, 23–29.
5. DoF (Department of Fisheries). *National Fish Week Compendium*; Department of Fisheries, Ministry of Fisheries and Livestock: Dhaka, Bangladesh, 2018; p. 160. (In Bengaliu)
6. Haque, M.M.; Rabbani, M.G.; Hasan, M.K. Efficiency of marine dry fish marketing in Bangladesh: A supply chain analysis. *Agriculturists* **2015**, *13*, 53–66. [CrossRef]
7. Afroz, Q.S. Small-Scale Fisheries of Bangladesh. *Small-Scale Fish. South Asia* **2018**, 67–80. Available online: https://www.researchgate.net/profile/Shiba-Giri/publication/333089404_Small-scale_Fisheries_in_South_Asia/links/5cdadb9aa6fdccc9ddabc5a0/Small-scale-Fisheries-in-South-Asia.pdf#page=76 (accessed on 15 May 2021).
8. FRSS (Fisheries Resources Survey System). *Fisheries Statistical Report of Bangladesh*; Department of Fisheries: Dhaka, Bangladesh, 2019; Volume 6, p. 129.
9. Allison, E.H.; Perry, A.L.; Badjeck, M.C.; Neil Adger, W.; Brown, K.; Conway, D.; Dulvy, N.K. Vulnerability of national economies to the impacts of climate change on fisheries. *Fish Fish.* **2009**, *10*, 173–196. [CrossRef]
10. Badjeck, M.C.; Allison, E.H.; Halls, A.S.; Dulvy, N.K. Impacts of climate variability and change on fishery-based livelihoods. *Mar. Policy* **2010**, *34*, 375–383. [CrossRef]

11. Merino, G.; Barange, M.; Blanchard, J.L.; Harle, J.; Holmes, R.; Allison, E.H.; Mullon, C. Can marine fisheries and aquaculture meet fish demand from a growing human population in a changing climate? *Glob. Environ. Change* **2012**, *22*, 795–806. [CrossRef]
12. Islam, M.M.; Barman, A.; Kundu, G.K.; Kabir, M.A.; Paul, B. Vulnerability of inland and coastal aquaculture to climate change: Evidence from a developing country. *Aquac. Fish.* **2019**, *4*, 183–189. [CrossRef]
13. Islam, M.M.; Sallu, S.; Hubacek, K.; Paavola, J. Vulnerability of fishery-based livelihoods to the impacts of climate variability and change: Insights from coastal Bangladesh. *Reg. Environ. Change* **2014**, *14*, 281–294. [CrossRef]
14. Mozumder, M.M.H.; Shamsuzzaman, M.; Rashed-Un-Nabi, M.; Harun-Al-Rashid, A. Socio-economic characteristics and fishing operation activities of the artisanal fishers in the Sundarbans Mangrove Forest, Bangladesh. *Turk. J. Fish. Aquat. Sci.* **2018**, *18*, 789–799. [CrossRef]
15. Billah, M.M.; Kader, M.A.; Siddiqui, A.A.M.; Shoeb, S. Studies on fisheries status and socio-economic condition of fishing community in Bhatiary coastal area Chittagong, Bangladesh. *J. Entomol. Zool. Stud.* **2018**, *6*, 673–679.
16. Almaden, C.R.C. A Case Study on the socio-economic conditions of the artisanal fisheries in the Cagayan de Oro River. *Int. J. Soc. Ecol. Sustain. Dev. (IJSESD)* **2017**, *8*, 14–30. [CrossRef]
17. Mondal, M.A.H.; Islam, M.K.; Islam, M.E.; Barua, S.; Hossen, S.; Ali, M.M.; Hossain, M.B. Pearson's correlation and likert scale based investigation on livelihood status of the fishermen living around the Sundarban estuaries, Bangladesh. *Middle-East J. Sci. Res.* **2018**, *26*, 182–190.
18. Roy, A.; Sharma, A.P.; Bhaumik, U.; Pandit, A.; Singh, S.R.K.; Saha, S.; Mitra, A. Socio-economic features of womenfolk of Indian Sunderbans involved in fish drying. *Indian J. Ext. Educ.* **2017**, *53*, 142–146.
19. Nath, K.D.; Nabadeep, S.; Pulakabha, C. Comparative study on quality of dry Fish (*Puntius* spp.) produce under Solar Tent Dryer and Open Sun Drying. *Agric. Ext. J.* **2017**, *1*, 88–91.
20. Zinnat, M.A. Dry Fish Trade Thriving in Cox's Bazar of Bangladesh. The Daily Star, 10 December 2015. Available online: https://www.thedailystar.net/backpage/dried-fish-trade-thriving-coxs-bazar-184966 (accessed on 11 October 2019).
21. Hossain, M.A.; Belton, B.; Thilsted, S.H. Preliminary rapid appraisal of dried fish value chains in Bangladesh. In *World Fish Bangladesh*; World Fish: Dhaka, Bangladesh, 2013.
22. Islam, M.M.; Aktar, R.; Nahiduzzaman, M.; Barman, B.K.; Wahab, M. Social considerations of large river sanctuaries: A case study from the Hilsa shad fishery in Bangladesh. *Sustainability* **2018**, *10*, 1254. [CrossRef]
23. Saaty, T.L. A scaling method for priorities in hierarchical structures. *J. Math. Physiol.* **1977**, *15*, 234–281. [CrossRef]
24. Hossain, M.S.; Rahman, M.F.; Thompson, S.; Nabi, M.R.; Kibria, M.M. Climate change resilience assessment using livelihood assets of coastal fishing community in Nijhum Dwip, Bangladesh. *Pertanika J. Sci. Technol.* **2013**, *21*, 397–422.
25. Saaty, T.L. *Multicriteria Decision Making: The Analytic Hierarchy Process: Planning, Priority Setting Resource Allocation*; RWS Publications: Pittsburgh, PA, USA, 1990; p. 287.
26. Kjærsgaard, J.; Andersen, J.; Mathiesen, C. Multiple objectives and perceptions of optimal management: The Danish industrial fishery in the North Sea. *Eur. Rev. Agric. Econ.* **2007**, *34*, 181–208. [CrossRef]
27. Sethi, S.A. Risk management for fisheries. *Fish Fish.* **2010**, *11*, 341–365. [CrossRef]
28. Sharifuzzaman, S.M.; Hossain, M.S.; Chowdhury, S.R.; Sarker, S.; Chowdhury, M.S.N.; Chowdhury, M.Z.R. Elements of fishing community resilience to climate change in the coastal zone of Bangladesh. *J. Coast. Conserv.* **2018**, *22*, 1167–1176. [CrossRef]
29. Kubra, K.; Hoque, M.S.; Hossen, S.; Husna, A.U.; Azam, M.; Sharker, M.R.; Ali, M.M. Fish drying and socio-economic condition of dried fish producers in the coastal region of Bangladesh. *Middle-East J. Sci. Res* **2020**, *28*, 182–192.
30. Touhiduzzaman, M.; Shamsunnahar, S.I.K.; Khan, A. Livelihood pattern of fishing communities in south-central coastal regions of Bangladesh. *Sch. J. Agric. Vet. Sci.* **2019**, *6*, 70–77.
31. Bhuyan, S.; Islam, S. Present status of socio-economic conditions of the fishing community of the Meghna River adjacent to Narsingdi district, Bangladesh. *J. Fish. Livest. Prod.* **2016**, *4*, 192.
32. Abdulla-Al-Asif, M.; Samad, A.; Rahman, M.H.; Almamun, M.; Farid, S.M.Y.; Rahman, B.S. Socio-economic condition of fish fry and fingerling traders in greater Jessore region, Bangladesh. *Int. J. Fish. Aquat. Stud.* **2015**, *2*, 290–293.
33. Siam, H.S.; Hasan, M.R.; Sultana, T. Socio-economic status of fisher community at Nijhum Dwip. *Asian J. Med. Biol. Res.* **2020**, *6*, 351–358. [CrossRef]
34. Singh, A.; Sahoo, P.K.; Srinath, K.; Kumar, A.; Tanuja, S.; Jeeva, J.C.; Nanda, R. Gender roles and livelihood analysis of women in dry fish processing: A study in coastal Odisha. *Fish. Technol.* **2014**, *51*, 267–273.
35. Rana, M.E.U.; Salam, A.; Shahriar Nazrul, K.M.; Hasan, M. Hilsa fishers of Ramgati, Lakshmipur, Bangladesh: An overview of socio-economic and livelihood context. *J. Aquac. Res. Dev.* **2018**, *9*, 2.
36. Omotayo, A.O.; Aremu, A.O. Evaluation of Factors Influencing the Inclusion of Indigenous Plants for Food Security among Rural Households in the North West Province of South Africa. *Sustainability* **2020**, *12*, 9562. [CrossRef]
37. Maharjan, K.L.; Joshi, N.P. Determinants of household food security in Nepal: A binary logistic regression analysis. *J. Mt. Sci.* **2011**, *8*, 403–413. [CrossRef]
38. Marimuthu, R. Socio-Economic and Livelihood Profile of Inland Fishermen Households in Theni District, Tamil Nadu. Ph.D. Thesis, Fisheries College and Research Institute, Tamil Nadu Fisheries University, Thoothukudi, India, 2015; pp. 1–110.

39. Haque, M.A.; Hossain, M.D.; Jewel, M.A.S. Assessment of fishing gears crafts and socio-economic condition of Hilsa (Tenualosa ilisha) fisherman of Padma River, Bangladesh. *Int. J. Fish. Aquat. Stud.* **2017**, *5*, 177–183.
40. Saberin, I.S.; Reza, M.S.; Mansur, M.A.; Kamal, M. Socio-economic status of fishers of the old Brahmaputra River in Mymensingh sadar upazila, Bangladesh. *Res. Agric. Livest. Fish.* **2017**, *4*, 229–235. [CrossRef]

International Journal of
*Environmental Research and Public Health*

*Article*

# Subsystem Hazard Analysis on an Offshore Waste Disposal Facility

**Sang-Ho Oh [1,*] and Seung-Woo Kim [2]**

1 Coastal Development and Ocean Energy Research Center, Korea Institute of Ocean Science and Technology, 385 Haeyang-ro, Yeongdo-gu, Busan 49111, Korea
2 Department of Risk Assessment, Risk Solutions Inc., Seoul 04701, Korea; seungwookim76@gmail.com
* Correspondence: coast.oh@gmail.com; Tel.: +82-51-664-3523

Received: 14 September 2020; Accepted: 14 October 2020; Published: 23 October 2020

**Abstract:** Offshore waste disposal facilities are unique marine infrastructures that exist only in a few countries. Although the existing facilities in Japan and Singapore have been successfully operated in general, there have been no investigations on the probable hazards they pose on the environment. Considering this, conceivable hazards were identified for an offshore waste disposal facility that has recently been proposed in Korea. The causes and consequences of each of the identified hazards were analyzed to seek countermeasures for reducing the environmental impact in advance. Hazards of waste disposal facilities can be classified according to their design, construction, maintenance, operation, and site utilization. For these areas, except for site utilization, subsystem hazard analysis was performed. In the initial assessment, seven elements were found to be in the extreme risk zone, 30 were in the high-risk zone, and six were in the moderate-risk zone. After applying the alternative mitigation methods, the final risk assessment resulted in 27 moderate-risk and 16 low-risk elements. Therefore, it was confirmed that the potential risks of the proposed offshore waste disposal facility were within acceptable ranges.

**Keywords:** offshore waste disposal facility; hazard analysis; risk matrix; subsystem; environmental impact

## 1. Introduction

Offshore waste disposal refers to the final landfilling of stabilized inorganic solid waste, such as land and marine waste incineration materials. Although the demand for new waste landfill sites continues, waste disposal space is insufficient because of the imminent end of life in current landfill sites, and the difficulty in securing new landfill sites. Thus, an offshore waste disposal facility as a final waste disposal space was proposed as an alternative landfill site [1]. In Japan and Singapore, offshore waste disposal facilities are being operated to solve the lack of space for land waste disposal, while also creating eco-friendly marine spaces [2–5].

Since the 1990s, Japan has been conducting a thorough reclamation plan with local environmental surveys in advance, based on experience in the construction of multiple offshore waste disposal facilities [6,7]. Furthermore, based on the survey results, efforts have been made to minimize the impact on the marine environment by predicting environmental impacts and establishing measures to mitigate environmental pollution. In general, offshore waste disposal facilities in Japan have been successfully operated in the long term, through continuous monitoring, without major environmental problems or damage [8].

In Singapore, meanwhile, the volume of urban waste started to increase when the economy grew rapidly from the 1970s [9]. Hence, Singapore promoted the construction of its Semakau Landfill offshore facility by enclosing the sea between two islands. Since then, efforts have been made to restore the environment through monitoring and management plans to minimize potential impacts. Examples include restoring the mangrove forests and coral reefs that were destroyed during the construction of the Semakau offshore waste disposal facility [10,11].

In Korea, the necessity of constructing an offshore waste disposal facility as an alternative to land-based landfills has steadily increased, and accordingly, studies on the development of the core technologies and necessary institutional standards have been conducted [12–14]. However, due to various environmental and social conflict factors, such as pollution and the opposition of local residents to the construction of waste treatment facilities nearby, to date there is no decision or plan for an offshore waste disposal facility construction [8]. Because the facility is environmentally and socio-economically sensitive and there have been no previous construction cases in Korea, risk analysis is required to understand the expected risks associated with the construction of this facility. By examining the factors that will cause risks in advance, it is helpful to increase the social acceptance of this facility and to eventually build safer structures.

Risk analysis is often carried out regarding structures that have not been built in the past or are considered to be highly hazardous [15]. Although such an analysis has not been generalized for coastal or offshore structures, hazard analysis from an accredited international organization is mandatory when constructing an offshore plant [16–20]. Usually, such a risk analysis is carried out in accordance with the international standard code ISO 31000:2009 procedure [21]. Moreover, various hazard assessment studies are being conducted in relation to natural disasters and various human activities occurring on the coast and in the sea [22–25]. Nevertheless, to the best of our knowledge, studies on hazard evaluation have not been conducted thus far for offshore waste disposal facilities.

However, a study was recently completed to identify hazards that could occur during the construction and operation of an offshore waste disposal facility, analyzing the causes and effects of identified hazards specific to Korea [26]. A preliminary hazard list analysis (PHL) and preliminary hazard analysis (PHA) on offshore waste disposal facilities were performed, and hazard risks were evaluated. However, the previous study [26] was an analysis at the conceptual design level of an offshore waste disposal facility, which did not reflect the regional characteristics of the facility. Accordingly, the study was limited in that it did not address the specific design of an offshore waste disposal facility.

To overcome this limitation, in the present study, the subsystem hazard analysis (SSHA) was performed using specific design data for an offshore waste disposal facility. SSHA is a procedure performed when detailed design is available as it provides a more in-depth analysis on the hazards previously identified by PHA [15]. Since SSHA is carried out on more detailed design information, hazard elements of PHA are inherited or eliminated, and new hazard elements are also discovered. Therefore, SSHA refines identifying hazards, their associated causal factors, level of risk, and mitigating design measures.

The location of the offshore waste disposal facility on which the SSHA in this study was performed is inside the extension section of the dredged soil disposal area close to Incheon Songdo International City. In comparison with the previous PHA, hazard items are more specifically identified, and the causes and effects of the identified hazard items become clearer. In particular, since a populated city under development is located near the offshore waste disposal facility, considerations for minimizing the environmental impact were taken seriously. However, detailed hazard elements to the residence around the facility were not directly dealt with because this study conducted a hazard analysis mainly focusing on the structural risks of the offshore waste disposal facility.

## 2. Outline of Subsystem Hazard Analysis

System safety identifies the potential hazards of the system and derives a measure to reduce each one. Hazard analysis (HA) focuses mostly on identifying hazards at the beginning of the entire process of system safety [15]. The identification of these hazards is carried out through the HA, and the SSHA is performed when basic and detailed design information is available. The purpose of hazard analysis at this stage is to accurately analyze the causes of the hazards identified in the PHA and to present specific risk mitigation measures. In addition, as the design progresses in detail, new hazards that were not recognized in the previous stage may be identified, and those previously identified may be removed. In other words, SSHA is a step that specifically expands PHA. Thus, the methodology is similar to PHA, however, more thorough results are obtained. Figures 1 and 2 are an overview and the conceptual diagram of SSHA, respectively. The causes and effects of hazards are accurately assessed by comparing them to the hazard checklist based on PHA results and the detailed subsystem design. The core content here confirms how hazards of the previously performed PHA changed during the detailed design process, specifying the causal factors of hazards and their mitigation measures. In this process, the system's top-level mishaps (TLMs), safety critical functions (SCFs), and system safety requirements (SSRs) are produced.

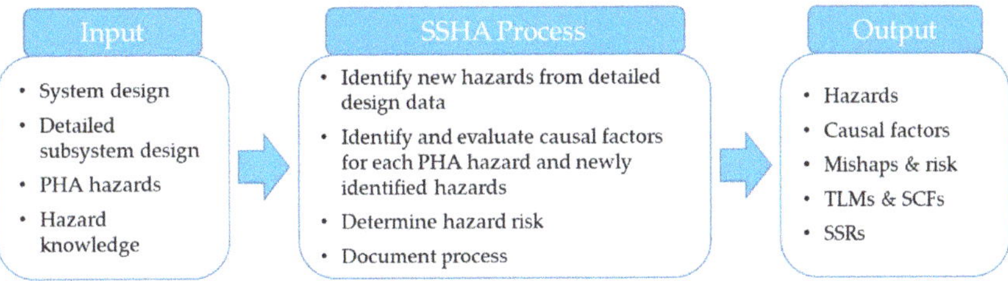


**Figure 1.** Overview of the subsystem hazard analysis (adapted and redrawn from [27]).

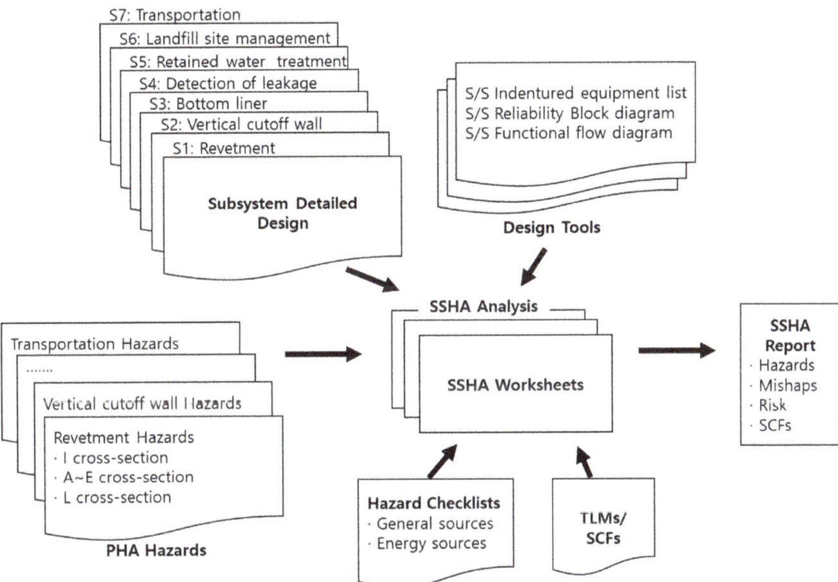


**Figure 2.** Conceptual diagram of the subsystem hazard analysis for the offshore waste disposal facility. PHA denotes preliminary hazard analysis, SSHA—subsystem hazard analysis, TLM—top-level mishaps, and SCFs—safety critical functions, respectively.

The detailed process of SSHA includes 10 steps. The first step is system definition, the second step is analysis planning, the third step involves the establishment of safety standards, the fourth step involves detailed designing and data collection (such as a hazard checklist), the fifth step requires the execution of SSHA, the sixth step is hazard evaluation, the seventh step is the proposal of a hazard reduction plan, the eighth step involves monitoring whether the presented hazard mitigation measures are effective as safety recommendations or system safety requirements, the ninth step is hazard tracking, and the tenth and final step is documentation.

The most important step in the above process is the fifth step of the SSHA. First, sub-divided subsystem elements are listed, elements of each list are evaluated, and the causal factors of any hazards are identified from the subsystem elements. At this time, the hazard risks identified in PHA are also analyzed in parallel. Furthermore, new hazards not identified in the PHA are identified through TLMs, SCFs, hazard checklists, and similar incident cases. When identifying all the hazards, the functional relationship, timing, and parallel functions of the subsystem elements should be understood.

## 3. The Offshore Waste Disposal Facility

### 3.1. The Facility of Investigation

An offshore waste disposal facility is a gravity-based structure that forms an outer revetment and stores waste inside it. The revetment protects the facility from maritime external forces such as waves and tsunamis. It should be a watertight structure to prevent any leaking of leachate. For this purpose, it is typically constructed on a seabed of clay deposit that effectively restricts any vertical flows through the seabed. Meanwhile, vertical cut-off barriers are formed by using the sheet piles that are installed along the inner side of the revetment to prevent any horizontal flows across the barrier. The space between the outer revetment structure and the sheet piles is typically filled with a backfill of rubble stones.

In this study, SSHA was carried out for the proposed project of constructing an offshore waste disposal facility within the dredged soil disposal pond that is being built in the course of constructing the new Incheon port [12]. Figure 3 shows the location of the pilot project, which is very close to the new Incheon port in the south and Songdo International City in the southeast. The planned waste disposal facility is 1.6 km long and 0.6 km wide, and it is divided into two zones of almost equal areas. Once the facility is built, Zone 1 will initially be used for waste disposal, following which Zone 2 would be used for the same purpose when Zone 1 is almost filled with disposed waste.

**Figure 3.** Location of the proposed offshore waste disposal facility.

By constructing a waste disposal facility here, construction costs can be reduced because a portion of the revetment of the disposal pond can be used as the outer revetment of the planned waste disposal facility. In addition, the sea transportation of waste may be unnecessary once this site is connected to the land in the future. On the other hand, there would be opposition from the residents because it is adjacent to Songdo International City, which is only 2.5 km from the residential town.

### 3.2. *Subsystem of the Offshore Waste Disposal Facility*

The SSHA of offshore waste disposal facilities must consider all the aspects of facility design, construction, maintenance, and operation. More specifically, hazards such as failures of the revetments, leakage of surface or vertical cut-off barriers, and the risks accompanied by waste transport should be carefully considered. Considering this, a hazard analysis was conducted by classifying the subsystem of the offshore waste disposal facility into the following seven major categories: revetment, vertical cut-off barrier, surface-cut-off barrier, leakage monitoring and detection, retained water treatment facility, landfill management, and waste transportation.

#### 3.2.1. Revetment

Figure 4 shows the plan view of the outer and inner revetments (marked using red lines) that constitute the proposed offshore waste disposal facility. Meanwhile, the black lines in Figure 4 indicate the revetment of the dredged soil disposal pond that is being built during the construction of the new Incheon port. Sections A to E are termed as barrier revetments because they act as a barrier that prevents the outflow of waste and retained water, rather than directly resisting the external forces from the ocean. The cross-section of these sections is composed of double vertical layers of sheet piles, as shown in Figure 5a.

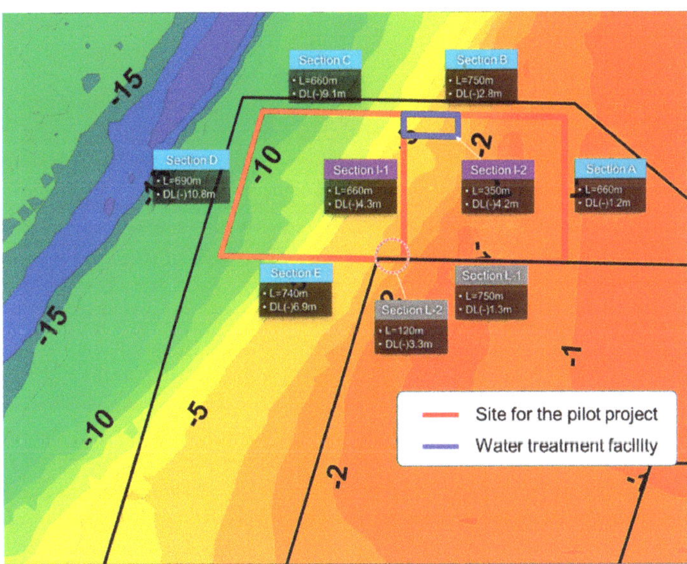


**Figure 4.** Details of the revetments constituting the proposed offshore waste disposal facility.

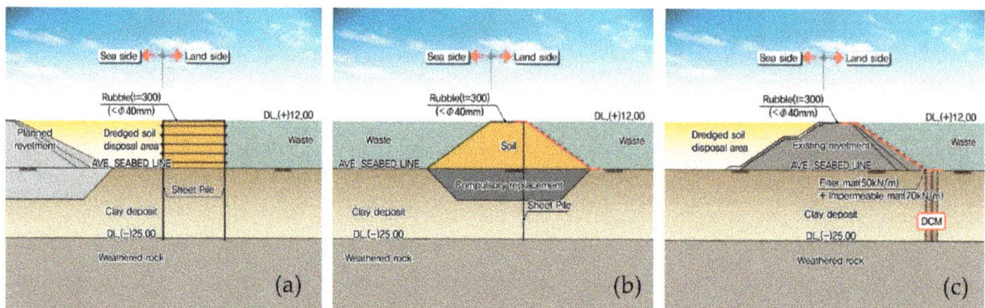


**Figure 5.** Cross-sections of the revetments constituting the offshore waste disposal facility: (**a**) the barrier revetment corresponding to Sections A–E; (**b**) the separating revetment corresponding to Section I; and (**c**) the connection revetment corresponding to Section L.

Meanwhile, Section I is termed as a separating revetment because its function is to divide the inner space of the waste disposal facility into parts depending on the usage of the facility. Section I-1 divides Zones 1 and 2, as illustrated in Figure 3, whereas Section I-2 separates the space for the water treatment facility from Zone 1. As shown in Figure 5b, the cross-section of the separating revetment is a newly constructed rubble mound structure that includes vertical sheet piles in the middle and water-proof sheets on the slope.

Finally, Section L, termed as a connection revetment, utilizes the revetment of the dredged soil disposal pond and supplements it with only an additional function as a barrier. Accordingly, the soil strength is improved by means of deep cement mixing (DCM) and waterproof sheets are installed on the slope of the existing revetment.

### 3.2.2. Vertical Cut-Off Barrier

For the barrier revetment (Sections A–E), the vertical cut-off barrier, which is composed of double-layer sheet piles, is applied as a means of preventing leachate movement. For the junction between the neighboring sheet piles, an expansive-type water stop material was used.

### 3.2.3. Surface and Floor Cut-Off Barrier

A surface cut-off barrier was applied to the separating revetment (Section I) and the connection revetment (Section L), marked in the red thick lines on the slope of the revetment in Figure 5b,c. The surface cut-off barrier is composed of a filter mat and a double-layer waterproof sheet. In general, the permeability coefficient of the waterproof sheet ranges from $1.0 \times 10^{-2}$ to $1.0 \times 10^{-4}$. For the connection revetment, the DCM method was applied to the clay deposit at the toe of the revetment as illustrated in Figure 5c. Then, the waterproofing performance is very important at the interface of the seabed where the revetment toe and the DCM meet.

On the other hand, the floor cut-off barrier is typically not considered if the sea bottom has a low permeable layer, which satisfies the conditions of the permeability coefficient and thickness standards required for the landfill facility. If there exists an insufficient low-permeable layer underneath the seabed, a separate floor cut-off barrier treatment is requested. For this purpose, an expandable particle-type barrier material can be applied to form the floor barrier.

### 3.2.4. Leakage Monitoring and Detection

Monitoring of water leakage in hydraulic structures or landfills is essential. In a waste landfill on land, the electric methods are commonly used for detecting damage to the barrier by installing electrodes on the waterproof sheet and measuring the electricity flow between two different electrodes [28]. If there is any leakage along the waterproof sheet, there should be significant electricity flow due to the insulation of the sheet being broken.

In the case of the offshore waste disposal facility in Tokyo Bay, Japan, the inspection of hazardous substances is carried out on a monthly basis at the wells around the revetment that are prepared for water quality testing. The leakage of retained water is also indirectly monitored by placing sand tubes between concrete blocks in the external revetment and monitoring whether there is any loss of sand from the sand tubes.

In this study, leakage monitoring and detection has been considered an independent subsystem of the offshore waste disposal facility. This deals with the methodology itself detecting a leakage when either the vertical barrier or surface and floor barrier are damaged. If damage occurs to a barrier and it is not detected, the hazard posed by the waste disposal facility may increase. Therefore, it is reasonable to be considered as an independent subsystem apart from the barrier structures.

#### 3.2.5. Retained Water Treatment Facility

The retained water treatment facility, represented using a blue box in Figure 4, was designed to be located in Zone 2. This facility treats the retained water in Zone 2 as well as Zone 1. The maximum capacity of the treatment facility is determined as 4000 $m^3$/day, taking into account the daily amount of carry-in waste. When only Zone 1 is used, it will treat up to 2000 $m^3$/day of the retained water. When Zone 2 is used in addition to Zone 1, the capacity will be extended to up to 4000 $m^3$/day.

#### 3.2.6. Landfill Management

Landfill management refers to how the waste disposal facility is managed while it is being landfilled with carry-in waste. As shown in Table 1, there exist various monitoring items that are required for efficient landfill management. In order to effectively monitor the items listed in Table 1, fixed or mobile sensing methods can be applied depending on the measurement purpose and preferences. A fixed sensing method is used when measuring the deformation of the revetment or change of water level inside and outside the disposal site. Meanwhile, a mobile sensing method is adequate for measuring parameters such as the landfill height or water quality, which are required to be measured in different locations over a wide area of the facility.

**Table 1.** Monitoring items and remarks for landfill management.

| Monitoring Items | Remarks |
| --- | --- |
| Changes in water level inside and outside the offshore disposal facility | Performance of the cut-off barrier can be checked by monitoring changes in water level inside and outside the facility. It is necessary to confirm that the water level inside the facility is not influenced by the adjacent tide level. |
| Revetment deformation due to ground settlement | It is necessary to measure the displacement of revetment on the ground as well as underwater. Monitoring ground settlement is also required by installing inclinometers and settlement gauges. |
| Landfill height change | Landfill height should be distributed as evenly as possible within the disposal facility for stable landfill management. |
| Water quality inside and outside the facility | Measurement of dissolved oxygen (DO), pH, and other water quality indices is requested at least once every 6 months over at least two locations. |

#### 3.2.7. Waste Transportation

The proposed site of the offshore waste disposal facility is located where access by land is possible. Accordingly, direct waste disposal to the waste disposal facility through land transportation can be carried out, which does not require a carry-in base for sea transportation. In this case, the hazards associated with water transportation can be eliminated. In addition, civil complaints due to scattering dust and odor, which typically take place in the vicinity of the carry-in base, can be significantly reduced.

## 4. Procedures of Subsystem Hazard Analysis

### 4.1. Selection of Relevant Hazard Checklists

A hazard checklist commonly includes elements such as energy sources, hazardous functions, hazardous operations, hazardous materials, undesired mishaps, and lessons learned from similar type systems [15]. A hazard checklist by itself is not complete and does not include all hazardous items. It helps in identifying potential hazards through past examples. Therefore, different hazard checklists can be established depending on the analysis. The bulleted list below shows the hazard checklists considered in this study, which are relevant to the hazard analysis of offshore waste disposal facilities:

- Structural damage or failure;
- Leakage;
- Contamination;
- Corrosion;
- Toxicity;
- Weather and environment.

### 4.2. Identification of TLMs

TLMs refer to an important hazardous accident caused by one or more causes. It is a concept used to collect various potential hazards that lead to the same TLM outcome. TLM defined in the previous PHA study [25] mostly focused on safety concerns on external structures and cut-off barriers. Among them, some TLMs such as the sliding and overturning of caisson are removed in the present SSHA study because caisson is no longer adopted as an external structure of the subsystem of the proposed offshore waste disposal facility. On the other hand, additional hazards such as suspended particles or traffic congestion need to be included considering that the proposed site for the pilot project is close to a populated city under development. Meanwhile, the hazards related to energy sources include mechanical and electrical equipment of the retained water treatment facility. However, the hazards from this equipment were not included in TLMs because they are of relatively low importance in the design. The bulleted list below illustrates the TLMs considered in this study:

- Corrosion of sheet piles;
- Deformation and damage at the joint of sheet piles;
- Insufficient embedded depth of sheet piles;
- Damage to the waterproof sheets of cut-off barriers;
- Difficulty in operating a land base into which waste is carried;
- Suspended particles in the air;
- Leakage of leachate to the sea.

### 4.3. Assessment of IMRI and FMRI

Table 2 shows a worksheet that is commonly used in SSHA [15]. In the table, ① is the number of subsystem hazards that are identified. Entry ② represents a specific hazardous element that is expected to occur, and ③ represents the initiating mechanism by which hazardous elements lead to mishaps. Entry ④ indicates the consequence when a hazard has occurred. Entry ⑤ is an initial mishap risk index (IMRI), which serves as a qualitative indicator of the importance of the identified hazards. Entry ⑥ indicates a possible preventive measure to eliminate or mitigate the identified hazard. Lastly, ⑦ is a final mishap risk index (FMRI), a qualitative indicator of the action taken to reduce the identified hazard.

**Table 2.** Worksheet used for the subsystem hazard analysis. IMRI—initial mishap risk index, FMRI—final mishap risk index.

| No. | Hazard | Causes | Consequences | IMRI | Mitigation | FMRI |
|---|---|---|---|---|---|---|
| ① | ② | ③ | ④ | ⑤ | ⑥ | ⑦ |

IMRI and FMRI use the US Department of Defense's system safety practice standard, MIL-STD-882D, which has widely been used internationally [27]. In MIL-STD-882D, the hazards are classified into four categories according to the degree of severity and five categories according to the probability of occurrence, as shown in Table 3. IMRI and FMRI are represented as a combination of a single index from each column in the table. For instance, a critical hazard that occasionally occurs is denoted as 2C.

**Table 3.** Mishap risk indices according to MIL-STD-882D [27].

| Severity | Probability |
|---|---|
| (1) Catastrophic | (A) Frequent |
| (2) Critical | (B) Probable |
| (3) Marginal | (C) Occasional |
| (4) Negligible | (D) Remote |
| | (E) Improbable |

When IMRI and FMRI are assessed in this study, the hazard severity was considered based on the importance and extent of the damage. Catastrophic damage corresponds to damage that leads to the long-term shutdown of the offshore waste disposal facility, such as the leakage of leachate to the marine environment due to the extensive damage of the external revetment. Critical damage refers to a case in which the operation of the offshore waste disposal facility is suspended for considerable time due to the severe degradation of the cut-off barrier performance caused by the deformation of the outer revetment, the breakage of vertical and surface barriers, or the failure of the retained water treatment facility. Marginal damage means a temporary suspension of the facility operation influenced by local damage to the cut-off barrier, problems with landfill management, or a shortage of capacity to process the retained water.

Meanwhile, hazard mitigation or reduction measures, corresponding to entry ⑥ in Table 2, are generally presented as forms of system improvement requirements. Hazard mitigation measures can be classified into four stages in terms of priority. First, it is ideal to remove hazards from the stage of system design. If it is difficult to change the system design, however, supplementary safety devices should be applied alternatively. Third, if the application of safety devices is unnecessary or their effects on improving system safety are limited, the introduction of an alarm system can be the next alternative. Finally, non-facility measures, such as education and training on safety, can be implemented in addition to the measures focusing on mitigating the hazards of facilities.

## 5. Results of Subsystem Hazard Analysis

### 5.1. Mishap Risk Indices and Methods of Mitigation

As a result of the hazard analysis on the seven subsystems of the offshore waste disposal facility, 43 potential hazardous elements were identified, as listed in Table 4. The severity and probability of each hazardous element were evaluated through a workshop in which the experts from design firms, research institutes, and relevant authorities have participated. Then, it was possible to determine the risk level of each hazardous element by following the risk assessment method described in MIL-STD-882E [29]. More specifically, the appropriate severity category for a given hazard element was determined by assessing the potential for death, injury, environmental impact, or monetary loss.

In addition, the appropriate probability level was determined by evaluating the likelihood of occurrence of a mishap. It is desirable to use relevant quantitative data as a basis for objectively assessing the severity and probability. When quantitative data are not available, a single opinion is obtained through the Delphi analysis or a second-round workshop after the initial evaluation by each expert.

**Table 4.** The list of severities and probabilities of all 43 identified hazards.

| Code | Hazard | Severity | Probability |
|---|---|---|---|
| S1-A | Cracking and collapse of the separate revetment | 3 | D |
| S1-B | Circular slip failure of the revetment | 2 | B |
| S1-C | Displacement of the revetment due to waste disposal | 3 | C |
| S2-A | Corrosion of barriers made of steel | 2 | B |
| S2-B | Deformation and damage of joints caused by an earthquake | 2 | D |
| S2-C | Deformation and damage of joints during landfill | 2 | D |
| S2-D | Deformation and damage of joints due to landfill pressure | 2 | B |
| S2-E | Insufficient stability of mortar at joints caused by poor filling | 3 | B |
| S2-F | Degradation of stability of mortar at joints caused by aging | 3 | B |
| S2-G | Poor construction of joints | 3 | C |
| S2-H | Insufficient depth of barriers due to poor construction | 2 | A |
| S2-I | Insufficient depth of barriers due to heterogeneity of ground | 2 | A |
| S2-J | Degradation of cut-off performance at joints | 2 | C |
| S3-A | Locally permeable ground | 3 | D |
| S3-B | Damage of waterproof sheets due to the poor fusion of adhesive parts | 3 | C |
| S3-C | Damage of waterproof sheets due to landfill weight | 3 | C |
| S3-D | Damage of waterproof sheets due to slip on the slope | 3 | C |
| S3-E | Degradation of cut-off performance of the separate revetment | 3 | C |
| S3-F | Degradation of cut-off performance of the connection revetment | 3 | C |
| S3-G | Floating of waterproof sheets due to lack of sufficient weight | 3 | C |
| S3-H | Degradation of cut-off performance of the barrier revetment | 3 | C |
| S4-A | Failure of detecting leakages of leachate | 3 | C |
| S4-B | Failure of confirming the location where leakages occur | 3 | C |
| S5-A | Failure of treating retained water | 2 | C |
| S5-B | Degradation of the capability of retained water treatment | 3 | D |
| S5-C | The outflow of retained water to the sea | 2 | C |
| S5-D | Exceedance of retained water greater than treatment capacity | 2 | C |
| S5-E | Fire | 2 | C |
| S5-F | Flooding | 2 | C |
| S5-G | Electric shock from high voltage | 2 | D |
| S5-H | Wave overtopping | 2 | C |
| S5-I | Deterioration of employee's health | 2 | C |
| S6-A | Malfunction of the sensors measuring water level | 3 | B |
| S6-B | Malfunction of the sensors measuring water quality | 3 | B |
| S6-C | Exceedance of measurement range of the sensors | 3 | B |
| S6-D | Power supply interruption to the sensors | 3 | B |
| S6-E | Malfunction of the movable sensing device | 3 | D |
| S6-F | Errors in the communication signal | 3 | C |
| S6-G | Inability of collecting sensing data | 3 | D |
| S6-H | Failure of operation and management system | 3 | D |
| S7-A | Collison of a ship to the revetment | 2 | D |
| S7-B | Suspended particles in the air | 2 | A |
| S7-C | Traffic congestion | 2 | A |

The results obtained from the above procedure are summarized in Figure 6 as a form of risk assessment matrix. As shown in Figure 6, seven out of 43 elements were classified into the extreme risk zone. Among the seven extremely risky hazardous elements, one is related to the subsystem of the revetment (S1), four to the vertical barriers (S2), and two to waste transportation (S7). Meanwhile, 30 elements were classified into the high-risk zone, and the remaining six were in the moderate risk zone.

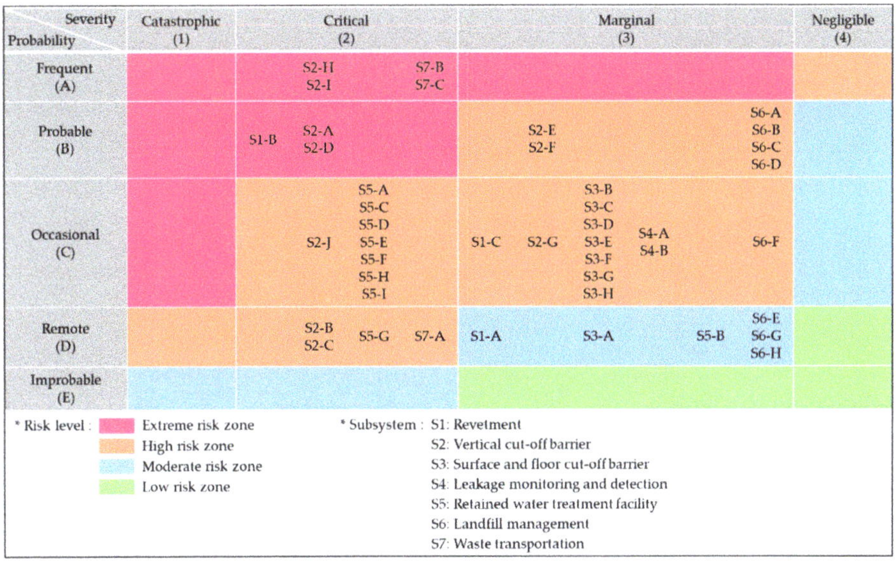

| Severity / Probability | Catastrophic (1) | Critical (2) | Marginal (3) | Negligible (4) |
|---|---|---|---|---|
| Frequent (A) | | S2-H S2-I S7-B S7-C | | |
| Probable (B) | | S1-B S2-A S2-D | S2-E S2-F S6-A S6-B S6-C S6-D | |
| Occasional (C) | | S2-J S5-A S5-C S5-D S5-E S5-F S5-H S5-I | S1-C S2-G S3-B S3-C S3-D S3-E S3-F S3-G S3-H S4-A S4-B S6-F | |
| Remote (D) | | S2-B S2-C S5-G S7-A | S1-A S3-A S5-B S6-E S6-G S6-H | |
| Improbable (E) | | | | |

* Risk level : Extreme risk zone; High risk zone; Moderate risk zone; Low risk zone

* Subsystem : S1: Revetment
S2: Vertical cut-off barrier
S3: Surface and floor cut-off barrier
S4: Leakage monitoring and detection
S5: Retained water treatment facility
S6: Landfill management
S7: Waste transportation

**Figure 6.** Risk assessment matrix of initial subsystem hazard analysis (SSHA) of the offshore waste disposal facility.

For every identified and classified hazard, necessary measures should be sought to eliminate or reduce the risk in advance. Table 5 summarizes a general principle for seeking a measure required for reducing or controlling the potential hazards depending on the associated risk level. The ultimate goal is always to find the most appropriate solution to eliminating the hazard. When it is impossible to completely eliminate a hazard, the assessed risk should be reduced to the lowest acceptable level with the approval or agreement of the authorities or parties, after taking account all of the relevant constraints such as cost, time, and other factors [29]. In other words, when removing a hazard is difficult, hazard management is required.

**Table 5.** Measures required for reducing or controlling the hazards according to the associated risk level [29].

| Risk Level | Measure |
|---|---|
| Extreme | Shall be eliminated |
| High | Shall only be accepted when risk reduction is impracticable and with the agreement of the authority |
| Moderate | Acceptable with adequate control and the agreement of the authority |
| Low | Acceptable with/without the agreement of the authority |

Table 6 shows the example worksheet presenting IMRIs, mitigation measures, and FMRIs of the seven hazardous elements that were classified to be extremely risky. For the remaining hazardous elements corresponding to high or moderate risk, similar worksheets were created, however, they were omitted as they were considered excessively lengthy to be included in this paper. Basically, the process of evaluating FMRI is similar to that of IMRI. For every hazard element evaluated as extreme or high risk in IMRI, a mitigation measure was suggested for reducing the associated risk to an acceptable range. Then, the severity and probability after applying the mitigation measure were evaluated by experts who conducted assessment of IMRIs by following the same procedure described in Section 5.1. If the risk of a hazard element was not reduced to the lowest acceptable level by the firstly proposed mitigation measure, this process was repeated until it was satisfied.

**Table 6.** Extremely hazardous elements identified by subsystem hazard analysis (SSHA) and measures to mitigate them. IMRI—initial mishap risk index, FMRI—final mishap risk index.

| No. | Hazard | Cause | Consequence | IMRI | Mitigation | FMRI |
|---|---|---|---|---|---|---|
| S1-B | Circular slip failure of the revetment | Excessive surface load on the separating revetment | Collapse of the revetment | 2B | Modify the design of the separating revetment to satisfy the safety standards by reinforcing the fore slope of the revetment | 2E |
| S2-A | Corrosion of barriers made of steel | Exposure to seawater | Cracks in the barrier and deterioration of cut-off performance | 2B | Apply cathodic protection method or use special steels for anti-corrosion | 2E |
| S2-D | Deformation and damage of joints due to landfill pressure | Pressures from the landfilled waste | Cracks in the barrier and deterioration of cut-off performance | 2B | Apply a double layer waterproof sheets on the joints or use sheet piles implementing fail-safe technology | 2E |
| S2-H | Insufficient depth of barriers due to poor construction | Uncertainty associated with the construction of vertical barriers | Failure of barriers or deterioration of cut-off performance | 2A | Monitor the driving depth of vertical barriers and check verticality during construction | 2E |
| S2-I | Insufficient depth of barriers due to heterogeneity of ground | Poor construction of vertical barriers influenced by the heterogeneity of ground | Failure of barriers or deterioration of cut-off performance | 2A | Apply longer barriers than required to have extra driving depth considering the associated uncertainty | 2E |

**Table 6.** *Cont.*

| No. | Hazard | Cause | Consequence | IMRI | Mitigation | FMRI |
|---|---|---|---|---|---|---|
| S7-B | Suspended particles in the air | Waste transportation and unloading to the disposal facility | Problems with the health and well-being of local residents | 2A | Apply a shield to the transportation vehicles and build a special device to minimize particle suspension during transportation and unloading | 2E |
| S7-C | Traffic congestion | Waste transportation on land | Inconvenience and uneasiness to local residents | 2A | Make a detour away from the residential area and transport waste avoiding crowded hours | 2E |

The comparison of IMRIs and FMRIs in Table 6 shows that the risk levels of the hazardous elements have been apparently reduced by applying the mitigation measures. This fact is more easily confirmed in Figure 7, which shows the modified risk assessment matrix of the offshore waste disposal facility after applying the mitigation measures. The items identified in the extreme risk zone and the high-risk zone in Figure 6 have been moved to the moderated risk zone and the low-risk zone in Figure 7, respectively. Hence, all the hazards associated with extreme or high risks in IMRIs have been evaluated as moderate or low risks in FMRIs. This indicates that the risks of the originally proposed design have been significantly mitigated through SSHA. By applying the mitigation measures to the original design, the overall stability of the design was improved, especially with regard to the safety of the revetment and the cut-off barriers, which are the core structures constituting the offshore waste disposal facility. Note that the FMRIs of all the identified hazardous elements were evaluated through a process similar to IMRIs, by holding a workshop with various members from relevant organizations.

*5.2. Design Modification by Reflecting on SSHA*

As described in Section 5.1, the risks of the offshore waste disposal facility were analyzed and measures to mitigate such risks were suggested through SSHA. Reflecting on this, the original design was subjected to a complete revision, from which some parts of the design were changed. Figure 8a–c show the modified cross-sections of the barrier revetment, separating revetment, and the connection revetment from the initial design shown in Figure 5a–c.

The major risks of barrier revetment (Sections A-E) were the poor construction of the barrier for some reasons and the deterioration of the cut-off performance that would lead to a mixture of waste and dredged soil. To mitigate these types of risks, additional waterproof sheets were placed along the interior sheet piles in the updated design as shown in Figure 8a.

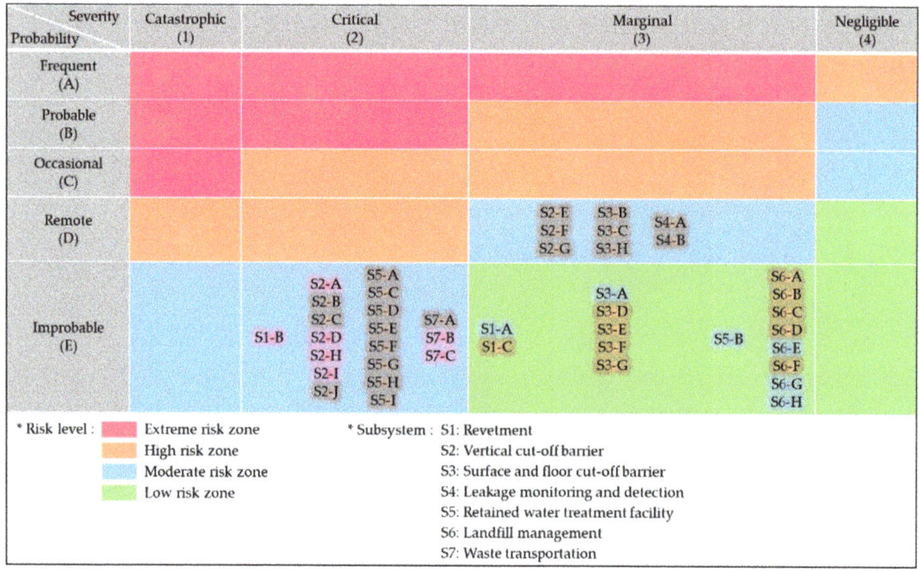


**Figure 7.** Risk assessment matrix of the final subsystem hazard analysis (SSHA) of the offshore waste disposal facility. Blurred background colors indicate the results of the initial SSHA.

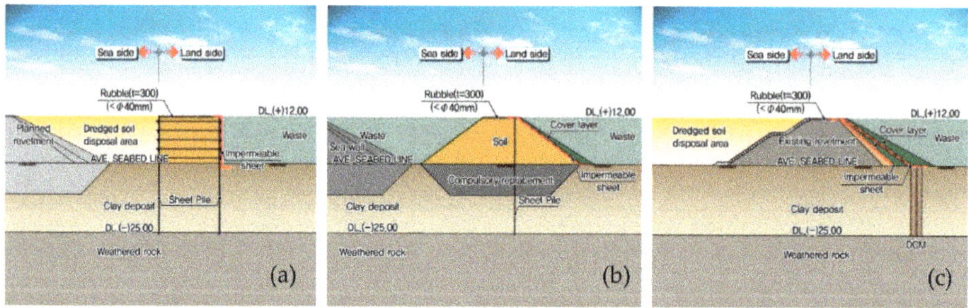


**Figure 8.** Cross-sections of the revetments constituting the offshore waste disposal facility that have been modified based on the subsystem hazard analysis. (**a**) The barrier revetment corresponding to Sections A–E; (**b**) the separating revetment corresponding to Section I; and (**c**) the connection revetment corresponding to Section L.

Meanwhile, the critical risks of the separating revetment (Section I) were the failure or collapse of the revetment due to the landfill overload or other unexpected external forces and exceedance of the capability of the waste treatment facility. Such risks have been reduced by placing a supplementary cover layer on the inner slope of the separating revetment and waterproof sheets underneath the cover layer. Figure 8b shows the modified design according to these changes.

Finally, the primary risk of the connection revetment (Section L) was the lack of continuity at the connection between the improved soil by DCM and the cut-off barrier on the revetment slope. If there exists any discontinuity, leachate shall be leaked to the dredged soil disposal pond through the revetment. To prevent this risk, a relatively thick cover layer was placed with waterproof sheets in the middle, as shown in the updated design in Figure 8c.

As illustrated in Figure 8, improved design outputs were obtained through SSHA by updating the original design where hazard elements not recognized by designers in the early stages of the design were included. In this context, the results presented in this study are meaningful in that they introduce a method to objectively recognize and manage the risks associated with the construction of the offshore waste disposal facility. In addition, the design drawings derived through this study were made at the conceptual design level but intrinsically are comparable to the basic design level. Therefore, they will be practically used as a key material for future detailed design when the construction of the facility is determined and a full-fledged demonstration project becomes visible.

## 6. Conclusions

This study provided the result of SSHA for an offshore waste disposal facility, which, to the best of our knowledge, has not previously been presented in any literature so far. The hazard analysis was performed for a detailed design of the proposed offshore waste disposal facility that was planned to be constructed near the new Incheon port, Korea. According to the IMRIs that were assessed by SSHA, seven hazard elements corresponded to extreme risks, 30 high risks, and seven moderate risks. After applying the risk mitigation measures, however, there were no hazard elements associated with extreme or high risks. All the hazard elements came into moderate or low risks, which can be controlled or managed adequately.

Therefore, this study shows that potential hazardous elements of the offshore waste disposal facility can be identified in advance through SSHA and countermeasures can be established to eliminate or mitigate the identified risks. However, it remains uncertain as to whether the proposed offshore waste disposal facility near Songdo International City can actually be built. Because such a facility has never been built in Korea, even if it can be constructed with only marginal or negligible influence on the environment, it may take a long time to obtain the consent and agreement of the local residents. In this context, when the actual construction plan is dealt with more specifically, it is necessary to perform an additional hazard analysis that focuses more on the risk of the residence.

**Author Contributions:** Conceptualization, S.-H.O.; methodology, S.-H.O. and S.-W.K.; software, S.-H.O. and S.-W.K.; validation, S.-H.O.; formal analysis, S.-W.K.; investigation, S.-H.O. and S.-W.K.; data curation, S.-W.K.; writing—original draft preparation, review, and edition, S.-H.O. and S.-W.K.; funding acquisition, S.-H.O. All authors have read and agreed to the published version of the manuscript.

**Funding:** This study is part of a research project titled "Development of Technology for Offshore Final Disposal Facility", which is funded by the Ministry of Oceans and Fisheries of Korea (Project No. 20160281/PM60660).

**Conflicts of Interest:** The authors declare no conflict of interest.

## References

1. Oh, M.H.; Kwon, O.S.; Kim, G.H.; Chae, K.S. Introduction on offshore waste landfill and potential sites. *KSCE J. Civ. Eng.* **2012**, *60*, 40–48.
2. Yamamura, K. Current status of waste management in Japan. *Waste Manag. Res.* **1983**, *1*, 1–15. [CrossRef]
3. Gotoh, S. Waste management and recycling trends in Japan. *Resour. Conserv.* **1987**, *14*, 15–28. [CrossRef]
4. Bai, R.; Sutanto, M. The practice and challenges of solid waste management in Singapore. *Waste Manag.* **2002**, *22*, 557–567. [CrossRef]
5. Tanaka, N.; Tojo, Y.; Matsuto, T. Past, present, and future of MSW landfills in Japan. *J. Mater. Cycl. Waste Manag.* **2005**, *7*, 104–111.
6. Hiroshima Prefecture. Environmental Impact Assessment Report on the Installation of Waste Disposal Plant in Dejima Landfill Area. Available online: https://www.pref.hiroshima.lg.jp/site/eco/i-i2-dejima-dejima-kizon.html (accessed on 21 October 2020).
7. Kitakyushu Prefecture. Preparations for Environmental Impact Assessment on the Construction of Waste Disposal Places in Eastern Hibikina. Available online: https://www.city.kitakyushu.lg.jp/kankyou/00600155.html (accessed on 21 October 2020).

8. Lee, H.; Son, M.; Kang, T.; Maeng, J. A study on environmental impact assessment guidelines for marine environments in construction projects of offshore waste disposal landfills. *J. Environ. Imp. Assess.* **2019**, *28*, 312–331. (In Korean)
9. Khoo, H.H.; Tan, L.L.Z.; Tan, R.B.H. Projecting the environmental profile of Singapore's landfill activities: Comparisons of present and future scenarios based on LCA. *Waste Manag.* **2012**, *32*, 890–900. [CrossRef] [PubMed]
10. Chan, J.K.H. The ethics of working with wicked urban waste problems: The case of Singapore's Semakau Landfill. *Lands. Urb. Plann.* **2016**, 123–131. [CrossRef]
11. Tan, S.K.; Yeo, R.K.H. The intertidal mollusks of Pulau Semakau: Preliminary results of "project Semakau". *Nat. Singap.* **2010**, *3*, 287–296.
12. Park, J.; Xu, X.; Oh, M.; Chae, K.; Kim, S.; Lee, K.; Hwang, Y. Development on the Technology for Offshore Waste Final Disposal in S. Korea. In *Environmental Geotechnology*; Springer: Singapore, 2019; pp. 17–42.
13. Xu, X.; Liu, X.; Oh, M.H.; Park, J. Development of a novel base liner material for offshore final disposal sites and the assessment of its hydraulic conductivity. *Waste Manage.* **2020**, *102*, 190–197. [CrossRef] [PubMed]
14. Chae, K.; Lee, J.; Jin, H.; Oh, M. Analysis of a new structure for offshore waste landfill: Contaminant transport characteristics affected by tidal fluctuations. In International Conference on Offshore Mechanics and Arctic Engineering. *Am. Soc. Mechan. Engin.* **2018**, *51227*, V003T02A060.
15. Ericson, C.A. *Hazard Analysis Technique for System Safety*; John Wiley & Sons, Inc.: Fredericksburg, VA, USA, 2005.
16. Det Norske Veritas (DNV) Integrity Management of Submarine Pipeline Systems, Recommended Practice DNV-RP-F116. Available online: https://rules.dnvgl.com/docs/pdf/DNV/codes/docs/2015-02/RP-F116.pdf (accessed on 21 October 2020).
17. American Bureau of Shipping (ABS) Guidance Notes on Risk Assessment Applications for the Marine and Offshore Oil and Gas Industries. Available online: https://books.google.co.kr/books/about/Guidance_Notes_on_Risk_Assessment_Applic.html?id=9ET6HAAACAAJ&redir_esc=y (accessed on 21 October 2020).
18. American Bureau of Shipping (ABS) Guide for Risk Evaluations for the Classification of Marine-Related Facilities. Available online: https://ww2.eagle.org/content/dam/eagle/rules-and-guides/current/other/117_riskevalforclassofmarinerelatedfacilities/pub117_riskeval.pdf (accessed on 21 October 2020).
19. American Bureau of Shipping (ABS). Guidance Notes on Accidental Load Analysis and Design for Offshore Structures. Available online: https://ww2.eagle.org/content/dam/eagle/rules-and-guides/current/other/197_accload/accidental_load_gn_e.pdf (accessed on 21 October 2020).
20. American Petroleum Institute (API). *Recommended Practice for Design and Hazards Analysis for Offshore Production Facilities*; American Petroleum Institute (API): Washington, DC, USA, 2019.
21. International Organization for Standardization (ISO) 31000. Risk Management—Principles and Guidelines. Available online: https://www.iso.org/obp/ui/#iso:std:iso:31000:ed-1:v1:en (accessed on 21 October 2020).
22. Yang, X.; Utne, I.B.; Holmen, I.M. Methodology for hazard identification in aquaculture operations (MHIAO). *Saf. Sci.* **2020**, *121*, 430–450. [CrossRef]
23. Sezen, H.; Hur, J.; Smith, C.; Aldemir, T.; Denning, R. A computational risk assessment approach to the integration of seismic and flooding hazards with internal hazards. *Nucl. Eng. Des.* **2019**, *355*, 110341. [CrossRef]
24. Chiri, H.; Abascal, A.J.; Castanedo, S. Deep oil spill hazard assessment based on spatio-temporal met-ocean patterns. *Mar. Poll. Bull.* **2020**, *154*, 111123. [CrossRef] [PubMed]
25. Xiao, Y.F.; Duan, Z.D.; Xiao, Y.Q.; Ou, J.P.; Chang, L.; Li, Q.S. Typhoon wind hazard analysis for southeast China coastal regions. *Struct. Saf.* **2011**, *33*, 286–295. [CrossRef]
26. Kim, S.W.; Oh, S.H.; Oh, M.; Seo, S.N. Preliminary hazard analysis based on a conceptual design of the offshore waste disposal facility. *J. Korean Soc. Hazard Mitig.* **2018**, *18*, 395–404. (In Korean) [CrossRef]
27. US Department of Defense. MIL-STD-882D, Standard Practice for System Safety. Available online: http://everyspec.com/MIL-STD/MIL-STD-0800-0899/MIL_STD_882D_934/ (accessed on 21 October 2020).

28. Oh, M.; Seo, M.W.; Lee, S.; Park, J. Applicability of grid-net detection system for landfill leachate and diesel fuel release in the subsurface. *J. Cont. Hydrol.* **2008**, *96*, 69–82. [CrossRef] [PubMed]
29. US Department of Defense. MIL-STD-882E, Standard Practice for System Safety. Available online: http://acqnotes.com/acqnote/tasks/mil-std-882e-system-safety (accessed on 21 October 2020).


**Publisher's Note:** MDPI stays neutral with regard to jurisdictional claims in published maps and institutional affiliations.

International Journal of
*Environmental Research and Public Health*

*Article*

# Transparency of Financial Reporting on Greenhouse Gas Emission Allowances: The Influence of Regulation

**Patricia Milanés Montero [1,*], Esteban Pérez Calderón [1] and Ana Isabel Lourenço Dias [2]**

1 Department of Accounting and Finance, University of Extremadura, Faculty of Economic and Business Sciences, 06006 Badajoz, Spain; estperez@unex.es
2 Department of Financial Accounting, Lisbon Polytechnic Institute/Lisbon Accounting and Business School, 1069-035 Lisbon, Portugal; aidias@iscal.ipl.pt
* Correspondence: pmilanes@unex.es; Tel.: +34-636324775

Received: 5 December 2019; Accepted: 22 January 2020; Published: 31 January 2020

**Abstract:** This study focuses on the transparency of financial reporting on emission allowances (EA) and greenhouse gas (GHG) emissions within the European Union Emissions Trading Scheme (EU ETS). In particular, the different accounting treatments adopted by standard setters and professionals were analyzed to evaluate the influence of regulation in the transparency of financial reporting on EA and GHG emissions. Based on a sample of 85 companies registered with the Portuguese, Spanish, and French National Plans of Allocation (NPAs), data collected from the annual reports were analyzed for the 2008–2014 period. The results were obtained based on descriptive, logistic regressions and panel data statistical techniques, and they show that better levels of transparency of financial reporting on EA and GHG emissions are conditioned by a variety of accounting policies, which compromises the comparability of the financial information. The adoption of the International Accounting Standards Board (IASB) standards set lead to a greater dispersion in the choice of the accounting approach and a higher probability of not disclosing any information, as well as adopting off-balance sheet policies. Therefore, the regulatory factor is a determinant of the level of transparency of financial reporting on EA and GHG emissions, contributing to reduce strategies of omission.

**Keywords:** EU ETS; Emission allowances; Greenhouse gas emissions; Transparency; Accounting regulation

---

## 1. Introduction

In the context of the EU ETS, a cap-and-trade scheme, the debate in the literature began on whether the company's exposure generated an accounting treatment for EA and GHG emissions [1,2]. The basis for the functioning of the EU ETS, until the end of the year 2012 and with some application in phase III, was the grandfathering system. This system incorporates the prediction of the tons of GHG emissions necessary to an entity, so that it may continue with its operational activities and, at the same time, comply with the commitment of reduction of GHG emissions. This would result in the free allocation of that number of EA, decided by national governments on a National Plan of Allocation (NPA). At the end of a given period, the entities have to surrender EA equivalent to tons of emitted GHG. In the case of emitting GHG that exceed the cap, they have to acquire the corresponding EA that is missing for compliance; if the GHG emissions are below the cap, the surplus can be sold. Further discussion in the literature agrees that these transactions should be included in financial reporting [3–9], as they are relevant, in nature and value. However, the history of the regulation of financial reporting regarding EA and GHG emissions has been marked by the consecutive postponements of the IASB that is still without an answer for an explicit treatment. This lack of international guidance has been subjected to extensive criticism in the literature, and it is viewed as the main reason to the omission of accounting information on EA and GHG emissions in the financial reports [10–15]. Nevertheless, the full International Financial Reporting Standards (IFRS) presents general principles that may be applied. Back in 2004, the International Financial Reporting Interpretations Committee (IFRIC) elaborated Interpretation 3 'Emission Rights', in accordance with the IASB Conceptual Framework (not reviewed at the date) and within the scope of International Accounting Standard (IAS) 38, IAS 20 and IAS 37. The absence of general support for IFRIC 3 led to its consequent withdrawal by the IASB, and allowed companies to base its financial reporting on the professional judgment—under the principles of IAS 8. As a reaction to the IASB position, some European national standard setters engaged on domestic solutions. Among others, it was the case of the Portuguese, Spanish and French standard setters: Comissão de Normalização Contabilística (CNC), Instituto de Contabilidad y Auditoría de Cuentas (ICAC) and Autorité des Normes Comptables (ANC), respectively. This scenario of duality between the international and domestic standard setters created some discrepancies in the adopted accounting treatments. Previous literature [3,11–16] identified different accounting policies, and even, the absence of disclosure. The purpose of the present study focuses that those practices may be compromising the transparency of financial reporting on EA and GHG emissions (not to discuss which accounting treatment is more suitable). The concept of transparency is not well defined in the literature, but it was understood as Barth and Schipper's [17] state "the extent to which financial reports reveal an entity's underlying economics in a way that is readily understandable by those using the financial reports." The main purpose of the study is to reinforce the necessity of regulation to increase transparency of the financial reporting on the EU ETS transactions within IFRS with two arguments: IFRS companies are allowed not to disclose information on real transactions that have a regulated market behind (EA) and a position by the IASB could decrease the differences that national regulators apply in their issued standards. This study is an additional contribution to prior empirical studies in the area with particular emphasis on the transparency purpose of IFRS Foundation which we believe it is a weak position for the EU ETS transactions. The IFRS Foundation [18] state "Our mission is to develop IFRS Standards that bring transparency ... ", although it is not explicit in the IASB Conceptual Framework that transparency is a purpose of the set of financial statements. In fact, IFRS companies may recognize and measure those transactions resorting IAS 38, or IFRS 9, for instance. So, although not completely disregarding the IFRS standards, there may be an omission of important financial information. The IASB position seems to be contrary to other regulators, such as European national regulators. This stronger position may contribute to increase the transparency of the financial reporting of the companies that are EU ETS participants. Therefore, this research distinguishes itself from previous studies because: (1) addresses the importance of transparency of financial reporting on EA and GHG emissions, and; (2) compares two scenarios of transparency within guidelines based on the (in)existence of specific accounting principles (domestic standards *versus* IFRS).

The selection of the sample was not made taking into consideration the level of pollution in each country. Instead of that, we selected three countries with interesting characteristics to study. For instance, the Portuguese companies showed a diversity of accounting policies both in terms of recognition and measurement of the EA and GHG emissions, while Spanish companies were more consensual on the recognition and more discordant on measurement. We believe that this is the consequence of the change in Portuguese legislation in 2010 (while Spanish Resolution of 2006 was maintained until 2016), as well as a more sense of compliance with environmental financial reporting regulation. These results (by country) were an initial justification for the role of regulation in the transparency of financial reporting. Therefore, the sample selection consists in 85 French, Spanish, and Portuguese companies listed in the NPAs of phase II and phase III, that have European operators exclusively in those three countries. This limitation was achieved by crossing the list of companies with available annual accounts with the operators listed in the report released by the European Commission on Verified Emissions for 2014 [19], and aggregate it to company-level. The individual financial statements were the first choice for a primary source. When not available, the information was collected from the consolidated financial statements. In these cases, an analysis of the perimeter of consolidation was the method to limit the sample to companies located in the three countries. Another condition to define the sample was the online availability of the financial reporting available in English, Spanish, Portuguese or French. The period of the study, from 2008 to 2014, represents the five years from phase II and the two initial years of phase III of the EU ETS. The implementation of the auction system was not seen as a major difference in the transparency of financial reporting as the surplus of EA back-loaded the distribution by auction [20]. After applying descriptive, logistic regression and panel data techniques, the results show that the transparency of financial reporting on EA and GHG emissions are negatively affected by the IFRS. Although the difference is not significantly different, there is more disparity when the IFRS are the basis of presentation.

This study is structured into five additional sections: In Section 2, we briefly present the literature review that enable the formulation of the hypothesis and describe the research method; the results are presented and discussed in Section 3, and the final considerations are presented in Section 4.

## 2. Materials and Methods

### 2.1. Development of the Hypothesis

#### 2.1.1. Financial Reporting on EA and GHG Emissions—An Overview

Back in 2005, the IASB released IFRIC 3 'Emission Rights' for application after the beginning of phase I of the EU ETS, just to six months later decided to withdraw the interpretation arguing that there had been a misinterpretation of its urgency. In line with EFRAG's negative advice, the IASB stated that the prescribed accounting treatment created mismatches in the financial statements. This was mainly due to the measurement discordant between the assets (the EA) and the liabilities (GHG emissions), that would result in artificial volatility of reported earnings and would not reflect the economic reality of the companies. The skepticism about the usefulness of the recognition of EA and GHG emissions [2,21] stemmed in financial reporting relatively neglected [22]. Previous descriptive studies [3,11–15], identified a multitude of accounting practices for the EU ETS transactions. Those can be distinguished into two basic approaches: Gross approaches (as prescribed by the IFRIC 3 or government grant approach identified by Ernst and Young [23], and net approaches (as prescribed by U.S. Generally Accepted Accounting Principles (GAAP). The U.S.A. regulation on EA is based on pronouncements issued by the Federal Energy Regulatory Commission (FERC)—Order 552 issued in 1993, codified in Uniform System of Accounts (USofA) 101.21. The prescribed accounting treatment may be summed in the following: (a) EA are not reflected in financial statements; (b) an asset is recorded as inventory for purchased emission certificates at its cost; (c) when an entity does not hold the estimated required amount of emission certificates, a liability is recognized to reflect the number and current price of lacking emission certificates. Those studies also identified relative high frequencies of non-disclosure that were associated with the IFRS lack of specific guidance. Giner [8] illustrates that although the financial reporting regarding EA and GHG emissions is not easily framed in conventional accounting categories, relying on the business model allows more discretion than imposing a common solution.

The role of the domestic standard setters was precisely to overcome the lack of international guidance and to find common ground for the accounting treatment regarding EA and GHG emissions. This was the case of: OIC—Organismo Italiano di Contabilità, DRSC—Deutsches Rechnungslegungs Standards Committee e.V.; AFRAG—Austrian Financial Reporting and Auditing Committee, DASB—Dutch Accounting Standards Board, CNC—Comissão de Normalização Contabilística, ICAC—Instituto de Contabilidad y Auditoría de Cuentas, ANC—Autorité des Normes Comptables. This study focuses on Portuguese, Spanish and French companies, and therefore it is presented in Table 1 a summary of the treatments prescribed by each national standard setter and by IFRIC 3.

**Table 1.** Summary of Accounting Approaches for Emission Allowances (EA) and Greenhouse Gas (GHG) Emissions.

| | IFRIC 3 | CNC—Portugal [1] | | ICAC—Spain [2] | | ANC—France [3] | |
|---|---|---|---|---|---|---|---|
| | | **Before 1 January 2010** | **After 1 January 2010** | **Before 1 January 2016** | **After 1 January 2016** | **Before 1 January 2013** | **After 1 January 2013** |
| **Recognition of EA** | Intangible assets/Government Grant (Deferred income) | Intangible assets/Government Grant (Deferred income) | Intangible assets/Government Grant (Owners' Equity) | Intangible assets/Government Grant (Owners' Equity) | Inventory [4]/Government Grant (Owners' Equity) | Intangible assets/Liability—*Quotas d'émission alloués par l'État* | Inventory |
| **Initial measurement of EA** | Market value | Fair value | Fair value | Market value | Market value (fair value) | Market value | Nil value |
| **Subsequent measurement of EA** | Cost model or Revaluation model (without amortizations) | Cost less accumulated impairments (extraordinary amortization) | Cost less accumulated amortizations | Cost less accumulated impairments | Cost less accumulated impairments | Cost or Acquisition cost of equivalents (best estimate) | - |
| **Recognition and measurement of GHG emissions** | Operational Loss/Provisions [1] using best estimate (usually the market price) | Operational Loss/Provisions using *First in First out* (FIFO) | Amortization loss/Accumulated Amortizations using *First in First out* (FIFO) | Operational Loss/Provisions using: 1) the carrying amount of the EA held (proportion of GHG emissions of the period to the total GHG emissions) 2) weighted average cost (WAC) of the remaining EA. | | Loss/Liability using the initial value of EA | Operational loss/Inventory using FIFO or WAC |
| | Recognize the government grant (as income) over the periods in which the related expenses are intended to offset | | | | | | |
| **GHG emissions above detained EA** | Loss/Provision using best estimate (when the estimate of GHG emissions exceeds the carrying amount of the EA held for purposes of compliance) (In the case of Portugal, before 1 January 2010, was prescribed the fair value for this measurement) | | | | | Loss/Liability using the close value of EA | Liability at acquisition cost for EA necessary to cover the GHG emissions |

[1] The Portuguese legislation **(Comissão de Normalização Contabilística-CNC)** is presented with two accounting treatments, due to the legislative amendment introduced for periods beginning on or after 1 January, 2010. The accounting treatment in force until 1 January, 2010, was introduced by Technical Interpretation (TI) 4. For periods after 1 January, 2010, Notice 15654/2009 of 7 September (modified by Notice 8256/2015 of July 29) changes the accounting treatment provided for in that TI.
[2] The Spanish legislation (Instituto de Contabilidad y Auditoría de Cuentas-ICAC) is presented with two accounting treatments, due to the legislative amendment introduced for periods beginning on or after 1 January, 2016. The accounting treatment in force until January 1, 2016, was introduced by the Resolution of 8 February, 2006, and by the Resolution of 23 May, 2013. Royal Decree 602/2016 changes the accounting treatment of the aforementioned Resolutions, for periods starting from 1 January 2016, but only as regards the nature of the EA held to comply with the obligation to settle the GHG emissions incurred.
[3] The French legislation (Autorité des Normes Comptables-ANC)is presented with two accounting treatments, due to the legislative amendment introduced for periods beginning on or after 1 January, 2013. The *Avis nº 2004-C du 23 mars du Comité de urgence* established the accounting treatment recommended until 1 January, 2013. *Réglement no. 2012-03 of 4 Octobre 2012* changes the accounting treatment of GHG emission allowances and assimilated units, but the comparison refers only to the production model and excludes the trading model.
[4] EA to be consumed after a period of one year shall be presented in a separated line.

**Source:** Own elaboratio.

#### 2.1.2. Transparency of Financial Reporting on EA and GHG Emissions and the Role of Accounting Regulation

One of the issues of introducing information on EA and GHG emissions in financial reporting has been the interpretation of the transactions from a market based on a cap-and-trade system, as materially relevant [24]. This is much due to the grandfathering system considered for allocation of EA in phases I, II and even phase III of the EU ETS. The surplus of EA from phase II amounted to around 2 billion EA at the beginning of phase III and increased to more than 2.1 billion in 2013. Consequently, the European Commission postponed the auction of 900 million EA until 2019-2020 [20]. Thus, to underline that information on EA and GHG emissions should be included in annual accounts, thus improving transparency, it is important to frame it within the concepts of relevance and faithful representation. The IASB Conceptual Frameworks (2010, 2018) expose them as qualitative characteristics. The concept of materiality is an entity-specific aspect of relevance based on nature or magnitude, or both, of the items to which the information relates in the context of an individual entity's financial report (Conceptual Frameworks, 2010, 2018). As such, relevance is not limited to value relevance; it includes nature relevance, which should cover the specific nature of EA and GHG emissions under the EU ETS. The concept of faithful representation, states that financial reports represent economic phenomena in words and numbers, in a way that (i)t would be complete, neutral and free from error (Conceptual Frameworks, 2010, 2018). These qualities should be maximized to the extent possible in financial reporting. This clarification provides the basis to include information on EA and GHG emissions in financial reporting, as the practice of non-disclosure under the scope of materiality issues, is not under the complete concept of relevance nor under faithful representation.

Transparency is not a principle established in the Conceptual Frameworks. Nevertheless, the concept of financial reporting transparency, as exposed by Barth and Schipper [17], includes the perspective of the underlying economics, closely related to faithful representation: ( ... ) [T]he extent to which financial reports reveal an entity's underlying economics in a way that is readily understandable by those using the financial reports. Interestingly, Lovell and Mackenzie [25] argue that with the IASB-FASB (Financial Accounting Standards Board) joint project on ETS, accountants of major companies-participants in the EU ETS have suggested a readiness for clear guidance from the standard setters, due to a strong desire of reducing choice and be fairly compared with their competitors. The outcome of the joint project was not as expected, and in December of 2012 the IASB reactivated this project as an IASB-only research project. In 2015 the project was renamed to Pollutant Price Mechanisms, to reflect a change in its scope, approach and direction and it is seen as a research pipeline project, waiting for the revised Conceptual Framework is closed to finalization. Hence, it is announced that by 2019/2020, the Board will recommence the work [26]. Lovell et al. [11] and Giner [8] underline the importance of specific accounting treatment for obtaining a fair and transparent comparison of the financial reporting of EA and GHG emissions. However, this relation between specific regulation and increased transparency of financial reporting is not supported in the literature. In fact, such topics related to environmental disclosures are related to strategies of disclosure even in the presence of mandatory disclosure [27]. Moreover, IFRS provide several answers to include EU ETS transactions in the financial statements that do not include its omission (within IAS 38, IAS 20, IAS 37 or IFRS 9). Nevertheless, as some IFRS reporting companies choose to report some type of information on the exposure to the EU ETS, we believe that regulation would increase the levels of transparency and particularly, the possibility of comparison of the accounting policies between companies. Hence, on the one hand, financial reporting transparency on EA and GHG emissions is a concept desired by professionals and academics, which can be provided by accounting standard setters. On the other hand, an identified threat for the transparency of environmental disclosures, is the adoption of strategies, even in the presence of mandatory standards, outlined as compliance, dismissal or concealment [27]. Holthausen [28] concludes that the influence of a common set of accounting standards across countries is unlikely to lead to similar financial reporting outcomes. Barbu et al. [29] state that compliance with the IASB standards may differ between entities, and even between countries, due to differences in reporting practices. Thus, there is a suggestion that adopting an explicit accounting treatment does not systematically ensure the transparency of financial reporting. Cowan and Gadenne [30] suggest that companies adopt different approaches to disclose environmental issues when they are potentially subjected to further scrutiny through legal requirements. As a possible concealment strategy, Llena et al. [31] concluded that companies disclose less 'bad news', as provisions and contingencies, than 'good news', such as investments and expenses. The conclusions of Llena et al. [31] are in line with Deegan and Gordon [32] but disagree with Larrinaga et al. [33] or Adams, et al. [34] that bad news would open companies to challenge. The evidence in the literature that supports that the disclosure of the underlying economics of the EU ETS transactions in the financial reporting—a higher level of transparency—is related to the existence of specific accounting principles is scarce. Based on this significant absence, the following non-directional hypothesis of investigation is formulated:

**H1:** *Lower levels of transparency of financial reporting on EA and GHG emissions are not associated with the adoption of International Financial Reporting Standards (IFRS).*

### 2.2. Sample and Data

The NPAs from Portugal, Spain and France were the starting point for determining the sample as it was the most complete source of installations subjected to the EU ETS. The selection of these countries has been motivated by the purpose of the study which consists of comparing two scenarios of transparency within guidelines based on the (in)existence of specific accounting principles, and taking into consideration the availability of comparable accounting information. Those installations were matched with the respective company-owner. The corporate websites were used to collect the annual financial statements, as they are considered a primary source of information for stakeholders [35]. The selection of the sample mostly depended on the availability of these, in English, Spanish, French or Portuguese. If accessible, the individual annual accounts were collected; in its absence, the consolidated annual accounts were also considered, as long as the company was either the parent company, or it was explicitly mentioned in the Notes as a subsidiary, included by full consolidation method. It was assumed that the parent company rely on the subsidiary to perform under the EU ETS and incorporates the accounting approaches that the subsidiary has on its individual annual accounts—mandatorily or optionally by a gap of the IFRS. The consolidated accounts that included companies that could have installations in other European NPAs were excluded from the sample to prevent the exposure to other domestic regulations that were not identified. Figure 1 presents a diagram for the sample selection.

The period of the study covers 2008 to 2014. Phase I was excluded as it was an experimental period of the EU ETS and also for companies to implement a number of procedures concerning financial reporting, that was much discussed. Previous literature [3,13] struggle to obtain financial data and therefore, it was chosen to commence the study with phase II, following the remaining descriptive studies [11,12,14,16]. It was considered all five years of phase II and the first two years of phase III, which includes the transition to auctioning. The distinction from 2012 to 2013 and 2014 was not made as there has been a postponement in the auctioning of 900 million EA until 2019–2020, due to the surplus of more than 2.1 billion EA in 2013 [20]. In addition, at the beginning of phase III, manufacturing industry received 80% of its EA for free, airlines also receive the large majority of their EA and some power generators, although do not receive any free EA by principle, if the country intends to modernize the sector, then the EU ETS have made some EA available [36].

The final sample is formed by 85 companies, but not for all seven year-period, as it was obtained 548 observations, an unbalanced panel. The representativeness of the sample may be questioned, as the data collected depended on the online availability of annual financial statements. The sample selection is greatly reduced, since a large number of installations that are listed in the NPAs belong to a transnational group of companies that incorporates companies that have installations in other European countries (not exclusively in Portugal, Spain or France), and therefore, apply other financial reporting regulations. This was more significant for French installations, which explains its lack of representativeness. For these companies, only 15 observations are regarding the use of French GAAP, and the results show that they are not disclosing information on EA or GHG emissions, disregarding the domestic regulation. However, this does not invalidate the analysis as it can be assumed that the results are applied to companies with the same characteristics of those in the sample [35].

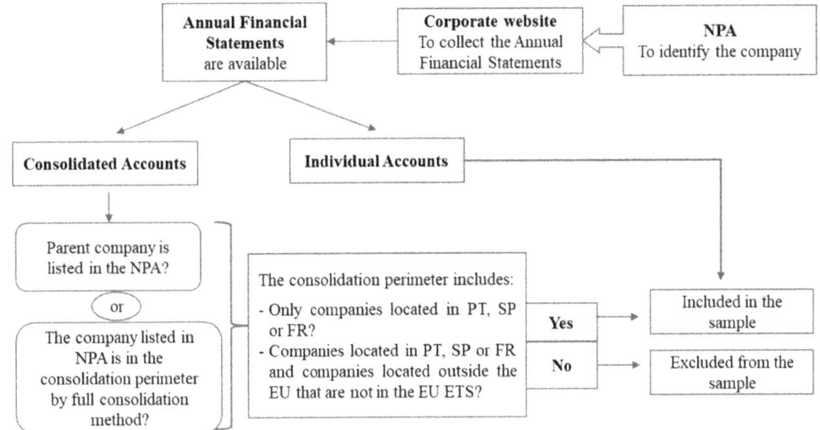


Source: Own elaboration

**Figure 1.** Summary of sample selection. FR, France; PT, Portugal; SP, Spain; NPA, National Plans of Allocation.

The content analysis technique was used to evaluate the accounting policies of EA and GHG emissions disclosed in the financial statements [12–15].

### 2.3. Variables

#### 2.3.1. Dependent Variables

Table 2 (panels A to C) describes the variables that were used as proxies for transparency of financial reporting on EA and GHG emissions. Panel A presents the eight accounting policies that were collected from the Notes to the annual accounts. These categorical variables are the basis for the formulation of the transparency variables: Accounting approaches and indexes of the disclosure, summarized in panels B and C, respectively.

**Table 2.** Variable definitions.

| **Panel A—Accounting policies for recognition and measurement of EA and GHG emissions** | |
|---|---|
| **Categorical variables** | **Abbreviation** |
| I1. Initial recognition of EA | Rec EA |
| I2. Counterpart on the initial recognition of EA | Count EA |
| I3. Initial measurement of EA | Initial Meas EA |
| I4. Subsequent measurement of EA | Subseq Meas EA |
| I5. Recognition of GHG emissions | Rec GHG emissions |
| I6. Measurement of GHG emissions | Meas GHG emissions |
| I7. Recognition of GHG emissions over-allocated EA | Rec GHGe over EA |
| I8. Recognition of EA acquired to cover GHG emissions over cap | Rec EA acq |

**Table 2.** *Cont.*

| **Panel B—Identification of Accounting Approaches** | | | | | | | |
|---|---|---|---|---|---|---|---|
| **Approach** | I1 | I2 | I3 | I4 | I5 | I6 | I7 |
| **IFRIC3_cost** | 1 | 3-5-6-7 | 3-4-5 | 3 | 3-6 | 4-7 | |
| **GovGrant** | 1 | 3-5-6-7 | 3-4-5 | 3 | 3-6 | 9-11-12-13 | |
| **NetLiab** | | | 1-6-7-99 | | 7-0 | | 1-2-3-4 |
| **Inventory** | 2 | | 1-6-7 | 1-6 | 2-3 | [a] 1-2-5-6-9-10-11-12-13 | |
| **ICAC** | 1 | 1-7 | 3-4-5 | 3 | 3-6 | 4-5-6-10 | |
| **AFRS26** [b] | 1 | 1-7 | 2-3-4-5 | 2-4 | | 2 | |
| **NoRec** | 99 | | 99 | | 7 | | 5 |
| **NoDisc** | 0 | | 0 | | 0 | | 0 |

The accounting approach is a dichotomous variable, that takes the value 1 if it is identified by the joint answers of each categorical variable (I1 to I7)—representing accounting policies, and 0 otherwise. Each categorical variable may present one or more possibilities to be considered that a certain accounting approach is being adopted. The numbers showed below each categorical variable correspond to the category expressed on Appendix A. The abbreviations for the identified accounting approaches are as follows: IFRIC 3—cost model (IFRIC3_cost), government grant approach (GovGrant) as exposed by Ernst and Young [37], net liability approach (NetLiab), the French inventory approach (Inventory) as exposed by ANC in 2012, the Spanish approach exposed in Resolution 8 of February of 2006 (ICAC), the Portuguese approach exposed in the appendix to Accounting and Financial Reporting Standard (AFRS) 26 (hereafter AFRS26), the approach of disclosing off-balance sheet policies (NoRec), and the approach of non-disclosure of EA or GHG emissions (NoDisc).
A I6 for inventory approach presents a variety of options in content analysis (1-2-5-6-9-10-11-12-13 in Appendix A), but there are only six observations that fulfil the previous criteria. It discloses I6 as the carrying amount of EA held.
B I7 was not considered to identify the approach because the disclosure may not have happened if GHG emissions are below EA allocated for free and held for compliance.

| **Panel C—Indexes of the Disclosure** | | | |
|---|---|---|---|
| **Variable** | **Type** | **Description** | **Measurement** |
| *ID* | Levels of transparency | Index of disclosure | Takes values from 0 to 8 (is the sum of I1.dummy to I8.dummy) |
| *ID.IFRS* | | Index of disclosure in an IFRS basis | |
| *ID.NatGAAP* | | Index of disclosure in a national GAAP basis | |
| *ID.EA* | Levels of transparency—Good *vs* Bad news | Index of disclosure of *good news* | Takes values from 0 to 4 (is the sum of I1.dummy to I4.dummy for EA; is the sum of I5.dummy to I8.dummy for GHG emissions) |
| *ID.IFRS.EA* | | Index of disclosure of *good news* in IFRS | |
| *ID.NatGAAP.EA* | | Index of disclosure of *good news* in national GAAP | |
| *ID.eGHG* | | Index of disclosure of *bad news* | |
| *ID.IFRS.eGHG* | | Index of disclosure of *bad news* in IFRS | |
| *ID.NatGAAP.eGHG* | | Index of disclosure of *bad news* in national GAAP | |

Source: Own elaboration.

The variables defined in panel A of Table 2 are a summary of the questions raised in the literature review for the recognition and measurement of EA and GHG emissions [11–14,38] supported with the issues prescribed by French, Spanish and Portuguese standard setters. Eight categorical variables were determined: I1. Initial recognition of free EA, I2. Counterpart on the initial recognition of free EA, I3. Initial measurement of free EA, I4. Subsequent measurement of EA, I5. Recognition of GHG emissions, I6. Measurement of GHG emissions, I7. Recognition of GHG emissions over-allocated EA, I8. Recognition of EA acquired to cover GHG emissions over cap. These categorical variables enable to identify the adopted accounting approach in each company-year observation, by the joint answers given to the accounting policies—I1 to I7 (panel B of Table 2). Each accounting approach is a dichotomous variable that takes the value 1 if it was identified and 0 otherwise. The following accounting approaches were identified: the IFRIC 3 cost model, the government grant approach as described by Ernst and Young [23] (an approach that is similar to the 'cost with balance at market value' mentioned by Lovell et al. [11], to 'modified IFRIC 3 approach' mentioned by Black [14], and also to ICAC Resolution of 8 of February as mentioned in Section 2.1.1—Table 1); the net liability approach; the French inventory approach defined by ANC in 2012; the Spanish approach (until 2016) defined by ICAC in Resolution 8 of February of 2006; the Portuguese approach defined by CNC in the appendix to AFRS 26; and the approach of disclosing off-balance sheet policies, and non-disclosure. Panel C of Table 2 shows self-elaborated indexes as proxies for the extent of disclosures that reflect aspects related to the EU ETS compliance transactions, as frequently used in the literature [12,37,39]. The indexes were based in the eight categorical variables, that were transformed into dichotomous variables (I1.dummy to I8.dummy) that take the value 1 if some categorical information was disclosed and 0 for the cases of non-disclosure and disclosure of off-balance sheet policies (coded as 0 and 99, respectively, in tables of Appendix A). The index of disclosure (ID) is the sum of the eight dummy variables (I1.dummy to I8.dummy) and takes values from 0 to 8. The level of transparency of EA and GHG emissions is also assessed by 'good news' and 'bad news': The ID related to EA (I1 to I4) is seen as a 'good news' once it gives the company free allowances and accordingly to market rules; the ID related to GHG emissions (I5 to I8) is seen as 'bad news' as it internalizes 'the cost of polluting'.

2.3.2. Independent Variables: Treatment and Control

The explanatory variable is the basis of presentation of financial reporting (BoPFS). The definition of the sample exposes companies to the preparation of financial reporting on accordance with IFRS, either mandatory or optional, or with national accounting standards. Previous literature [13,35,40–42] state that accounting practices are distinguished by the standard set that includes specific regulation on the subject versus those leading to professional judgment. Thus, the variable basis of presentation of financial statements under IFRS (BoPFS.IFRS) is a dummy variable that takes the value 1 for observations in IFRS, and the value 0 to represent observations in national standards (Portuguese, Spanish or French). Some caution has to be considered because the consolidated annual accounts of the parent company may be prepared within IFRS, although national regulation obligates the subsidiary to prepare individual annual accounts within national standards (this is the case of Spanish and French regulation, but not Portuguese). Considering that the IFRS have a specific regulatory gap, but the Conceptual Frameworks of Portuguese, Spanish and French standard setters do not contradict the IASB's, there is an argument for the potential influence of the subsidiary financial reporting on the consolidated. Allini et al. [9] present a similar argument to assess whether domestic standard setters have influence in the consolidated accounts. It is therefore important to consider that the majority of observations are based in IFRS reporting and that the use of the French standards is residual (Table 3).

**Table 3.** Frequencies on the basis of presentation of financial reporting (BoPFS) per country.

| BoPFS | Spain | | France | | Portugal | | Total | |
|---|---|---|---|---|---|---|---|---|
| | **Freq** | **Perc** | **Freq** | **Perc** | **Freq** | **Perc** | **Freq** | **Perc** |
| IFRS | 215 | 72,88 | 73 | 82,95 | 68 | 41,21 | 356 | 64.96 |
| Spanish GAAP | 80 | 27,12 | | | | | 80 | 14.60 |
| French GAAP | | | 15 | 17,05 | | | 15 | 2,74 |
| Portuguese GAAP | | | | | 97 | 58,79 | 97 | 17.70 |
| **Total** | **295** | **100,00** | **88** | **100,00** | **165** | **100,00** | **548** | **100.00** |

Regarding control variables, we expect that size, profitability, financial risk, external supervision and industry provide good factors to control the transparency of financial reporting on EA and GHG emissions. Size is considered as a positive influence on the disclosure of information related to GHG emissions [39,43–45], since it emphasizes that large companies are politically more sensitive, facing pressures that smaller ones do not have. There are several measures to proxy size, and we choose the logarithm of total assets as Giner [40] and Gallego-Álvarez et al. [35]. Companies that are more profitable may be more transparent about their activities, and therefore, are expected to present more disclosures to justify good performance. Several studies simultaneously use both return on assets (ROA) and return on equity (ROE) to control for profitability (most observations are from unlisted companies, and therefore, a proxy for profitability could not be market-based), as ROA reflects an operational performance and ROE a financial performance [40,44,46]. We also introduced return on sales (ROS) as it represents the short-term return [47]. Leverage ratios are used to inform about default risks that may lead to avoiding investment and financing. They are often used as a covenant, which leads managers to adopt accounting policies that avoid an inconvenient outcome [44]. There are also several measures used in the literature, as we choose to include debt-to-equity ratio [35,44] and debt-to-assets ratio [24,45]. These control variables (size, ROA, ROE, ROS; debt-to-equity and debt-to-assets) were categorized into '1—small', '2—medium' or '3—large'. Class 1 is up to 25th percentile, class 2 is between the 25th percentile (inclusive) and the 75th percentile (inclusive), and from the 75th percentile, it was categorized as class 3.

The presentation of an independent auditor report certifies the reliability and relevance of accounting practice, thus providing more credibility to financial statements [48]. It is expected that the presentation of an independent auditor report has a positive influence on transparency. The variable Audit is a dichotomous variable that takes the value 1 if the annual report incorporates the auditor's report and the value 0 if not. We realize that unlike the remaining control variables, the presence of auditors' report as a dummy does not capture changes between years.

The literature review provided arguments that practices regarding financial reporting may vary across the industry. On another perspective, it has to be considered that the legal requirements for the financial reporting regarding EA and GHG emissions are similar for all companies in the sample. An initial analysis considered industry as a control variable, coded as a dummy for each industry. Nevertheless, the multivariate analysis showed multicollinearity issues, which led us to omit it from the model. Table 4 presents the frequencies for these variables.

**Table 4.** Frequencies for Industry.

| Industry | Freq. | Percent |
|---|---|---|
| Aviation | 9 | 1.64 |
| Cement | 56 | 10.22 |
| Combustion | 231 | 42.15 |
| Pulp and paper | 102 | 18.61 |
| Production of electricity | 67 | 12.23 |
| Oil refineries | 6 | 1.09 |
| Steel | 42 | 7.66 |
| Glass | 35 | 6.39 |
| | **548** | **100.00** |

### 2.3.3. Model and Econometric Analysis

The proposed econometric model intends to assess if the exposure to IFRS harms the transparency of financial reporting on EA and GHG emissions:

**Model 1:** $\text{LevelTransparency}_{it} = \int (\text{Basis of presentation of financial reporting, control variables})_{it}$

Equations (1) and (2) were designed to verify the influence of the IFRS in the transparency of financial reporting.

$$IDit = \beta 0 + \beta 1 \text{ BoPFS.IFRS}it + \varnothing \text{X}it + \varepsilon it, \tag{1}$$

$$AAit = \beta 0 + \beta 1 \text{ BoPFS.IFRS}it + \varnothing \text{X}it + \varepsilon it, \tag{2}$$

where i and t are, respectively, company and year; ID represents the indexes of the disclosure: ID, ID.EA, ID.eGHG; AA embodies the identified accounting approach; BoPFS.IFRS is the adoption of the IFRS as BoPFS; X corresponds to control variables: Size, Profitability, Leverage and Audit; ε is the error term.

In order to assess the transparency of the accounting policies disclosed by BoPFS, a univariate analysis was carried out, mainly focused on measures of frequency and summary statistics.

For Equation (1), the developments in panel data techniques allow to independently treat the dynamics of the individuals in a period of time and eliminates the bias of the aggregation, which is known as individual effects or unobservable heterogeneity [49,50]. The consistency of parameter estimators, and consequently, the validity of the economic interpretation, rely on the correct functional form specification and controlling for unobserved heterogeneity [51]. The Hausman test compares the use of two estimators (fixed or random effects model), which depends on whether there are (or not) individual effects. However, Wald's' modified test indicated heteroscedasticity, and therefore, standard error estimates had to be corrected for the homoscedasticity condition. It was used a Panel Corrected Standard Errors (PCSE) regression. The use of this estimator also allowed to correct issues of first order autocorrelation, whenever detected, by Wooldridge test. When it was verified heteroscedasticity and first-order autocorrelation, the correction was made through Prais–Winsten estimator. Moundigbaye et al. [52] state that PCSE preserves the (Prais–Winsten) weighting of observations for autocorrelation, but uses a sandwich estimator to incorporate cross-sectional dependence when calculating standard errors.

Equation 2 is based on a model of binary choice. It expresses the likelihood of the company adopting an accounting approach, depending on the regulatory BoPFS. This is a logistic regression model in which the normal distribution is replaced by the logistic [53]. However, the identification of certain accounting approaches (IFRIC 3 cost model, net liability approach) ischaracterized a rare event (see panel A of Appendix B). In order to minimize the bias of the maximum likelihood estimation against the reduced sample, Firth's [54] Penalized Maximum Likelihood Estimation was used, because in cases of 'separation', it allows convergence for finite estimates.

The next chapter presents the results and discussion.

## 3. Results and Discussion

### 3.1. Univariate Analysis

The characterization of the sample by country and by year suggests that throughout the years, there is consistency on the disclosure of the accounting policies, as it is supposed. It should be noted that both Portuguese and French companies may present less consistency, due to alterations in accounting regulation. More specifically, the majority of Spanish companies disclose the recognition of the EA, while French companies are those that present a large frequency of lack of disclosures of such information (for both disclosures on EA and GHG emissions). Portuguese companies are present more discretion, which is less evident from the year of 2010 on regarding GHG emissions, when the recognition of it as the amortization of the EA, using FiFo, becomes part of accounting regulation.

Appendix A presents the frequencies of each accounting policy for the recognition and measurement of EA and GHG emissions. It is assessed that within the application of the Spanish standards, 98,75% disclose the recognition of EA, but in the IFRS adoption, only 68,82% do it. Whilst, 18,26% of the IFRS observations do not disclose information about EA and 9,83% disclose the adoption of off-balance sheet policies. The analysis shows the following tendencies for EA: Recognition as an intangible asset with a counterpart in owners' equity or deferred income, at market value; subsequently, EA are preferably measured at cost fewer impairment losses. PWC and IETA [3], Steenkamp et al. [12] and Black [14] concluded that most companies initially measure the EA at nil value, which is not validated in this study. One possible explanation for recognition in owner's equity and initial measurement at market value is the influence of Portuguese and Spanish standards. Regarding GHG emissions, in line with what was verified for EA, almost 20% of the observations in IFRS discloses the adoption of off-balance sheet policies. The results for the basis of measurement of the GHG emissions identify 13 bases. Similarly, Warwick and Ng [13] found 11 bases of measurement for GHG emissions. Nevertheless, the most common bases are the cost and the carrying amount of the EA, as PWC and IETA [3] and Ayaz [15] also conclude. It may be related to the high frequencies of initial measurement for EA at market value. It should be noted that there is more diversity within the use of the IFRS and that, for all accounting policies, the observations based on French standards only report non-disclosure.

Appendix B (panels A and B) presents the mean and standard deviation for the accounting approaches and indexes of the disclosure. The most used accounting approach under the IFRS is government grant—as identified by Ernst and Young [23], followed by not disclosing information on EA or GHG emissions. Under the adoption of the IFRS and Portuguese standards, the report of off-balance sheet policies suggests the use of strategies of concealment. The index of disclosure presents a mean of approximately five items, increased when it is based in national standards, particularly Spanish, and decreased when the bases are the IFRS. The standard deviation regarding the IFRS indexes provides evidence of the disparity in the levels of transparency. These results are in line with Larrinaga [21] that argued the lack of guidance from the IASB as a motivating factor for the adoption of discretionary views. The companies that present their annual accounts using the Spanish standard are the most transparent, as well as those with less discrepancy in accounting policies for EA and GHG emissions. The results also suggest that whether companies apply IFRS or national standards, the tendency is to disclose more information on EA than on GHG emissions. However, this does not mean dismissal of the standard or a strategy of concealment. The index of 'bad news' includes the recognition of GHG emissions above EA held and the acquisition of EA, which may not be a necessity of disclosure, although it would be more transparent to present an accounting policy for it. The difference is more significant for the Portuguese standard, which we attribute to the prescribed accounting treatment for GHG emissions—the amortization of the EA and not the recognition of a liability.

As expected, when specific accounting treatment is missing—in an IFRS context—more approaches are allowed, thus improving the level of transparency of financial reporting of EA and GHG emissions. However, even transparency under specific accounting regulations is not consistent (in an analysis across time). The changes on national regulations in the period 2008-2014 may have contributed to this situation (as companies that apply Spanish GAAP—a steady regulation in such period, present a lesser number of adopted accounting approaches, and even so, government grant and ICAC approaches are similar). Discretion is, therefore, a situation that the existence of regulation may minimize, but to stabilize the levels of transparency of financial reporting, thus reducing discretion, we believe to be necessary a common approach. Without that, the companies that are under unspecific accounting treatments will have the possibility to present the transactions with different levels of transparency (even across time); and those with specific accounting treatments, will also benefit with the doubt created by the international community. Given this situation, the IASB should create a more stable situation, that we hope it comes to support a full recognition of the assets and liabilities, hence ensuring more transparency and less discretion in the financial reports.

### 3.2. Multivariate Analysis

Appendix D presents the results for Equation (1). It should be noted that industry variables presented a variance inflation factor (VIF) above 10 (excluding aviation). As dummy variables, it is understandable that they present correlation among them (see Appendix C). These results do not affect the coefficient of the explanatory variable, as it was decided to remove them from the models. For models 1 and 3 it was used a linear regression heteroscedastic PCSE and for model 2, to deal with simultaneous heteroscedastic and first order autocorrelation, it was used a Prais–Winsten regression, heteroscedastic PCSE. The following post-estimation tests are presented: Wald tests to verify the significance of time dummy variables—identified as i.Year; the Wald modified test to assess the presence of heteroscedasticity—identified as Waldchi2; and the Wooldridge test verifies the presence of first-order autocorrelation—identified as AR(1).

The results show that the application of the IFRS has a negative effect on the indexes of the disclosure. That effect is stronger regarding disclosures of EA than GHG emissions, suggesting that there is a tendency to incorporate less information about the EA in IFRS reporting than information on GHG emissions. Control variables, such as ROS, debt-to-equity, or auditors' report have a positive influence on the indexes of the disclosure.

Appendix E shows the results for Equation (2) using the Penalized Maximum Likelihood Estimation of Firth's [54]. Models 1 and 4 on the probability of disclosure the IFRIC 3—cost model and inventories approach are not statistically significant. Model 5 regarding the probability of disclosing the ICAC model (from 2006), although statistically significant, presents no significant probability of being influenced by the use of the IFRS. Somewhat similarly, Moneva and Llena [55] provide empirical evidence that, in the Spanish context, there are no significant differences regarding environmental disclosures between large quoted and non-quoted companies. The probability of reporting the government grant and the net liability approaches is positively and strongly associated with the adoption of the IFRS. The same positive direction applies to the disclosure of off-balance sheet policies and also, for non-disclosure. The probability of adopting the AFRS 26 approach is negatively associated with the IFRS, suggesting that only under the adoption of Portuguese standards this approach has been disclosed. Monteiro and Vilas Boas [39] that concluded that although AFRS 26 is legally imposed for unlisted companies, listed companies also presented a higher index of disclosures on EA. Therefore, it may be implied that listed companies disclose information on EA and GHG emissions although not under the direct influence of the Portuguese standard, are concerned in presenting financial information on the participation on the EU ETS. The probability of reporting the government grant and the net liability approaches is positively and strongly associated with the adoption of the IFRS. The same positive direction applies to the disclosure of off-balance sheet policies, and also, for non-disclosure

The results obtained suggest that the transparency of financial reporting on EA and GHG emissions are negatively influence by the IFRS. Hence, H1 is rejected because there is an association regarding the indexes of the disclosure and concerning four specific approaches: Non-disclosure, off-balance sheet recognition, government grant approach and net liability approach. This is due to the fact that the

IASB does not specifically state its position on the EU ETS transactions, but the IFRSs provide several answers to include EU ETS transactions in the financial statements that do not include its omission (within IAS 38, IAS 20, IAS 37 or IFRS 9). The levels of transparency of financial reporting on EA and GHG emissions are lower in companies that adopt the IFRS. The use of the IFRS is positively related to the probability to use accounting approaches that minimize transparency, such as net liability approach, off-balance sheet policies or non-disclosure. However, it is also positively associated with the government grant approach, a gross approach that provides more information in financial reporting. Hence, there is an association between transparency and accounting regulation, but there is no evidence of a significant reduction of the levels of transparency of financial reporting prepared under the IFRS. The existence of mandatory regulation seems to be sufficient to provide a relative increase in the transparency of financial reporting. This is in contrast with Giner [40] conclusion that legislation appears to produce a strong increase in disclosure, even before being compulsory. Barbu et al. [29] stated that compliance (with IASB standards) might differ between companies and even between countries, due to differences in reporting practices. So, even in environments where the standard setter did not issue specific regulation, the transparency of financial reporting on EA and GHG emissions may be positively influenced by national regulation, but not sufficient to disregard other influencing factors. The suggestion that companies could use strategies, such as disclose more 'good news', related to EA, than with 'bad news', regarding GHG emissions [31,33], was not verified. Therefore, the lack of specific legislation does not influence companies to disclose more accounting policies related to EA than with GHG emissions. Therefore, companies are disclosing accounting practices on both "the asset" as "the liability", possibly to compensate the 'cost of polluting' with the recognition of licenses that permit to emit a given number of GHG for free. Generally, it can be argued that lower levels of transparency regarding EA and GHG emissions are associated with this lack of specific regulation, but is not as significant as it could be expected.

In summary, we believe that as regulation may not be decisive in the disclosure process, it could increase the levels of transparency and particularly the possibility of comparison amongst companies. The results show that the many approaches taken by IFRS companies include mainly approaches of non-disclosure, off-balance sheet recognition and a net liability approach (although some choose government grant disclosure). A clear position by the IASB would help to reduce that discretion and to increase the transparency of the financial reporting that represents EU ETS transactions. The next chapter presents the conclusions and contributions of the study, as well as its limitations.

## 4. Conclusions

### 4.1. Conclusions and Contribution

This study aims to provide evidence of the influence of accounting regulation in the transparency of financial reporting on EA and GHG emissions. It was considered the perspectives from the IASB, national standard setters and the reporting practices commonly adopted by companies, accordingly to previous literature. This issue arises from the lack of specific guidance in the IFRS, which motivated European national standard setters to overcome it, and led professionals to exercise their professional judgment. The Portuguese and Spanish standard setters adopted a gross approach, similar to the IFRIC 3—although this is not a condition to adopt a gross approach; the French standard setter changed the fundamentals of the prescribed approach in 2012 (for periods beginning at 1/1/2013 or after), from a gross to a net approach. We trust that gross approaches provide better levels of transparency and provide stakeholders with relevant and faithfully represented information on the companies' exposure to the EU ETS. Notwithstanding, as a high-quality standard set, the IFRS provide a basis to adopt policies that meet those characteristics.

Multiple accounting approaches are identified, mainly within the adoption of IFRS. The transparency of financial reporting based in IFRS is positively influenced by the probability of adopting approaches such as the government grant (gross approach) and net liability approach; and negatively influenced by the probability of disclosing off-balance sheet and practices of non-disclosure. Hence, the companies that adopt the IFRS tend to disclose less information than companies that apply regulations with specific accounting principles. Moreover, it is within the IFRS that it is verified more dispersion in the reported information, implying that there is not a commonly accepted framework for disclosures. As a conclusion, this study provides evidence that the existence of specific mandatory regulation is sufficient to provide a relative increase on the transparency of financial reporting, but other factors have to be considered for a more comprehensive analysis. Companies reporting on the basis of the IFRS have looked forms of reporting to their stakeholders the exposure to the EU ETS; however, we do not see it as sufficient. The lack of specific regulation prejudices transparency (allowing the not disclosing or the disclosure of off-balance sheet policies) but also leads to total discretion on the accounting approaches that are adopted. The results suggest that such discretion is a situation that the existence of consistent regulation may minimize, but to stabilize the levels of transparency of financial reporting between companies, thus reducing discretion, we believe to be necessary a common approach. Without that, the companies that are under unspecific accounting treatments will have the possibility to present the transactions with different levels of transparency; and those with specific accounting treatments will also benefit with the doubt created by the surrounding community. This is a fundamental reason that urges the regulation of IFRS transactions, in line with the international harmonization process. The IASB is, ultimately, responsible for creating a more stable situation, that we hope it supports a full recognition of the assets and liabilities, hence ensuring more transparency and less discretion in the financial reports.

Given the reduced number of studies concerning such matters, this research highlighted that transparency in financial reporting should be focused in the underlying economics of the EU ETS transactions: The full recognition of the related assets and liabilities. The lack of comparability between companies, the strategies adopted for disclosing some information (such as off-balance sheet policies) and the lack of disclosure, harm the desired transparency for financial reporting.

### 4.2. Limitations

A note of caution is needed because of the inherent limitations. The first is related to the availability of annual financial statements. This led us not to consider important determinants as listed/unlisted companies as some of the annual reports were only available for consolidated reporting. In such cases, the individual company, that was the receiver of the EA was not listed, although the parent company (and the reporting entity) was, . Hence, the consolidated financial disclosures were considered as the accounting treatment adopted by the individual company, but the inherent characteristics (as size or profitability) were regarding the individual company. Second, the use of content analysis may result in bias, as sometimes it is necessary to make some subjective judgements. This is regarding the analysis of the perimeter of consolidation, to include a wider scope of companies, and to assess the proxies of transparency. Finally, we acknowledge that the use of all companies listed in the NPAs with available annual financial statements may potentially include information within materiality issues that increments the adoption of off-balance sheet policies and non-disclosure.

**Author Contributions:** For research articles with several authors, a short paragraph specifying their individual contributions must be provided. The following statements should be used "Conceptualization, P.M.M., E.P.C. and A.I.L.D.; Data curation, E.P.C. and A.I.L.D.; Formal analysis, P.M.M., E.P.C. and A.I.L.D.; Funding acquisition, P.M.M.; Investigation, P.M.M., E.P.C. and A.I.L.D.; Methodology, P.M.M., E.P.C. and A.I.L.D.; Project administration, P.M.M.; Resources, A.I.L.D.; Software, A.I.L.D.; Supervision, P.M.M.; Validation, P.M.M., E.P.C. and A.I.L.D.; Visualization, P.M.M.; Writing—original draft, E.P.C. and A.I.L.D.; Writing—review & editing, E.P.C. and A.I.L.D. All authors have read and agreed to the published version of the manuscript.

**Funding:** This research was funded by European Regional Development Fund, European Union "A way of making Europe", and by Council of Economy and Infrastructure, Regional Government of Extremadura (Spain), grant number GR18128. It was also funding by Junta of Extremadura, grant 2017 for the mobility of teaching and research staff at the University of Extremadura.

**Acknowledgments:** We acknowledge the support given by the Cardiff Business School during the academic stay developed from July to September 2017.

**Conflicts of Interest:** "The authors declare no conflict of interest." "The funders had no role in the design of the study; in the collection, analyses, or interpretation of data; in the writing of the manuscript, or in the decision to publish the results".

## Appendix A. Frequencies of the Accounting Policies Regarding EA and GHG Emissions

See Table 2 for variable descriptions.

| | IFRS | | Spanish GAAP | | French GAAP | | Portuguese GAAP | |
|---|---|---|---|---|---|---|---|---|
| | Freq | Perc | Freq | Perc | Freq | Perc | Freq | Perc |
| **Are EAs' recognized?** | | | | | | | | |
| Yes | 245 | 68.82 | 79 | 98.75 | | | 85 | 87.63 |
| No | 35 | 9.83 | | | | | 2 | 2.06 |
| Only when GHG emissions exceed EA held | 9 | 2.53 | | | | | | |
| Do not have EA | 2 | 0.56 | | | | | 1 | 1.03 |
| Not disclosed | 65 | 18.26 | 1 | 1.25 | 15 | 100.00 | 9 | 9.28 |
| **I1. Rec EA** | | | | | | | | |
| 1—Intangible Asset | 222 | 62.36 | 79 | 98.75 | | | 82 | 84.54 |
| 2—Inventory | 22 | 6.18 | | | | | | |
| 3—Financial Asset | | | | | | | 3 | 3.09 |
| 4—Other current asset | 5 | 1.40 | | | | | | |
| 0—Not disclosed | 70 | 19.66 | 1 | 1.25 | 15 | 100.00 | 9 | 9.28 |
| 99- Not applied | 37 | 10.39 | | | | | 3 | 3.09 |
| **I2. Count EA** | | | | | | | | |
| 1—Owners' Equity | | | 65 | 81.25 | | | 64 | 65.98 |
| 2—Other non-current liability | 7 | 1.97 | | | | | | |
| 3—Liability | | | | | | | 4 | 4.12 |
| 4—Financial liability | | | | | | | 3 | 3.09 |
| 5—Liability—EA allocated | 4 | 1.12 | | | | | | |
| 6—Deferred income | 145 | 40.73 | | | | | 9 | 9.28 |
| 7—Government grant | 45 | 12.64 | 7 | 8.75 | | | | |
| 8—Grant reduced to asset carrying amount | 11 | 3.09 | | | | | | |
| 9—Income—Government grant | 5 | 1.40 | | | | | | |
| 10—Operational income | 7 | 1.97 | | | | | | |
| 0—Not disclosed | 95 | 26.69 | 8 | 10.00 | 15 | 100.00 | 14 | 14.43 |
| 99—Not applied | 37 | 10.39 | | | | | 3 | 3.09 |

| | | | | | | | | |
|---|---|---|---|---|---|---|---|---|
| **I3. Initial Meas EA** | | | | | | | | |
| 1—Acquisition cost | 6 | 1.69 | 2 | 2.50 | | | | |
| 2—Fair value | | | | | | | 15 | 15.46 |
| 3—Market value | 175 | 49.16 | 64 | 80.00 | | | 64 | 65.98 |
| 4—Price | 7 | 1.97 | | | | | | |
| 5—Quoted average price | 7 | 1.97 | | | | | | |
| 6—Zero | 41 | 11.52 | | | | | | |
| 7—Symbolic value | 6 | 1.69 | 7 | 8.75 | | | | |
| 0—Not disclosed | 77 | 21.63 | 7 | 8.75 | 15 | 100.00 | 15 | 15.46 |
| 99—Not applied | 37 | 10.39 | | | | | 3 | 3.09 |
| **I4. Subseq Meas EA** | | | | | | | | |
| 1—Historical cost | 58 | 16.29 | 4 | 5.00 | | | 6 | 6.19 |
| 2—Cost less accumulated amortisations | | | | | | | 15 | 15.46 |
| 3—Cost less accumulated impairment losses | 120 | 33.71 | 43 | 53.75 | | | 5 | 5.15 |
| 4—Cost model | | | | | | | 25 | 25.77 |
| 5—*First in First Out* | 2 | 0.56 | | | | | | |
| 6—Cost or net realisable value, the lower | 6 | 1.69 | | | | | | |
| 7—Fair value | 12 | 3.37 | | | | | 20 | 20.62 |
| 8—Revaluation model | | | | | | | 5 | 5.15 |
| 9—*Net liability approach* | | | 7 | 8.75 | | | 1 | 1.03 |
| 10—*Net asset approach* | 7 | 1.97 | | | | | | |
| 0—Not disclosed | 114 | 32.02 | 26 | 32.50 | 15 | 100,00 | 17 | 17.53 |
| 99—Not applied | 37 | 10.39 | | | | | 3 | 3.09 |
| **I5. Rec GHG emissions** | | | | | | | | |
| 1—Amortization expenses | | | | | | | 39 | 40.21 |
| 2—Cost of consumed materials | 2 | 0.56 | | | | | | |
| 3—Other operational losses | 14 | 3.93 | | | | | 6 | 6.19 |
| 4—Liability | 11 | 3.09 | | | | | 4 | 4.12 |
| 5—Financial Liability | | | | | | | 3 | 3.09 |
| 6—Provision | 182 | 51.12 | 79 | 98.75 | | | 14 | 14.43 |
| 7—No recognition | 70 | 19.66 | | | | | 18 | 18.56 |
| 0—Not disclosed | 77 | 21.63 | 1 | 1.25 | 15 | 100,00 | 13 | 13.40 |
| **I6. Mens GHG emissions** | | | | | | | | |
| 1—Cost | | | 26 | 32.50 | | | | |
| 2—*First in First Out* | 2 | 0.56 | | | | | 32 | 32.99 |
| 3—Initial price | 7 | 1.97 | | | | | | |
| 4—Average price of EA | 7 | 1.97 | | | | | 3 | 3.09 |
| 5—Average cost of EA + Acq. Cost EA | 6 | 1.69 | | | | | | |
| 6—Average cost of EA + Best estimate to acquire | 7 | 1.97 | 7 | 8.75 | | | | |
| 7—Fuel consumption + average price of EA | | | 3 | 3.75 | | | | |
| 8—Cancelation of the obligation order | | | 7 | 8.75 | | | | |

| | | | | | | | | |
|---|---|---|---|---|---|---|---|---|
| 9—Carrying amount (CA) EA | 68 | 19.10 | 11 | 13.75 | | | 1 | 1.03 |
| 10—CA EA + average price of EA + best estimate | 14 | 3.93 | 12 | 15.00 | | | | |
| 11—CA EA + Best estimate to acquire | 40 | 11.24 | | | | | | |
| 12—CA EA + price at closing | 17 | 4.78 | | | | | | |
| 13—CA EA + market value | 4 | 1.12 | | | | | | |
| 0—Not disclosed | 117 | 32.87 | 14 | 17.50 | 15 | 100.00 | 43 | 44.33 |
| 99—Not applied | 67 | 18.82 | | | | | 18 | 18.56 |
| **I7. Rec GHGe over EA** | | | | | | | | |
| 1—Expense accrual | | | | | | | 2 | 2.06 |
| 2—Liability | 19 | 5.34 | | | | | | |
| 3—Financial Liability | 14 | 3.93 | | | | | 14 | 14.43 |
| 4—Provision | 109 | 30.62 | 28 | 35.00 | | | 44 | 45.36 |
| 5—No recognition | 28 | 7.87 | | | | | 11 | 11.34 |
| 0—Not disclosed | 186 | 52.25 | 52 | 65.00 | 15 | 100.00 | 26 | 26.80 |
| **I8. Rec EA acq** | | | | | | | | |
| 1—Intangible Asset | 185 | 51.97 | 67 | 83.75 | | | 45 | 46.39 |
| 2—Inventory | 27 | 7.58 | | | | | | |
| 3—Never acquired EA | | | | | | | 14 | 14.43 |
| 0—Not disclosed | 144 | 40.45 | 13 | 16.25 | 15 | 100.00 | 38 | 39.18 |

## Appendix B. Levels of Transparency of EA and GHG Emissions Financial Reporting

See Table 2 for variable definitions. Std. Dev. is the standard deviation.

**Panel A—Descriptive summary of accounting approaches by BoPFS.**

| Accounting Approach | Full sample (Obs: 548) | | IFRS (obs = 356) | | Spanish (obs = 80) | | French (obs = 15) | | Portuguese (obs = 97) | |
|---|---|---|---|---|---|---|---|---|---|---|
| | Mean | Std. Dev. | Mean | Std. Dev. | Mean | Std. Dev. | Mean | Std. Dev. | Mean | Std. Dev. |
| IFRIC3_cost | 0,0128 | 0,1124 | 0,0197 | 0,1390 | 0,0000 | 0,0000 | 0,0000 | 0,0000 | 0,0000 | 0,0000 |
| GovGrant | 0,1569 | 0,3641 | 0,2809 | 0,4501 | 0,0625 | 0,2436 | 0,0000 | 0,0000 | 0,0000 | 0,0000 |
| Inventory | 0,0511 | 0,2204 | 0,0787 | 0,2696 | 0,0000 | 0,0000 | 0,0000 | 0,0000 | 0,0000 | 0,0000 |
| *NetLiab* | 0,0109 | 0,1042 | 0,0169 | 0,1289 | 0,0000 | 0,0000 | 0,0000 | 0,0000 | 0,0000 | 0,0000 |
| ICAC | 0,0438 | 0,2048 | 0,0393 | 0,1946 | 0,0875 | 0,2843 | 0,0000 | 0,0000 | 0,0309 | 0,1740 |
| AFRS26 | 0,0511 | 0,2204 | 0,0000 | 0,0000 | 0,0000 | 0,0000 | 0,0000 | 0,0000 | 0,2887 | 0,4555 |
| NoRec | 0,0420 | 0,2007 | 0,0590 | 0,2359 | 0,0000 | 0,0000 | 0,0000 | 0,0000 | 0,0206 | 0,1428 |
| NoDisc | 0,1369 | 0,3440 | 0,1433 | 0,3508 | 0,0125 | 0,1118 | 1,0000 | 0,0000 | 0,0825 | 0,2765 |

**Panel B—Descriptive summary of indexes of disclosure by BoPFS**

| Variable | Full sample (Obs: 548) | | IFRS GAAP (Obs: 356) | | Spanish GAAP (Obs: 80) | | Portuguese GAAP (obs = 97) | | Min | Max |
|---|---|---|---|---|---|---|---|---|---|---|
| | Mean | Std. Dev. | Mean | Std. Dev. | Mean | Std. Dev. | Mean | Std. Dev. | | |
| ***I1.dummy*** | **0.7442** | 0.4367 | 0.6994 | 0.4591 | 0.9875 | 0.1118 | 0.8763 | 0.3310 | 0 | 1 |
| *I2.dummy* | 0.6762 | 0.4683 | 0.6292 | 0.4837 | 0.9000 | 0.3019 | 0.8247 | 0.3822 | 0 | 1 |
| *I3.dummy* | 0.7102 | 0.4541 | 0.6798 | 0.4672 | 0.9125 | 0.2843 | 0.8144 | 0.3908 | 0 | 1 |
| *I4.dummy* | 0.6064 | 0.4890 | 0.5758 | 0.4949 | 0.6750 | 0.4713 | 0.7938 | 0.4067 | 0 | 1 |
| *I5.dummy* | 0.6386 | 0.4808 | 0.5871 | 0.4931 | 0.9875 | 0.1118 | 0.6804 | 0.4687 | 0 | 1 |
| *I6.dummy* | 0.4955 | 0.5004 | 0.4831 | 0.5004 | 0.8250 | 0.3824 | 0.3711 | 0.4856 | 0 | 1 |
| *I7.dummy* | 0.4132 | 0.4929 | 0.3989 | 0.4904 | 0.3500 | 0.4800 | 0.6186 | 0.4883 | 0 | 1 |
| *I8.dummy* | 0.6100 | 0.4882 | 0.5955 | 0.4915 | 0.8375 | 0.3712 | 0.6082 | 0.4907 | 0 | 1 |
| ID | 4.9543 | 2.9162 | 4.6489 | 3.0743 | 6.4750 | 1.3685 | 5.5876 | 2.3352 | 0 | 8 |
| ID.EA | 2.7719 | 1.6418 | 2.5843 | 1.7235 | 3.4750 | 0.8711 | 3.3093 | 1.3099 | 0 | 4 |
| ID.eGHG | 2.1823 | 1.4773 | 2.0646 | 1.5305 | 3.0000 | 0.9808 | 2.2784 | 1.2726 | 0 | 4 |

## Appendix C. Correlation Matrix

See Table 2 for variable definitions.

**Panel A—Spearman's correlation matrix (Equation (1)).**

| | (1) | (2) | (3) | (4) | (5) | (6) | (7) | (8) | (9) | (10) | (11) | (12) | (13) | (14) | (15) | (16) | (17) | (18) | (19) |
|---|---|---|---|---|---|---|---|---|---|---|---|---|---|---|---|---|---|---|---|
| (1) ID | 1.0000 | | | | | | | | | | | | | | | | | | |
| (2) ID.EA | 0.8487 * | 1.0000 | | | | | | | | | | | | | | | | | |
| (3) ID.eGHG | 0.9542 * | 0.6954 * | 1.0000 | | | | | | | | | | | | | | | | |
| (4) BoPFS.IFRS | −0.0898 ** | −0.1404 * | −0.0970 ** | 1.0000 | | | | | | | | | | | | | | | |
| (5) Size | −0.0179 | −0.0335 | −0.0048 | 0.2434 * | 1.0000 | | | | | | | | | | | | | | |
| (6) ROA | 0.1340 * | 0.0957 ** | 0.1385 * | −0.1406 * | −0.0693 | 1.0000 | | | | | | | | | | | | | |
| (7) ROE | 0.0888 ** | 0.0226 | 0.1103 * | −0.0541 | −0.0328 | 0.6971 * | 1.0000 | | | | | | | | | | | | |
| (8) ROS | 0.1167 * | 0.1427 * | 0.1083 | −0.2164 * | −0.2445 * | 0.1241 * | 0.0073 | 1.0000 | | | | | | | | | | | |
| (9) DebtE | −0.0190 | −0.1252 * | 0.0277 | 0.0433 | 0.0876 | −0.1314 * | 0.0109 | −0.1277 * | 1.0000 | | | | | | | | | | |
| (10) DebtTA | −0.0791 *** | −0.1915 * | −0.0270 | 0.0865 | 0.1095 | −0.2044 * | 0.0365 | −0.1131 * | 0.8905 * | 1.0000 | | | | | | | | | |
| (11) Audit | 0.1128 * | 0.0548 | 0.1524 * | 0.1774 * | 0.0851 | −0.0851 | −0.0568 | −0.1774 * | −0.1064 | −0.0568 | 1.0000 | | | | | | | | |
| (12) Aviation | −0.0967 ** | −0.1461 * | −0.0625 | −0.0255 | 0.0203 | −0.0203 | −0.0203 | 0.1827 * | −0.0406 | 0.1218 *** | 0.0558 | 1.0000 | | | | | | | |
| (13) Cement | −0.0203 | 0.0085 | −0.0439 | −0.0174 | 0.0426 | −0.0426 | −0.1193 * | −0.1874 * | −0.1193 * | −0.1193 * | 0.1456 * | −0.0436 | 1.0000 | | | | | | |
| (14) Combustion | −0.1062 ** | −0.0894 ** | −0.0816 *** | −0.1477 * | 0.1045 ** | 0.0523 | 0.0470 | 0.2143 * | −0.0366 | −0.0366 | −0.2413 * | −0.1103 * | −0.2880 * | 1.0000 | | | | | |
| (15) PulpPaper | −0.0249 | 0.0588 | −0.0673 | 0.0269 | −0.2984 * | 0.0265 | −0.0530 | 0.0199 | −0.0663 | −0.0928 *** | −0.0644 | −0.0618 | −0.1613 * | −0.4082 * | 1.0000 | | | | |
| (16) P.Electricity | 0.0967 ** | −0.0482 | 0.1230 * | −0.1229 * | 0.0079 | 0.0788 *** | 0.1891 * | −0.2521 * | 0.2757 * | 0.2757 * | 0.0845 * | −0.0482 | −0.1259 * | −0.3186 * | −0.1785 * | 1.0000 | | | |
| (17) OilRef. | 0.0640 | 0.0448 | 0.0804 *** | 0.0773 *** | 0.0000 | −0.0496 | −0.0992 ** | 0.0000 | −0.0496 | −0.0496 | 0.0454 | −0.0136 | −0.0355 | −0.0898 * | −0.0503 | −0.0393 | 1.0000 | | |
| (18) Steel | 0.0749 *** | 0.0694 | 0.0566 | 0.2116 * | 0.2037 * | −0.1067 ** | −0.0097 | 0.0291 | 0.0679 | 0.0388 | 0.1243 * | −0.0372 | −0.0972 * | −0.2459 * | −0.1378 * | −0.1075 * | −0.0303 | 1.0000 | |
| (19) Glass | 0.0912 ** | 0.1225 * | 0.0982 ** | 0.1918 * | −0.0317 | −0.0528 | −0.0528 | −0.0211 | −0.0739 *** | −0.0844 ** | 0.1127 * | −0.0338 | −0.0881 * | −0.2230 * | −0.1249 * | −0.0975 * | −0.0275 | −0.0753 ** | 1.0000 |

*, **, *** represent significance levels (two-tailed) at 1%, 5% and 10%, respectively.

**Panel B—Spearman's correlation matrix (Equation (2)).**

| | (1) | (2) | (3) | (4) | (5) | (6) | (7) | (8) | (9) | (10) | (11) | (12) | (13) | (14) | (15) | (16) | (17) | (18) | (19) | (20) | (21) | (22) | (23) | (24) |
|---|---|---|---|---|---|---|---|---|---|---|---|---|---|---|---|---|---|---|---|---|---|---|---|---|
| (1) IFRIC3_cost | 1.0000 | | | | | | | | | | | | | | | | | | | | | | | |
| (2) GovGrant | −0.0491 | 1.0000 | | | | | | | | | | | | | | | | | | | | | | |
| (3) NetLiab | −0.0264 | −0.1001 ** | 1.0000 | | | | | | | | | | | | | | | | | | | | | |
| (4) Invent | −0.0120 | −0.0454 | −0.0244 | 1.0000 | | | | | | | | | | | | | | | | | | | | |
| (5) ICAC | −0.0243 | −0.0923 ** | −0.0497 | −0.0225 | 1.0000 | | | | | | | | | | | | | | | | | | | |
| (6) AFRS26 | −0.0264 | −0.1001 ** | −0.0538 | −0.0244 | −0.0497 | 1.0000 | | | | | | | | | | | | | | | | | | |
| (7) NRec | −0.0238 | −0.0903 ** | −0.0486 | −0.0220 | −0.0448 | −0.0486 | 1.0000 | | | | | | | | | | | | | | | | | |
| (8) NDisc | −0.0453 | −0.1718 * | −0.0924 ** | −0.0419 | −0.0852 ** | −0.0924 ** | −0.0833 *** | 1.0000 | | | | | | | | | | | | | | | | |
| (9) BoPFS.IFRS | 0.0835 *** | 0.3169 * | 0.1704 * | 0.0773 *** | −0.0297 | −0.3160 * | 0.1156 * | 0.0253 | 1.0000 | | | | | | | | | | | | | | | |
| (10) Size | 0.0000 | 0.1703 * | 0.0703 | −0.1488 * | 0.0000 | −0.1641 * | −0.1158 * | 0.0225 | 0.2434 * | 1.0000 | | | | | | | | | | | | | | |
| (11) ROA | −0.0689 | 0.1064 ** | −0.0586 | 0.0496 | 0.1009 ** | −0.1406 * | 0.0901 ** | −0.0676 | −0.1406 * | −0.0693 | 1.0000 | | | | | | | | | | | | | |
| (12) ROE | −0.0689 | 0.1419 * | 0.0586 | 0.0000 | 0.0504 | −0.1524 * | 0.0386 | −0.0601 | −0.0541 | −0.0328 | 0.6971 * | 1.0000 | | | | | | | | | | | | |
| (13) ROS | 0.1609 * | −0.0922 ** | −0.1406 * | 0.0992 ** | 0.1135 * | −0.0352 | −0.1158 * | 0.0000 | −0.2164 * | −0.2445 * | 0.1241 * | 0.0073 | 1.0000 | | | | | | | | | | | |
| (14) DebtE | 0.0689 | 0.0922 ** | 0.1641 * | −0.1488 * | −0.0883 ** | 0.0469 | 0.0772 *** | −0.1352 * | 0.0433 | 0.0876 ** | −0.1314 * | 0.0109 | −0.1277 * | 1.0000 | | | | | | | | | | |
| (15) DebtTA | 0.0460 | 0.0709 *** | 0.1524 * | −0.1488 * | −0.1135 * | 0.0469 | 0.0515 | −0.0375 | 0.0865 ** | 0.1095 | −0.2044 * | 0.0365 | −0.1131 * | 0.8905 * | 1.0000 | | | | | | | | | |
| (16) Audit | 0.0491 | −0.0621 | 0.1001 ** | 0.0454 | −0.0793 *** | −0.0594 | 0.0903 ** | −0.3099 * | 0.1774 * | 0.0851 | −0.0851 | −0.0568 | −0.1774 * | −0.1064 | −0.0568 | 1.0000 | | | | | | | | |
| (17) Aviation | −0.0147 | −0.0558 | −0.0300 | −0.0136 | −0.0277 | −0.0300 | −0.0270 | 0.1992 * | −0.0255 | 0.0203 | −0.0203 | −0.0203 | 0.1827 * | −0.0406 | 0.1218 *** | 0.0558 | 1.0000 | | | | | | | |
| (18) Cement | −0.0384 | −0.1456 * | −0.0783 *** | −0.0355 | −0.0722 *** | 0.3321 * | −0.0706 *** | −0.1343 * | −0.0174 | 0.0426 | −0.0426 | −0.1193 * | −0.1874 * | −0.1193 * | −0.1193 * | 0.1456 * | −0.0436 | 1.0000 | | | | | | |
| (19) Combustion | −0.0971 ** | −0.2565 * | 0.1376 * | 0.1233 * | 0.2507 * | −0.1981 * | −0.1787 * | 0.2084 * | −0.1477 * | 0.1045 ** | 0.0523 | 0.0470 | 0.2143 * | −0.0366 | −0.0366 | −0.2413 * | −0.1103 * | −0.2880 * | 1.0000 | | | | | |
| (20) PulpPaper | −0.0544 | 0.1288 * | −0.1110 * | −0.0503 | −0.1023 * | 0.0381 | 0.2974 * | −0.0267 | 0.0269 | −0.2984 * | 0.0265 | −0.0530 | 0.0199 | −0.0663 | −0.0928 *** | −0.0644 | −0.0618 | −0.1613 * | −0.4082 * | 1.0000 | | | | |
| (21) P.Electricity | −0.0425 | 0.1606 * | −0.0866 ** | −0.0393 | −0.0799 *** | 0.0652 | 0.0886 ** | −0.0514 | −0.1229 * | 0.0079 | 0.0788 *** | 0.1891 * | −0.2521 * | 0.2757 * | 0.2757 * | 0.0845 * | −0.0482 | −0.1259 * | −0.3186 * | −0.1785 * | 1.0000 | | | |
| (22) Oil Refiner | −0.0120 | −0.0454 | −0.0244 | −0.0111 | −0.0225 | −0.0244 | −0.0220 | −0.0419 | 0.0773 *** | 0.0000 | −0.0496 | −0.0992 ** | 0.0000 | −0.0496 | −0.0496 | 0.0454 | −0.0136 | −0.0355 | −0.0898 * | −0.0503 | −0.0393 | 1.0000 | | |
| (23) Steel | 0.3948 * | −0.1243 * | 0.1201 * | −0.0303 | −0.0617 | −0.0669 | −0.0603 | −0.1147 * | 0.2116 * | 0.2037 * | −0.1067 ** | −0.0097 | 0.0291 | 0.0679 | 0.0388 | 0.1243 * | −0.0372 | −0.0972 * | −0.2459 * | −0.1378 * | −0.1075 * | −0.0303 | 1.0000 | |
| (24) Glass | −0.0297 | 0.4618 * | 0.0072 | −0.0275 | −0.0559 | −0.0606 | −0.0547 | −0.1040 ** | 0.1918 * | −0.0317 | −0.0528 | −0.0528 | −0.0211 | −0.0739 *** | −0.0844 ** | 0.1127 * | −0.0338 | −0.0881 * | −0.2230 * | −0.1249 * | −0.0975 * | −0.0275 | −0.0753 ** | 1.0000 |

*, **, *** represent significance levels (two-tailed) at 1%, 5% and 10%, respectively.

## Appendix D. Statistic Models: the Influence of Accounting Regulations on the Indexes of the Disclosure

See Table 2 for variable definitions.

This table shows the results for Equation (1), described in Section 2.3.3, using heteroscedastic Panel Corrected Standard Errors (PCSE). Model 1 presents the results for the dependent variable that represents the level of transparency of financial reporting on EA and GHG emissions, model 2 presents the results for the dependent variable that represents the level of transparency of financial reporting on EA using Prais–Winsten regression that corrects first order autocorrelation issues and model 3 presents the results for the dependent variable that represents the level of transparency of financial reporting on GHG emissions. [(1)] Wald test for time dummies is not significant.

| | (1) ID | | (2) ID.EA | | (3) ID.eGHG | |
|---|---|---|---|---|---|---|
| | *Coef.* | *Std. Err.* | *Coef.* | *Std. Err.* | *Coef.* | *Std. Err.* |
| **BoPFS.IFRS** | −0.8402 * | 0.2541 | −0.5416 ** | 0.2424 | −0.3228 ** | 0.1329 |
| **Size** | 0.1711 | 0.1855 | −0.0345 | 0.1211 | 0.0975 | 0.0949 |
| **ROA** | 0.0314 | 0.2605 | 0.0282 | 0.0600 | 0.1219 | 0.1337 |
| **ROE** | 0.3676 | 0.2508 | 0.0253 | 0.0705 | 0.1755 | 0.1272 |
| **ROS** | 0.5978 * | 0.1811 | 0.1357 *** | 0.0751 | 0.3069 * | 0.0872 |
| **DebtE** | 1.4218 * | 0.4364 | 0.3590 ** | 0.1450 | 0.7047 * | 0.2148 |
| **DebtA** | −1.5074 * | 0.4473 | −0.1524 | 0.1470 | −0.6186 * | 0.2204 |
| **Audit** | 1.7479 * | 0.4013 | 0.6490 * | 0.2499 | 0.9450 * | 0.1938 |
| **_cons** | 1.8618 ** | 0.9113 | 1.7397 * | 0.4320 | 0.2075 | 0.4506 |
| **i.Year** | No[(1)] | | Yes | | Yes | |
| *N obs/N groups* | 548/85 | | 548/85 | | 548/85 | |
| | $chi^2$ (8) = 60.36 * | | $chi^2$ (14) = 35.24 * | | $chi^2$ (14) = 64.04 * | |
| *Hausman chi²* | 26.94 ** | | 33.06 * | | 31.37 * | |
| *AR(1)* | 0.330 | | rho=0.835 | | 0.371 | |
| *Wald chi²* | $1.4 \times 10^8$ * | | $3.8 \times 10^5$ * | | $3.9\times 10^5$ * | |
| *LM chi²* | 953.52 * | | 947.92 * | | 925.01 * | |

*, **, *** represent levels of significance (two-tailed) at 1%, 5% and 10%, respectively.

## Appendix E. Logistic Regressions: the Influence of Accounting Regulations on the Disclosure of Accounting Approaches Statistic Models: the Influence of Accounting Regulations on the Indexes of the Disclosure

See Table 2 for variable definitions.

This table shows the results for Equation (2), described in Section 2.3.3, through Firth's (1993) Penalized Maximum Likelihood Estimation. *, **, *** represent levels of significance of 1%, 5% and 10%, respectively.

| | (1) IFRIC3_cost | | (2) GovGrant | | (3) NetLiab | | (4) Inventory | | (5) ICAC | | (6) AFRS26 | | (7) NRec | | (8) NDiv | |
|---|---|---|---|---|---|---|---|---|---|---|---|---|---|---|---|---|
| | *Coef.* | *Std. Err.* | *Coef.* | *Std. Err.* | *Coef.* | *Std. Err.* | *Coef.* | *Std. Err.* | *Coef.* | *Std. Err.* | *Coef.* | *Std. Err.* | *Coef.* | *Std. Err.* | *Coef.* | *Std. Err.* |
| **BoPFS.IFRS** | −0.8978 | 2.8800 | 5.3734 * | 1.5752 | 3.0022 ** | 1.4854 | 4.9124 ** | 2.2842 | 0.4792 | 0.4764 | −8.4853 * | 2.8815 | 2.5282 * | 0.8312 | 0.7992 ** | 0.3409 |
| **Size** | −2.8876 | 1.9260 | 2.6690 * | 0.4795 | −0.5220 | 0.3442 | −4.8690 ** | 2.1321 | 0.0126 | 0.3311 | 1.6844 | 1.2739 | −1.0232 ** | 0.4004 | −0.0830 | 0.2254 |
| **ROA** | −1.8704 | 1.5785 | 0.3796 | 0.4705 | −0.5879 | 0.5307 | 1.1698 | 1.9157 | 0.5006 | 0.5279 | −0.9742 | 0.9509 | 1.3884 ** | 0.6107 | −0.1607 | 0.3495 |
| **ROE** | 0.9363 | 1.6338 | 1.2157 ** | 0.5012 | 0.5529 | 0.4257 | −1.1375 | 1.9803 | −0.1213 | 0.4985 | −1.2298 | 0.8326 | −0.7003 | 0.6201 | −0.5054 | 0.3426 |
| **ROS** | 3.6618 ** | 1.6906 | 0.6545 *** | 0.3958 | −0.9763 ** | 0.4223 | 0.4112 | 1.0079 | 0.3548 | 0.3427 | 1.0470 | 1.0798 | −1.0024 ** | 0.4429 | −0.5643 ** | 0.2240 |
| **DebtE** | 1.0276 | 1.8870 | 0.2645 | 0.6306 | 1.3149 ** | 0.6443 | −1.1202 | 2.8090 | 0.4784 | 0.8797 | −2.2846 | 1.6745 | 2.5496 *** | 1.3017 | −1.5548 * | 0.4578 |
| **DebtA** | 0.0293 | 2.5682 | 0.1770 | 0.6407 | −0.3339 | 0.6669 | −1.7385 | 3.9064 | −0.9422 | 0.8947 | 4.0748 ** | 2.0740 | −1.8880 | 1.2806 | 0.6681 | 0.4773 |
| **Audit** | 0.1271 | 2.4471 | −2.2061 * | 0.5353 | 2.5268 *** | 1.4688 | −1.3267 | 2.0880 | −0.0725 | 0.4990 | −0.0733 | 1.1126 | 3.2177 ** | 1.6331 | −2.2944 * | 0.3548 |
| **Aviation** | −4.7746 | 3.4028 | −6.4056 * | 2.0933 | 1.3466 | 1.9044 | 5.0777 | 5.1772 | 2.0857 | 2.1055 | −7.1057 *** | 4.2955 | 3.3988 | 2.3195 | 5.8969 * | 1.7164 |
| **Cement** | −4.4409 | 4.4055 | −7.8923 * | 1.6211 | −1.7274 | 1.6184 | 3.0690 | 2.9951 | −0.1952 | 2.0264 | −1.4140 | 2.4352 | −0.2303 | 2.0437 | −0.6079 | 2.0258 |
| **Combustion** | −4.2134 | 3.2466 | −6.8711 * | 0.9024 | 1.0934 | 0.7474 | 2.1054 | 2.1110 | 2.2567 | 1.4638 | −12.8288 ** | 5.4975 | −1.2894 | 2.0162 | 2.9376 ** | 1.4486 |
| **PulpPaper** | −5.4676 *** | 3.3006 | −2.9223 * | 0.6310 | −2.4641 | 1.6056 | −0.8796 | 2.3527 | −0.9251 | 2.0180 | −3.3876 | 2.4631 | 3.0432 ** | 1.4676 | 2.1672 | 1.4697 |
| **P.Electricity** | −5.0719 *** | 2.8624 | −2.1843 * | 0.8293 | −2.3606 | 1.6142 | 1.7172 | 2.9785 | 0.0393 | 2.0513 | −4.8502 | 3.2045 | 2.0424 | 1.5550 | 2.7249 *** | 1.5182 |
| **Oil Ref.** | 2.3537 | 2.2461 | −3.7216 ** | 1.5504 | 0.8134 | 1.6637 | 4.2271 | 3.1434 | 1.6912 | 2.0525 | 0.9334 | 2.0857 | 1.9094 | 2.1255 | 1.4649 | 2.0486 |
| **Steel** | 2.1221 | 2.3374 | −9.0969 * | 1.7550 | 1.0417 | 0.8914 | 3.5759 | 3.2872 | 0.0125 | 2.0271 | −1.8213 | 2.4410 | 0.3770 | 2.0392 | 0.2953 | 2.0248 |
| **Glass** | – | – | – | – | – | – | – | – | – | – | – | – | – | – | – | – |
| **cons** | −5.9999 | 5.5174 | −11.7422 * | 2.6302 | −7.3692 * | 2.6402 | 2.2574 | 5.0230 | −5.3300 ** | 2.2466 | −1.0186 | 3.9577 | −8.8620 * | 2.9971 | 1.1811 | 1.7927 |
| **i.Ano** | Não | | Não | | Não | | Não | | Não | | Não | | Não | | Não | |
| **N** | 548 | | 548 | | 548 | | 548 | | 548 | | 548 | | 548 | | 548 | |
| **chi$^2$ (15)** | 17.53 | | 17.44 * | | 37.55 * | | 12.46 | | 24.14 *** | | 31.40 * | | 40.44 * | | 77.50 * | |
| **Penalized Log likelihood** | −9.9687 | | −69.9257 | | −60.2749 | | −9.8626 | | −69.0688 | | −28.437 | | −42.937 | | −136.3877 | |

## References


1. Wambsganss, J.R.; Sanford, B. The Problem with Reporting Pollution Allowances. *Crit. Perspect. Account.* **1996**, *7*, 643–652. [CrossRef]
2. Gibson, K. The Problem with Reporting Pollution Allowances: Reporting Is Not the Problem. *Crit. Perspect. Account.* **1996**, *7*, 655–665. [CrossRef]
3. PWC & IETA. Trouble-Entry Accounting-Revisited, 2007, 1–48. Available online: https://www.ieta.org/resources/Resources/Reports/trouble_entry_accounting.pdf (accessed on 28 January 2020).
4. Ragan, J.M.; Stagliano, A.J. Cap and Trade Allowance Accounting: A Divergence Between Theory and Practice. *Int. J. Econ. Bus. Res.* **2007**, *5*, 47–58. [CrossRef]
5. Bebbington, J.; Larrinaga-González, C. Carbon trading: Accounting and reporting issues. *Eur. Account. Rev.* **2008**, *17*, 697–717. [CrossRef]
6. KPMG. *Accounting for Carbon: The Impact of Carbon Trading on Financial Statements*. 2008, p. 22. Available online: http://content.ccrasa.com/library_1/3201%20-%20Accounting%20for%20carbon%20The%20impact%20of%20carbon%20trading%20on%20financial%20statements.pdf (accessed on 28 January 2020).
7. Cook, A. Emission rights: From costless activity to market operations. *Account. Organ. Soc.* **2009**, *34*, 456–468. [CrossRef]
8. Giner, B. Accounting for Emission Trading Schemes: A Still Open Debate. *Account. Organ. Soc.* **2014**, *34*, 45–51. [CrossRef]
9. Allini, A.; Giner, B.; Caldarelli, A. Opening the black box of accounting for greenhouse gas emissions: The different views of institutional bodies and firms. *J. Clean. Prod.* **2018**, *172*, 2195–2205. [CrossRef]
10. Inchausti, B.G. La contabilidad de los derechos de emisión: Una perspectiva internacional. *Revista Espanola de Financiacion y Contabilidad* **2007**, *36*, 175–193. [CrossRef]
11. Lovell, H.; Aguiar, T.S.; de Bebbington, J.; Larrinaga-González, C. Monitoring, Accounting for Carbon: Reporting and Verifying Emissions in the Climate Economy. In *ACCA Research Report*; The Association of Certified Chartered Accountants (ACCA): London, UK, 2010. [CrossRef]
12. Steenkamp, N.; Rahman, A.; Kashyap, V. Recognition, measurement and disclosure of carbon emission allowances under the EU ETS—An exploratory study. In Proceedings of the 10th CSEAR Australasian Conference, Launceston Campus, University of Tasmania, Hobart, Australia, 5–7 December 2011.
13. Warwick, P.; Ng, C. The "Cost" of Climate Change: How Carbon Emissions Allowances are Accounted for Amongst European Union Companies. *Aust. Account. Rev.* **2012**, *22*, 54–67. [CrossRef]
14. Black, C. Accounting for Carbon Emission Allowances in the European Union: In Search of Consistency. *Account. Eur.* **2013**, *10*, 223–239. [CrossRef]
15. Ayaz, H. Analysis of Carbon Emission Accounting Practices of Leading Carbon Emitting European Union Companies. *Athens J. Bus. Econ.* **2017**, *3*, 463–486. [CrossRef]
16. Ernst & Young. *Carbon Market Readiness*. 2010. Available online: http://globalsustain.org/files/Carbon_market_readiness.pdf. (accessed on 28 January 2020).
17. Barth, M.E.; Schipper, K. Financial Reporting Transparency. *J. Account. Audit. Finance* **2008**, *23*, 173–190. [CrossRef]
18. IFRS–Home. Available online: ifrs.org. (accessed on 11 January 2020).
19. European Commission, Union Registry. Available online: ec.europa.eu/clima/policies/ets/registry_en#tab-0-1. (accessed on 26 June 2019).
20. European Commission, Market Stability Reserve. Available online: ec.europa.eu/clima/policies/ets/reform_en. (accessed on 26 June 2019).
21. Larrinaga, C. Carbon accounting and carbon governance. *Soc. Environ. Account. J.* **2014**, *34*, 1–5. [CrossRef]
22. Lovell, H.; Bebbington, J.; Larrinaga, C.; Sales de Aguiar, T.R. Putting carbon markets into practice: A case study of financial accounting in Europe. *Environ. Planning C: Govern. Policy* **2013**, *31*, 741–757. [CrossRef]
23. Ernst & Young. Accounting for Emission Reductions and Other Incentive Schemes. 2009, p. 12. Available online: https://www.ey.com/Publication/vwLUAssets/Accounting_for_emission_reductions_and_other_incentive_schemes/%24FILE/Accounting_emission_reductions_July09.pdf. (accessed on 28 January 2020).
24. Busch, T.; Hoffmann, V.H. How Hot Is Your Bottom Line? Linking Carbon and Financial Performance. *Bus. Soc.* **2011**, *50*, 233–265. [CrossRef]

25. Lovell, H.; Mackenzie, D. Accounting for Carbon: The Role of Accounting Professional Organisations in Governing Climate Change. *Antipode* **2011**, *43*, 704–730. [CrossRef]
26. IFRS–Research Programme. Available online: ifrs.org/projects/work-plan/research-programme/#pipeline. (accessed on 28 January 2020).
27. Criado-Jiménez, I.; Fernández-Chulián, M.; Husillos-Carqués, F.J.; Larrinaga-González, C. Compliance with mandatory environmental reporting in financial statements: The case of Spain (2001–2003). *J. Bus. Ethics* **2008**, *79*, 245–262. [CrossRef]
28. Holthausen, R.W. Accounting Standards, Financial Reporting Outcomes and Enforcement. *J. Account. Res.* **2009**, *47*, 447–458. [CrossRef]
29. Barbu, E.M.; Dumontier, P.; Feleagă, N.; Feleagă, L. Mandatory environmental disclosures by companies complying with IASs/IFRSs: The cases of France, Germany, and the UK. *Inter. J. Account.* **2014**, *49*, 231–247. [CrossRef]
30. Cowan, S.; Gadenne, D. Australian corporate environmental reporting: A comparative analysis of disclosure practices across voluntary and mandatory disclosure systems. *J. Account. Organ. Change* **2005**, *1*, 165–179. [CrossRef]
31. Llena, F.; Moneva, J.M.; Hernandez, B. Environmental disclosures and compulsory accounting standards: The case of Spanish annual reports. *Bus. Strategy Environ.* **2007**, *16*, 50–63. [CrossRef]
32. Deegan, C.; Gordon, B. A Study of the Environmental Disclosure Practices of Australian Corporations. *Account. Bus. Res.* **1996**, *26*, 187–199. [CrossRef]
33. Larrinaga, C.; Carrasco, F.; Correa, C.; Llena, F.; Moneva, J. Accountability and accounting regulation: The case of the Spanish environmental disclosure standard. *Eur. Account. Rev.* **2002**, *11*, 723–740. [CrossRef]
34. Adams, C.A.; Coutts, A.; Harte, G. Corporate Equal Opportunities (Non-) Disclosure. *Br. Account. Rev.* **1995**, *27*, 87–108. [CrossRef]
35. Gallego-Álvarez, I.; Martínez-Ferrero, J.; Cuadrado-Ballesteros, B. Accounting Treatment for Carbon Emission Rights. *Systems* **2016**, *4*, 12. [CrossRef]
36. European Commission, Free Allocation. Available online: ec.europa.eu/clima/policies/ets/allowances_en. (accessed on 9 July 2019).
37. Matsumura, E.M.; Prakash, R.; Vera-Muñoz, S.C. Firm-value effects of carbon emissions and carbon disclosures. *Account. Rev.* **2014**, *89*, 695–724. [CrossRef]
38. Elfrink, J.; Ellison, M. Accounting for Emission Allowances: An Issue in Need of Standards. *CPA J.* **2009**, *79*, 30. Available online: https://www.journalofaccountancy.com/issues/2009/jul/20081312.html. (accessed on 28 January 2020).
39. Monteiro, S.; Vilas Boas, S. O Relato sobre Licenças de Emissão de CO2: O Caso das Empresas Portuguesas do PNALE I. In XV Encuentro AECA, 2012. Available online: http://www.aeca1.org/xvencuentroaeca/cd/72h.pdf. (accessed on 28 January 2020).
40. Giner, B. The influence of company characteristics and accounting regulation on information disclosed by Spanish firms. *Eur. Account. Rev.* **1997**, *6*, 45–68. [CrossRef]
41. Mateos-Ansótegui, A.I.; Bilbao-Estrada, I. Accounting Planning and Greenhouse Gas Emission Rights. *Revista Universo Contábil* **2007**, *3*, 101–122.
42. Fornaro, J.M.; Winkelman, K.A.; Glodstein, D. Accounting for Emissions—Emerging issues and the need for global accounting standards. *J. Account. July* **2009**, 1–7. Available online: https://search.proquest.com/openview/bad6d41e993f45c93972168f8f822c37/1?pq-origsite=gscholar&cbl=41065 (accessed on 28 January 2020).
43. Clarkson, P.M.; Li, Y.; Richardson, G.D.; Vasvari, F.P. Revisiting the relation between environmental performance and environmental disclosure: An empirical analysis. *Account. Organ. Soc.* **2008**, *33*, 303–327. [CrossRef]
44. Prado-Lorenzo, J.-M.; Rodríguez-Domínguez, L.; Gallego-Álvarez, I.; García-Sánchez, I.-M. Factors influencing the disclosure of greenhouse gas emissions in companies world-wide. *Manag. Decis.* **2009**, *47*, 1133–1157. [CrossRef]
45. Luo, L.; Lan, Y.C.; Tang, Q. Corporate Incentives to Disclose Carbon Information: Evidence from the CDP Global 500 Report. *J. Int. Financial Manag. Account.* **2012**, *23*, 93–120. [CrossRef]
46. Qiu, Y.; Shaukat, A.; Tharyan, R. Environmental and social disclosures: Link with corporate financial performance. *Br. Account. Rev.* **2016**, *48*, 102–116. [CrossRef]

47. Liu, Y.S.; Zhou, X.; Yang, J.H.; Hoepner, A.G.F. Corporate Carbon Emission and Financial Performance: Does Carbon Disclosure Mediate the Relationship in the UK? Disccusion paper ICM-2016-03, junio 2016. Available online: https://pdfs.semanticscholar.org/217e/7771283088f61d895e6225663e175c371b2f.pdf (accessed on 28 January 2020).
48. Dechow, P.; Ge, W.; Schrand, C. Understanding earnings quality: A review of the proxies, their determinants and their consequences. *J. Account. Econ.* **2010**, *50*, 344–401. [CrossRef]
49. Labra, R.; Torrecillas, C. Guía Cero para datos de panel. Un enfoque práctico. UAM-Accenture Working Papers. 2014; 16, 57.
50. Pindado, J.; Requejo, I. 10th Course: Data Management and Panel Data Models, Salamanca. 11th to 15th of July 2016. Universidad de Salamanca, 2016. Available online: https://fundacion.usal.es/es/images/stories/pdf/Syllabus_Data_Management_and_Panel_Data_Models_2016.pdf (accessed on 22 January 2020).
51. Lee, Y. Diagnostic Testing for Dynamic Panel Data Models, 2006. Available online: http://citeseerx.ist.psu.edu/viewdoc/download?doi=10.1.1.466.8012&rep=rep1&type=pdf. (accessed on 28 January 2020).
52. Moundigbaye, M.; Rea, W.S.; Reed, W.R. Which Panel Data Estimator Should I Use? A Corrigendum and Extension. *Economics* **2018**, *12*, 1–31. [CrossRef]
53. Verbeek, M. *A Guide to Modern Econometrics*, 5th ed.; Wiley: Hoboken, NJ, USA, 2017.
54. Firth, D. Bias Reduction of Maximum Likelihood Estimates Author. *Biometrika* **1993**, *80*, 27–38. Available online: http://www.jstor.org/stable/2336755 (accessed on 28 January 2020). [CrossRef]
55. Moneva, J.M.; Llena, F. Environmental disclosures in the annual reports of large companies in Spain. *Eur. Account. Rev.* **2000**, *9*, 7–29. [CrossRef]

International Journal of
*Environmental Research and Public Health*

*Article*

# Wave Energy Assessment at Valencia Gulf and Comparison of Energy Production of Most Suitable Wave Energy Converters

**Raúl Cascajo [1], Emilio García [2], Eduardo Quiles [2,*], Francisco Morant [2] and Antonio Correcher [2]**

[1] Área de Ingeniería Naval y Oceánica, Universidad Politécnica de Madrid, 28040 Madrid, Spain; rcascajo@gmail.com

[2] Instituto de Automática e Informática Industrial, Universitat Politècnica de València, 46022 Valencia, Spain; egarciam@isa.upv.es (E.G.); fmorant@isa.upv.es (F.M.); ancorsal@ai2.upv.es (A.C.)

* Correspondence: equiles@isa.upv.es; Tel.: +34-96-387-7007

Received: 24 September 2020; Accepted: 13 November 2020; Published: 16 November 2020

**Abstract:** Seaports' energy strategy should rely on the use of renewable energy. Presently, the share of renewable energy used by many of the ports worldwide is negligible. Some initiatives are in the process of implementation to produce some of the energy used by the Port of Valencia, one the largest ports in the Mediterranean Basin. Among these initiatives, a photovoltaic plant with an installed capacity of 5.5 MW is under a tendering process and the assessment studies for the deployment of three to five windmills are close to being finished. However, this is not enough to make it a "zero emissions port" as some of the energy demand would still be covered by fossil fuels. Therefore, we should consider clean alternative energy sources. This article analyses the wave energy resources in the surroundings of the Port of Valencia using a 7-year series of data obtained from numerical modelling (forecast). The spatial distribution of wave power is analysed using data from 3 SIMAR points at Valencia Bay and is compared to the data obtained by the Valencia Buoy I (removed in 2005). The obtained results are used to estimate the power matrices and the average energy output of two wave energy converters suitable to be integrated into the port's infrastructure. Finally, the wave energy converters' production is compared to the average amount of energy that is forecast to be obtained from other renewable sources such as solar and wind. Due to the nature of the Gulf's wave climate (mostly low waves), the main conclusion is that the energy obtainable from the waves in the Valencia Gulf will be in correlation with such climate. However, when dealing with great energy consumers every source of production is worthwhile and further research is needed to optimize the production of energy from renewable sources and its use in an industrial environment such as ports.

**Keywords:** wave energy; energy production; renewable energy; zero emissions port; wave energy converter

## 1. Introduction

Some of the major seaports worldwide are developing energy policies aimed at the reduction of their greenhouse gas emissions. These policies consist of the reduction of the ports' dependency from fossil fuels and the generation of energy from renewable sources, such as photovoltaic, wind, tidal or waves.

The level of maturity of wind and solar energy technologies makes these two energy sources the most feasible for installation in ports. However, if we consider solar energy, it must be taken into account that most of the port areas are very busy and subjected to the action of meteorological elements.

Wave energy seems to be one of the most promising alternatives to fossil fuels, the use of wave energy is estimated to increase significantly over the next few decades as wave energy converter (WEC) technology matures [1]. Moreover, according to a study by the European Technology and Innovation Platform for Ocean Energy, which was published in May 2020: "Europe's ocean energy resource is considerable. By 2050, ocean energy can deliver 100GW of capacity- equivalent to 10% of Europe's electricity consumption today. Flexible and predictable, ocean energy complements variable renewables such as wind or solar, that will dominate Europe's electricity system in 2050. Ocean energy will play an important role in smoothing production peaks and balancing Europe's electricity grid. By 2050, the ocean energy sector will employ 400,000 Europeans, ensuring a just transition to a decarbonized economy. Europe's technological advantage in ocean energy will ensure European companies a large share of a strong global market, as they do on offshore wind. With zero carbon emissions, ocean energy will help tackle climate change and achieve a cleaner, more sustainable and more prosperous Europe."

While ocean wave energy conversion is still unproven in a commercial scale, significant advances in research, design and testing continue to be made [2] and a number of wave energy projects, almost all still in a pilot stage, are currently underway around the world [3,4].

According to some estimates [5], the potential of wave energy in the region of Valencia is 3.64 TWh/year gross or 0.58 TWh/year net.

Some WECs can be integrated into port infrastructure [6], thus obtaining significant savings from the power take off systems (PTO) as well as installation, operation and maintenance costs, while providing erosion protection to the marine structure. This means that wave energy could be part of the energy mix aiming to cover most of the energy needs of the port [7].

Research on wave energy resource assessment has been carried out in countries where great wave energy potential is expected [1]. Wave energy is particularly interesting for islands, which are usually highly dependent on external energy supply. Taking this into account, one could say that the ideal place for the exploitation of wave energy potential are islands located in large oceans.

Furthermore, ports need great amounts of energy from the shore and the fact that they have provided energy supplying infrastructure allows us to consider them as "energy islands", making wave energy a good candidate to cover part of the ports' energy needs.

Some ports worldwide have already integrated renewable energy generators in their facilities. Rotterdam, Antwerp, Amsterdam [8–10], amongst others, are examples of the integration of wind energy in ports, while Naples, Gibraltar, Mutriku and Civitavecchia [11,12], are examples of locations that have integrated wave energy converters into their infrastructure. At present, only at the location of the Port of Civitavecchia is the system not running.

In relation to the Port of Valencia there are studies being developed by the Port Authority of Valencia (PAV) on the estimated energy production from renewable sources. The preliminary estimations provided a result of 8 GWh/year considering a total of 12,650 photovoltaic panels of 435 W each (size of each panel is around 2.1 square meters in size, resulting in overall space of 26,565 square meters required for the solar power station) on the roof of a new vehicle storage silo, covering almost 10% of the total consumption of the port of Valencia. In the case of the wind, the result obtained for this resource in the studied area is 5.19 m/s, with the predictions from the Spanish Energy Institute (IDAE) at the Valencia location being between 5.0 and 5.5 m/s at 100 m above the sea level [13]. Considering three wind turbines of 4.8 MW of nominal power, a total energy production of 26.63 GWh/y is estimated. This estimation would cover around 50% of the energy consumed by the Port of Valencia [7].

The aim of this paper is the evaluation of the wave energy resource at the Bay of Valencia, near the Port of Valencia. Once the wave energy resource has been estimated, two suitable WECs for ports are selected in order to estimate the energy production simulating their installation within the port's existing infrastructure and general comparison to the production to be achieved if given the same space, as required by 12,650 solar panels. With the results obtained, the share of energy demand covered by the WECs is estimated comparing it to that generated by other renewable energy sources.

Section 2 describes the study area. Section 3 shows the available data and the methodology used to quantify the wave resource. Section 4 estimates the average energy power (AEP) for two different WEC at the chosen locations nearby the Port of Valencia and Section 5 discusses the results obtained in Section 4. Finally, Section 6 presents the conclusions of the paper.

## 2. Data and Methodology

### 2.1. Study Area

The study area is located in the western Mediterranean Sea (39.45° N—39.40° N, 0.31° W—0.20° W) (Figure 1).

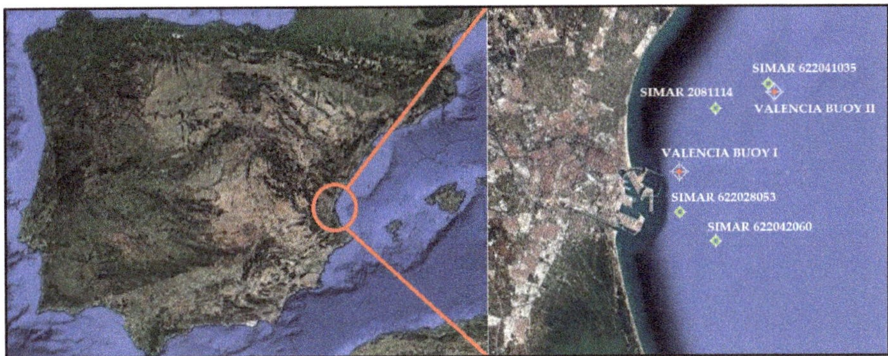


**Figure 1.** Location of the Port of Valencia in the western Mediterranean Sea (**left**). Location of the analysed buoys and points (**right**).

The Port of Valencia has a total surface of 5,626,534 $m^2$ [14]. The main cargo handled within the Port of Valencia is containerized, with over 5 million (Mio) TEU (Twenty feet Equivalent Units) handled in 2018. This means that the majority of the port's surface is dedicated to container handling. The Port of Valencia has three container terminals covering a total surface of over 2 million $m^2$. These three terminals together with a terminal for refrigerated cargo are the main electric energy consumers of the port, with over 80% of the consumption overall [15].

The climate in the Bay of Valencia is influenced by the semi-arid climate of mid-latitudes. This means that it is dominated by extra-tropical cyclones formed through baroclinic instability, which is at its highest during the winter season. In addition, it is affected by depressions in movement generated in the Atlantic Ocean or in north-western Europe [16]. According to [17], Mediterranean storms are usually shorter and less intense than those in northern Europe, with many sub-regional and mesoscale effects that produce great spatial and seasonal variability [18]. The reduced scale, together with the peculiar characteristics of the basin (the complex orography and the humidity of a relatively large body of water) makes the Mediterranean climate more difficult to predict than climates elsewhere [19]. During the summer season, thermal and orographic effects play a more important role in the genesis and maintenance of cyclones. In the north-western Mediterranean, the main cyclone centres are located again in the Gulf of Genoa and in the Iberian Peninsula, the latter caused by the contrasts of temperature between land and sea [18]. During the warm periods, the Mediterranean is also exposed to tropical systems [20] as a result of its location in a transition zone between humid mountains in the north and arid regions in the south. Finally, spring and autumn can be considered periods of transition between the contrasting patterns of winter and summer [20].

The Mediterranean Sea represents, in terms of wave energy power availability, an intermediate level between the open ocean and the enclosed small-fetch basins such as the Black Sea or the Baltic Sea.

Hence, wave energy exploitation seems to be promising even if the net quantities are not as significant as in the open ocean [21].

### 2.2. Available Wave Data

Wave Modelling (WAM) is a third generation model that integrates the basic transport equation that describes the evolution of a two-dimensional ocean wave spectrum without additional unplanned assumptions regarding spectral shape. There are three explicit source functions that describe wind input, non-linear transfer and white-capping dissipation. There is an additional source function for background dissipation and the refraction terms are included in the finite depth version of the model. The model runs in a spherical grid of latitude and longitude and can be used in any ocean region.

WAM predicts the directional spectra along with wave properties such as significant wave height, wave direction and mean wave frequency, wave height and mean wave direction, and corrected wind stress fields including wave-induced stress and drag coefficient at each point in the grid at the chosen output times.

WAM can be coupled to a range of other models. Examples include the Southeast Asian Ocean Model (SEAOM), the Coastal Ocean Modelling System of the Proudman Oceanographic Laboratory (POLCOMS), the Core Model for European Ocean Modelling (NEMO), the High-Resolution Limited Area Model (HIRLAM), and the Regional Climate Model (RegCM) of the National Centre for Atmospheric Research (NCAR) [22].

The Spanish Ports Authority (Puertos del Estado) has developed and implemented a two-way nesting procedure in the model for the Spanish coast. Using this system, the equation is integrated at the same time for all points. Since it is possible to define the spacing depending on the grid point location, it works as a variable spacing schema. The resolution is enhanced using intermediate grids, which are placed between the coarse and the fine grids.

In the particular case of the Mediterranean domain, the shallow water version of the WAM model is used, therefore, refraction and attenuation effects are considered for those (few) grid points located in shallow waters [23].

This model produces the wave directional spectra for each grid point. Then, it is used to obtain further information, i.e., Hs (significant wave height), Tp (peak wave period), Tm (mean wave period), mean direction, wind sea and swell components, etc.

The WAM model obtained information that is made available for the Bay of Valencia from the wave climate database stored by Puertos del Estado. This database contains data of 7-year hindcast wave climate using the WAM model and forced by the wind output of the HIRLAM regional atmospheric model. Data are taken from three points within the Bay of Valencia. However, a comparison between the data given by these points, obtained from a numerical simulation, with the real data obtained by the Buoy of Valencia II (0.20° W—39.51° N) from the same time period must be made in order to validate the data given by the WAM model.

### 2.3. Methodology

First of all, the validation of the SIMAR data must be granted. This validation lies in the work of Puertos del Estado [24,25]. In addition, the similarity of the data can be shown by looking at the Figure 2 below that is from a period of time in which both systems were working (2012–2013). This Figure 2 shows the wave rose according to both the Valencia buoy II and the SIMAR 622041035.

The period of 2012–2013 has been selected because it is the only period in which data from both the Valencia Buoy II and the SIMAR 622041035 are recorded in the Puertos del Estado database. After 2013 the Valencia Buoy II was dismantled.

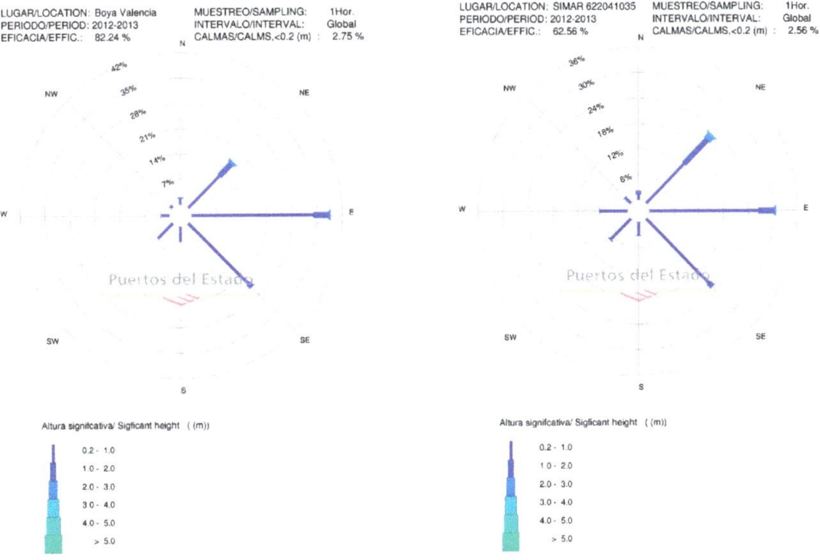


**Figure 2.** Wave rose: Valencia Buoy II (**left**). SIMAR 622041035 (**right**). Period 2012–2013.

As a preliminary conclusion, the data from both the buoy and the SIMAR 622041035 are alike and not much difference is noted between them. From now onwards, the available wave resource will be estimated by using the SIMAR points when we refer to the area around the Port of Valencia, as there are no available data from any buoys.

Both the Valencia Buoy and the SIMAR 6220241035 show a predominant wave direction from the E, together with other contributions from the NE and SE.

Considering Tables 1 and 2, it is deduced that for the Valencia Buoy II, 86% of the waves have a Hs of 1 m or less while 85% of them have a Tp of 7 s or less. In the case of the SIMAR, there is the same range of Hs and Tp, 86% and 76% of frequency are obtained, respectively. These results represent no deviation for the Hs and a 9% for the Tp, which could be acceptable according to [25–28].

**Table 1.** Hs (significant wave height) vs. Tp (peak wave period) table Valencia Buoy II.

| Year 2005–2013 | | Tp (s) | | | | | | | | | | | |
|---|---|---|---|---|---|---|---|---|---|---|---|---|---|
| | | **≤1.0** | **2.0** | **3.0** | **4.0** | **5.0** | **6.0** | **7.0** | **8.0** | **9.0** | **10.0** | **>10.0** | **TOTAL** |
| Hs (m) | ≤0.5 | – | — | 3.870 | 8.883 | 9.467 | 8.688 | 6.514 | 2.029 | 0.267 | 0.160 | 0.066 | 39.946 |
| | 1.0 | — | — | 3.747 | 8.944 | 12.058 | 7.647 | 6.834 | 5.199 | 1.171 | 0.237 | 0.015 | 45.854 |
| | 1.5 | — | — | — | 0.360 | 1.940 | 2.158 | 2.040 | 1.880 | 0.911 | 0.533 | 0.047 | 9.869 |
| | 2.0 | | | | 0.002 | 0.103 | 0.481 | 0.697 | 0.694 | 0.360 | 0.405 | 0.059 | 2.798 |
| | 2.5 | — | — | — | — | 0.003 | 0.056 | 0.246 | 0.302 | 0.160 | 0.142 | 0.042 | 0.952 |
| | 3.0 | — | — | — | — | — | 0.002 | 0.033 | 0.145 | 0.053 | 0.070 | 0.035 | 0.337 |
| | 3.5 | — | — | — | — | — | — | 0.012 | 0.098 | 0.035 | 0.023 | 0.006 | 0.174 |
| | 4.0 | — | — | — | — | — | — | — | 0.029 | 0.017 | 0.011 | 0.003 | 0.059 |
| | 4.5 | — | — | — | — | — | — | — | — | 0.003 | 0.005 | 0.005 | 0.012 |
| | 5.0 | — | — | — | — | — | — | — | — | — | — | — | 0.000 |
| | >5.0 | — | — | — | — | — | — | — | — | — | — | — | 0.000 |
| | TOTAL | — | — | 7.617 | 18.188 | 23.571 | 19.031 | 16.377 | 10.376 | 2.977 | 1.585 | 0.278 | 100% |

**Table 2.** Hs vs. Tp Table SIMAR 622041035.

| Year 2005–2013 | | Tp (s) | | | | | | | | | | | |
|---|---|---|---|---|---|---|---|---|---|---|---|---|---|
| | | **≤1.0** | **2.0** | **3.0** | **4.0** | **5.0** | **6.0** | **7.0** | **8.0** | **9.0** | **10.0** | **>10.0** | **TOTAL** |
| Hs (m) | ≤0.5 | — | 4.931 | 9.735 | 10.245 | 6.878 | 7.461 | 6.405 | 1.710 | 1.137 | 0.801 | 0.391 | 49.695 |
| | 1.0 | — | — | 1.456 | 5.350 | 6.296 | 5.714 | 7.679 | 4.176 | 2.184 | 1.410 | 2.156 | 36.421 |
| | 1.5 | — | — | — | 0.127 | 1.065 | 1.092 | 1.219 | 1.337 | 1.638 | 0.901 | 1.747 | 9.126 |
| | 2.0 | — | — | — | — | 0.064 | 0.209 | 0.191 | 0.209 | 0.582 | 0.464 | 1.028 | 2.748 |
| | 2.5 | — | — | — | — | — | 0.009 | 0.146 | 0.064 | 0.155 | 0.246 | 0.355 | 0.974 |
| | 3.0 | — | — | — | — | — | — | 0.082 | 0.027 | 0.191 | 0.200 | 0.045 | 0.546 |
| | 3.5 | — | — | — | — | — | — | 0.018 | 0.073 | 0.082 | 0.100 | 0.100 | 0.373 |
| | 4.0 | — | — | — | — | — | — | — | — | 0.009 | — | 0.073 | 0.082 |
| | 4.5 | — | — | — | — | — | — | — | — | — | — | 0.036 | 0.036 |
| | 5.0 | — | — | — | — | — | — | — | — | — | — | — | 0.000 |
| | >5.0 | — | — | — | — | — | — | — | — | — | — | — | 0.000 |
| | TOTAL | — | 4.931 | 11.191 | 15.722 | 14.303 | 14.485 | 15.740 | 7.597 | 5.978 | 4.122 | 5.932 | 100% |

Once the origin of the data is defined, using the following equations, the power resource can be estimated.

All points considered are located in deep water (50 m), therefore the propagation processes such as refraction and diffraction can be neglected. Wave power estimation can be calculated with the following deep-water equation:

$$P = \frac{\rho g^2}{64\pi} Hs^2 Te \cong 0.491 Hs^2 Te \tag{1}$$

where $P$ is the wave power per unit of crest length (kW/m), $Hs$ is the significant wave height, $Te$ is the energy period, $\rho$ is the density of seawater (assumed to be 1025 kg/m$^3$) and $g$ is the gravitational acceleration.

As pointed out in [29], one approach when $Tp$ is known is to assume the following:

$$Te = \alpha Tp \tag{2}$$

A value of $Te = 0.9\ Tp$ [30] was used to assess the wave energy resource, estimating $P$ using Equation (2).

$$P \cong 0.442 Hs^2 Tp \tag{3}$$

## 3. Analysis of Wave Energy Resource

The aim of this research is, apart from estimating the available wave resource in the study area, to propose different alternatives of marine energy generators that can meet the infrastructural requirements of a port such as the Port of Valencia. Special focus is made on those able to produce energy from waves. In this case, it is considered that the data showing the state of the sea of the SIMAR 622028053 provides more information about the actual sea state at the Port of Valencia than the data given by the other possible SIMAR points, as its location is a better match with the potential location of the WECs. Therefore, from now on the data obtained from this SIMAR 622028053 will be considered for the purpose of this study.

Using the previous Equation (2), and the data given by the Puertos del Estado database [24], for the SIMAR 622042060, SIMAR 2081113 and SIMAR 622028053 during the period of 2012–2019, the results shown in Table 3 can be obtained.

**Table 3.** Average values of Tp and Hs and estimation of the Pw for the SIMAR points around the Port of Valencia.

| Data Source | Tpav (s) | Hsav (m) | Pw (kW/m) |
|---|---|---|---|
| SIMAR 622042060 | 5.94 | 0.93 | 5.67 |
| SIMAR 2081113 | 6.85 | 0.85 | 5.12 |
| SIMAR 622028053 | 5.96 | 0.84 | 4.99 |
| Valencia buoy I (*) | 5.98 | 0.82 | 4.85 |
| Valencia buoy II (**) | 5.58 | 1.10 | 5.51 |

(*) Valencia buoy I was working during the 1985–2005 period. (**) Valencia buoy II was working during the 2005–2013 period.

Where Tpav, Hsav and Pw are the average values for Tp, Hs and Pw for the 2012–2019 period at the selected SIMAR points and buoys and Pw the wave power.

The seasonal wave power fluctuation at the Valencian Bay will now be analysed. With the purpose of having more detailed information about the variability of the resource, the seasonal variability (SV) index is introduced and defined as the difference of the most energetic season minus the lesser one divided by the yearly average value evaluated throughout the whole data set [31]:

$$SV = \frac{P_{Smax} - P_{Smin}}{P_{Year}} \tag{4}$$

where $P_{Smax}$ is the mean wave power for the highest-energy season and $P_{Smin}$ is the mean wave power for the lowest energy season, and $P_{year}$ is the annual mean wave power. The greater the value of $SV$ the larger the seasonal variability, with values lower than 1 indicating moderate seasonal variability.

We will now analyse the SV for the two boys close to the study area as there are no data available for the SIMAR 622028053.

In our case, considering the information from Puertos del Estado for the Valencia Buoy II, available at [24], $P_{Smax}$ (winter) = 6.62 kW/m and $P_{Smin}$ (autumn) = 5.79 kW/m, we obtain, thus, a result of $SV = 0.17$. However, if we take the same data from the Valencia Buoy I, $P_{Smax}$ (winter) = 7.04 kW/m and $P_{Smin}$ (spring) = 4.79 kW/m, we obtain, thus, a result of $SV = 0.46$. As we have considered for the purpose of this paper the data obtained from SIMAR 622028053 whose location is closer to the Valencia Buoy I, we should consider the seasonal variability of the latter instead.

The above results mean that there is very low seasonal variation for the wave resource in the study area, which would be a good starting point if the monthly variation of energy demand is considered at the Port of Valencia (Figure 3).

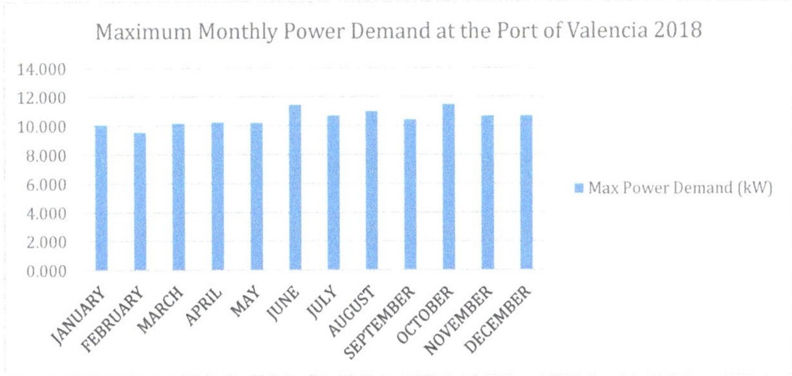


**Figure 3.** Maximum monthly power demand at the Port of Valencia in 2018 (Source: Port Authority of Valencia).

Only data from a two-year interval have been provided by the Puertos del Estado database. There are no data for longer than two years. The variation of Hs, Tp and direction for 2017–2019 period is shown in Figure 4, Figure 5, and Figure 6.

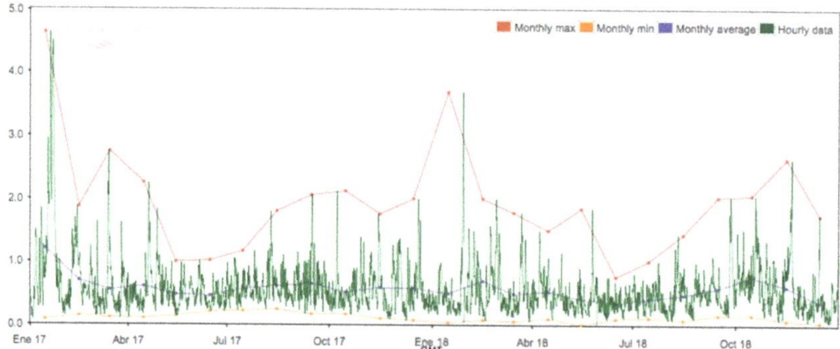


**Figure 4.** SIMAR 622028053 wave significant height 2017–2019.

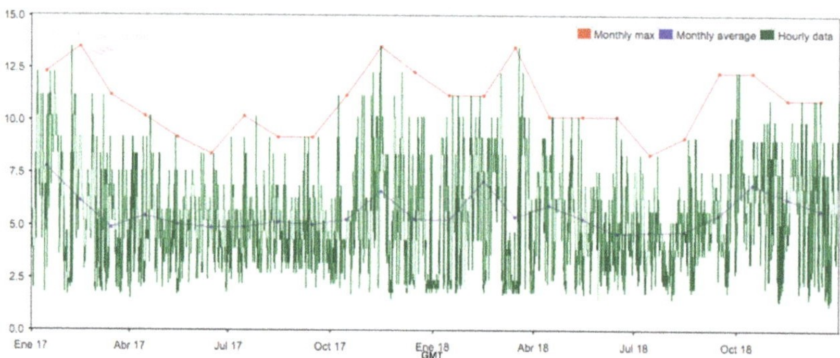


**Figure 5.** SIMAR 622028053 peak period 2017–2019.

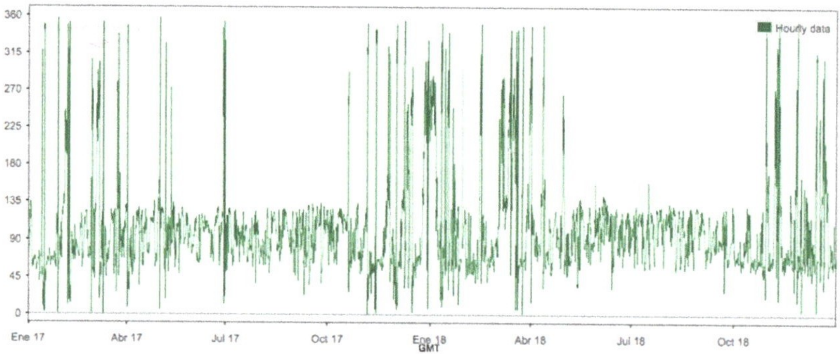


**Figure 6.** SIMAR 622028053 predominant wave direction 2017–2019.

Several studies have provided analysis of potential production and estimated performance of different WEC technologies [32–38].

For a sea state like that described in Table 4 the average energy production (AEP) could be estimated for two types of WEC suitable to be integrated into the port's breakwater [39,40]. The first system is the Eco Wave Power and the second is the Overtopping Breakwater (OTD) for Wave Energy

Conversion (OBREC). This election is based on the discussion shown in [7], as these two systems are easily embedded in the port's breakwaters.

**Table 4.** Hs vs. Tp Table SIMAR 622028053.

| Year 2012–2019 | | Tp (s) | | | | | | | | | | | |
|---|---|---|---|---|---|---|---|---|---|---|---|---|---|
| | | ≤1.0 | 2.0 | 3.0 | 4.0 | 5.0 | 6.0 | 7.0 | 8.0 | 9.0 | 10.0 | >10.0 | TOTAL |
| Hs (m) | ≤0.5 | — | 3.170 | 8.712 | 10.982 | 8.249 | 8.606 | 6.023 | 1.794 | 1.226 | 0.857 | 0.432 | 50.051 |
| | 1.0 | — | — | 1.397 | 5.184 | 6.962 | 6.641 | 8.340 | 3.504 | 2.420 | 2.159 | 1.441 | 38.048 |
| | 1.5 | — | — | — | 0.107 | 0.897 | 1.271 | 1.199 | 1.009 | 1.075 | 1.110 | 1.497 | 8.165 |
| | 2.0 | — | — | — | — | 0.030 | 0.243 | 0.320 | 0.184 | 0.316 | 0.434 | 0.740 | 2.267 |
| | 2.5 | — | — | — | — | — | 0.012 | 0.149 | 0.068 | 0.093 | 0.173 | 0.255 | 0.750 |
| | 3.0 | — | — | — | — | — | — | 0.024 | 0.072 | 0.094 | 0.084 | 0.061 | 0.335 |
| | 3.5 | — | — | — | — | — | — | 0.003 | 0.049 | 0.072 | 0.070 | 0.040 | 0.234 |
| | 4.0 | — | — | — | — | — | — | — | 0.002 | 0.014 | 0.042 | 0.028 | 0.086 |
| | 4.5 | — | — | — | — | — | — | — | — | 0.007 | 0.005 | 0.044 | 0.056 |
| | 5.0 | — | — | — | — | — | — | — | — | — | 0.005 | 0.003 | 0.008 |
| | >5.0 | — | — | — | — | — | — | — | — | — | — | — | 0.000 |
| | TOTAL | — | 3.170 | 10.109 | 16.273 | 16.138 | 16.773 | 16.058 | 6.682 | 5.317 | 4.939 | 4.541 | 100% |

The choice of these systems is based on: (a) the development stage of both technologies, running real tests, one of them being grid connected, (b) the manufacturers of both technologies have made their forecasted power matrices available for the purpose of this study, (c) the WECs have different operating principles and (d) both technologies are suitable for the integration of the devices into a port's infrastructure.

The characteristics of these two systems are explained below.

The Eco Wave Power located in Gibraltar (Figure 7) comprises 8 floaters that draw energy from incoming waves by converting the rising and falling motion of the waves into a clean energy generation process. Each floater is linked by a robust arm to a jetty and has the shape of a rectangular box and a surface area of about 4.68 m$^2$.

**Figure 7.** Eco Wave Power system in Gibraltar. Source: www.ecowavepower.com.

The second device is OBREC. This system is integrated in a traditional breakwater and can be considered an innovative non-conventional breakwater (Figure 8). It fulfils the same functions as traditional breakwaters with the added value of power generation. This technology captures part of the wave run-up on the slope of the device. The potential energy of the stored water in the reservoir is converted into kinetic energy, flowing through low head turbines located in an engine room behind the reservoir. The energy is, thus, converted into electrical energy by means of generators coupled to turbines. It fulfils the same functions as traditional breakwaters in terms of coastal protection with the added value of power generation. It is the first overtopping WEC totally integrated into an existing breakwater.

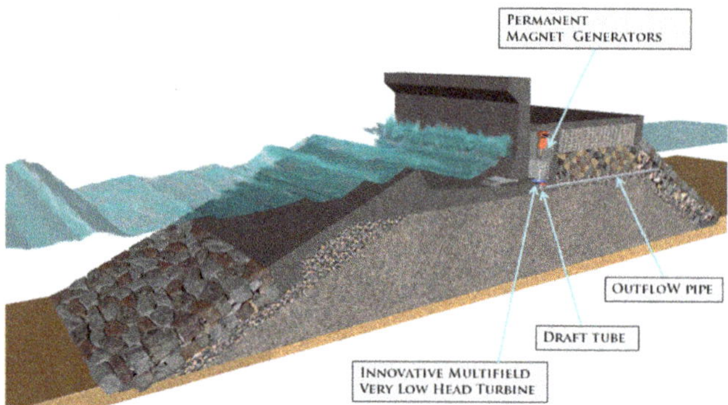


**Figure 8.** Lateral view of the OBREC device in Italy (Reprinted from Sustainability, Contestabile et al., 2016).

## 4. Results

As mentioned in the previous section, the AEP of the WEC devices can be calculated by combining the wave energy resource characterization matrix in the study area with the power matrices of the devices.

According to the Eco Wave Power calculation method, wave heights and period on the site, and considering a potential location on the Valencia breakwaters, Eco Wave Power has forecasted the possibility of installation of up to 70 floaters on the external side of the breakwater, with 0.5 spacing, and with a surface area of 20 $m^2$ for each floater. This kind of arrangement will enable an installed capacity of 300 kW, with an AEP of 0.8 GWh/year and with a capacity factor of 28.8 %, which is higher than the capacity factor available by solar energy and equal to the capacity factor to be achieved by installing wind turbines (see Table 5, showing the capacity factor for solar is 18%). However, considering the scaling potential of the Eco Wave Power system and an available breakwater length of 1000 m the AEP would be in the region of 2.6 GWh/year. In comparison with that, the photovoltaic plant would need 26,565 square meters for an installed power of 5.5 MW with an estimate of production of 10.1 GWh/year considering a capacity factor of 21%, which is 18.7% of the energy consumption by the Port of Valencia.

**Table 5.** Capacity factors in 2013 for various UK power plants. Data from the Department of Energy and Climate Change (DECC).

| Plant Type | Capacity Factor (%) |
|---|---|
| Nuclear power plants | 73.8 |
| Combined cycle gas turbine stations | 27.9 |
| Coal-fired power plants | 58.4 |
| Hydroelectric power stations | 31.7 |
| Wind power plants | 32.3 |
| Photovoltaic power stations | 10.2 |
| Marine (wave and tidal power stations) | 9.7 |
| Bioenergy power stations | 58.0 |

On the other hand, a definite power matrix of the OBREC is not available yet, since the first full-scale prototype installed at Naples Harbour (Italy) is currently under monitoring (Figure 9);

therefore, for the purpose of this paper and based on previous research [40], considering that the device could be integrated along the designed length of the new breakwater (500 m), and based on the aforementioned research, the AEP that could be provided by the OBREC would be in the region of 2.5 GWh [40–43]. However, this type of technology cannot be installed on each meter as some spacing and energy lost in conversion should be considered. In that case the capacity factor of the OBREC would be estimated at around 15% [44], much less compared to that from Eco Wave Power.

**Figure 9.** Lateral view of the OBREC device in Italy (Reprinted from Sustainability, Contestabile et al., 2016).

In the case of Valencia, however, with a breakwater length of 500 m, calculations for the OBREC gave an estimation of around 1.60 GWh/year of energy production (EnerOcean, personal communication, 26 July 2019), a little lower than the estimates given by previous research that reached AEPs of 3.5 GWh for Imbituba, Brazil and 2.5 GWh for San Antonio, Chile [45] but in line with the expected production considering the poorer wave resource of Valencia compared to both aforementioned locations.

As some researchers have already stated [46], OBREC devices give competitive energy outputs when $0.5 < Hs < 1$ and $7 < Tp < 8$ [47] which, apparently, is a better match with the wave resource available in the study area.

This condition occurs only 20% of the year in the case of the Bay of Valencia. This fact would be an advantage for the Eco Wave Power System as it could work at any wave height and period.

For the Valencia sea conditions previously set and the simulated performance of the two selected WECs, 0.8 GWh/year and 1.6 GWh/year could be expected, which means, according to the electricity consumed by the Port of Valencia in 2016 given by [48], around 3% of the electricity needs overall. However, and due to the scaling potential of the Eco Wave Power technology, if an array of 1000 on the Port of Valencia's breakwater is installed, an energy production of 2.6 GWh/year is estimated.

Considering that the $CO_2$ equivalent to the electricity used by the Port of Valencia in 2016 was 18,392 $tCO_2eq$, using the WEC technologies, around 500 $tCO_2eq$ could be avoided per year in the worst scenario.

## 5. Discussion

Extracting energy from waves is a promising solution to cover part of a port's energy demand. In addition, this solution is carbon-free and can be combined with other carbon-free energy sources such as wind or solar.

Port infrastructure could be designed to accommodate new facilities able to generate energy from renewable sources such as the ones mentioned above. Considering the wave energy converters, we should choose those capable of performing two tasks at the same time: producing energy and protecting the coastline from the effects of climate change.

When assessing the possibility to use a port's infrastructure for the installation of wave energy converters, some of the topics related to the climate conditions at the test site should be considered: the orientation of the breakwater with respect to the incident waves, geographical location and wave stability. Other issues to be considered are technical issues: PTO efficiency, cost effectiveness, scalability, grid connectivity, environmental impact and logistics. In addition, when considering a project as a whole, all costs and risks involved should be taken into account and the efficiency is only part of the equation while other factors, such as maintenance and operating costs or environmental issues, among others, could make one project a better choice than another.

Some of these technologies are not at an optimal technology readiness level (TRL) yet, however, we have experiences of facilities presently working close to the study area giving promising results. For example, Eco Wave Power has had an installed and grid connected power station in Gibraltar since 2016 that is not far from feasible TRL, and the OBREC system is installed in a breakwater at the Port of Naples. The architecture of some of these WECs make them capable of giving reasonable protection to coastal environments, as these devices could provide both power generation and coastal protection, one could be compensated by the other [4].

As for coastal protection, a terminator WEC, such as the OBREC, would be the best option, as it can better dissipate the energy of the incoming waves [40], although for the purpose of this paper a point absorber near-shore system is also considered, especially given the pointed design of the Eco Wave Power floaters, which cut through the waves, thus decreasing some of the erosive effects of the waves.

By deploying WECs the significant wave heights breaking at a certain location can be modified [40], and this is a positive factor that could support the success of a project. Therefore, when assessing the commissioning project of a WEC array, it is important to assign a monetary value to the ecological effects, the coastal protection action and the emissions of greenhouse gases and other air pollutants that have been avoided.

Two WEC systems particularly capable of being embedded in a port's breakwater have been chosen to evaluate the AEP from the available wave resource at the Valencia Bay.

Both systems are scalable but different in their architecture, the OBREC being part of the breakwater infrastructure while the Eco Wave Power system must be linked to it. The results obtained for these two systems for the sea condition of the Valencia Bay are that far in preliminary assumptions. In terms of efficiency, the Eco Wave Power's efficiency is close to 50%, and the capacity factor is almost 29% in the aforementioned configuration, while the efficiency of the OBREC is around 15% according to the estimates in the case study.

The results previously obtained for the WECs can be compared to the results of the evaluation studies of energy production of wind and solar origin, carried out by the Valencia Port Authority.

In the case of the wind energy assessment, and according to the previous studies, the estimated installed power in the port of Valencia would be around 15 MW. To meet this requirement, various wind turbines between 3 MW and 4.8 MW nominal power could be placed. According to the data obtained from the meteorological network located in the Port of Valencia, the average wind resource available would be about 5.2 m/s. By introducing this data in the different performance curves of the suitable wind turbines together with the aforementioned limitations, a capacity factor between 26% and 30% would be obtained. Considering the most pessimistic estimates, the operating hours of the wind turbine would be around 2300 h, obtaining an AEP of approximately 27 GWh/year.

On the other hand, when considering the installation of solar panels, the port of Valencia has free surfaces on the roof of a new vehicles storage silo capable of holding approximately 12,650 panels. Considering a panel type of 435 W/unit, its capacity factor in the latitude of the Port of Valencia is about 18%, and therefore the estimate AEP would be in the order of 8 GWh/year.

There are references to the capacity factor of different technologies, including those that produce energy from renewable sources in Table 5 [49].

Most of the estimated CF for the three evaluated technologies in the study area are in line with those shown in Table 5 for the UK, except for the solar technology which is a bit higher for the location in Valencia, and wave energy, which is at 29% in the case of the Eco Wave Power technology.

Consequently, the installation of renewable energy sources (including the ocean) is feasible if the amount of energy production is not the key factor for the decision-making process. All the three potential energy sources considered above supply an amount of energy capable of covering more than 50% of the current energy needs of the Port of Valencia [7].

The great advantage of the Eco Wave Power system in contrast to the OBREC one is that the former is highly scalable, the cost is much lower, and the maintenance operations are quite simple while the latter is fully integrated into the breakwater and has less margin in case of failure or scaling needs.

Further research has to be done to improve the performance of the WEC devices. For example, other authors have designed a hybrid system that combines the OTD and OWC technologies, that could eventually improve the system performance in terms of energy production, however there are no available data in this respect [50].

## 6. Conclusions

It is a fact that for some years now ports have been adapting their infrastructures to face the challenges posed by the energy transition and climate change. There is an increasing number of examples of ports that are progressively integrating renewable energy (and marine energies) into their facilities with the double objective of producing clean energy to meet their needs and protecting their infrastructure from the effects of climate change.

This paper evaluates the potential integration of renewable marine energies in port infrastructure, with special focus on the estimation of the wave energy resource at the Bay of Valencia, near the Port of Valencia. The purpose of this paper is to estimate the wave energy resource by simulation tools using the WAM model forced by the wind output of the HIRLAM regional atmospheric model, and to study of the performance of two WECs capable of being integrated into the port's shelter infrastructure.

Among the different technologies of WEC, those categorized as near-shore would be the best option for the purpose of this paper. This assumption is reached considering as primary selection criteria the existence of a relatively exposed deep-water breakwater, such as those existing in ports. Due to the early stage of development, a reliable system should be implemented, that can be granted in ports, where a direct-current transmission can be obtained and low maintenance costs are involved, compared to those for offshore WECs.

Research has been undertaken among the different technological developers and two potential WECs have been identified: the OBREC and the Eco Wave Power systems, respectively.

The OBREC estimated AEP is relatively low, and the Eco Wave Power's AEP is comparable to solar, due to the fact that the available wave resource in the study area is poor, but this could be compensated by the fact that logistics, operation and maintenance costs are much lower, and coastal protection performance is much higher than those of the offshore WECs and other power generators based on different energy sources such as wind or solar.

At this point, a decision must be made considering the installation of renewable energy generators in ports. Although ports have large surfaces, these surfaces are not really made to accommodate energy generators as they do not produce large profits. On the other hand, a price should be set on the new climate conditions created by climate change. We put in the same equation all these variables; cost of energy, cost of missing biodiversity, higher flood risks and consequently economic loss in port terminals, revenues by exploiting the port infrastructure, etc.

In addition, although there is no ideal technology, we should opt for those that allow a greater use of the resource, adjusting more to the wave regime that is most often present in the study area. More pilot tests would be needed with the acquired knowledge from previous locations and the technological development from these experiences.

It is a fact that wind and solar energy technologies are at a very high level of maturity although there is still room for further improvement. By contrast, further research should be conducted in the case of wave power generators capable of being integrated into ports' infrastructure. This research should focus on hybrid solutions that combine several renewable sources, such as wind and waves or tides and waves or wave and solar, such as the new patent submitted by Eco Wave Power, which incorporates solar panels on top of the floaters, thus saving space, while providing an additional energy source. In this case, the design of these energy devices should be considered when designing ports' shelter structures.

**Author Contributions:** Conceptualization, E.G. and R.C.; methodology, R.C. and E.Q.; validation, A.C. and R.C.; data curation, E.G. and R.C.; writing—original draft preparation, R.C.; writing—review and editing, R.C. and E.Q.; supervision, E.G. and F.M.; project administration, E.G. and E.Q.; funding acquisition, A.C. and F.M. All authors have read and agreed to the published version of the manuscript.

**Funding:** This research received no external funding.

**Acknowledgments:** Special thanks should be given to the Valencia Port Authority, EnerOcean, S. L. and Eco Wave Power for their contribution to the article by providing essential data not openly accessible. Also, thanks should be given to the research team from the Catania University for giving their permission to publish some of the pictures.

**Conflicts of Interest:** The authors declare no conflict of interest whatsoever.

## Abbreviations

| | |
|---|---|
| AEP | Annual Energy Production |
| AWP | Autonomous Wave Prediction System |
| CF | Capacity Factor |
| HIRLAM | High Resolution Limited Area Model |
| OBREC | Overtopping Breakwater for wave Energy Conversion |
| OTD | Overtopping Devices |
| OWC | Oscillating Water Column |
| PAV | Port Authority of Valencia |
| PTO | Power Take Off |
| SV | Seasonal variability |
| TEU | Twenty Feet Equivalent Unit |
| TRL | Technology Readiness Level |
| WAM | Wave prediction Model |
| WEC | Wave Energy Converters |

## References

1. Liberti, L.; Carillo, A.; Sannino, G. Wave energy resource assessment in the Mediterranean, the Italian perspective. *Renew. Energy* **2013**, *50*, 938–949. [CrossRef]
2. Lenee-Bluhm, P.; Paasch, R.; Özkan-Haller, H.T. Characterizing the wave energy resource of the US Pacific Northwest. *Renew. Energy* **2011**, *36*, 2106–2119. [CrossRef]
3. Hughes, M.G.; Heap, A.D. National-Scale wave energy resource assessment for Australia. *Renew. Energy* **2010**, *35*, 1783–1791. [CrossRef]
4. Manasseh, R.; Sannasiraj, S.A.; McInnes, K.L.; Sundar, V.; Jalihal, P. Integration of wave energy and other marine renewable energy sources with the needs of coastal societies. *Int. J. Ocean Clim. Syst.* **2017**, *8*, 19–36. [CrossRef]
5. Evolution of Wave Energy Potential (Evolución del Potencial de la Energía de Las Olas). IDEA. 2011. Available online: https://www.idae.es/uploads/documentos/documentos_11227_e13_olas_b31fcafb.pdf (accessed on 11 May 2020).
6. Mustapa, M.; Yaakob, O.; Ahmed, Y.M.; Rheem, C.K.; Koh, K.; Adnan, F.A. Wave energy device and breakwater integration: A review. *Renew. Sustain. Energy Rev.* **2017**, *77*, 43–58. [CrossRef]
7. Cascajo, R.; García, E.; Quiles, E.; Correcher, A.; Morant, F. Integration of marine wave energy converters into seaports: A case study on the port of Valencia. *Energies* **2019**, *12*, 787. [CrossRef]

8. Port of Rotterdam. Available online: https://www.portofrotterdam.com/en/doing-business/setting-up/existing-industry/energy-industry/sustainable-energy (accessed on 12 September 2020).
9. Port of Antwerp. Available online: www.portofantwerp.com/en/focus-renewable-energy-development (accessed on 13 September 2020).
10. Port of Amsterdam. Available online: https://www.portofamsterdam.com/en/port-amsterdam/energy-transition-towards-international-port-renewable-fuels-and-energy (accessed on 12 September 2020).
11. Marine Biotech Project Website. Available online: http://www.marinebiotech.eu/wiki/Wave_energy_converters_in_coastal_structures (accessed on 12 September 2019).
12. Vicinanza, D.; Di Lauro, E.; Contestabile, P.; Gisonni, C.; Lara, J.; Losada, I. Review of innovative harbor breakwaters for wave-energy conversion. *Waterw. Port Coast. Ocean Eng.* **2019**, *145*, 1–18. [CrossRef]
13. Available online: https://www.idae.es/uploads/documentos/documentos_11227_e4_atlas_eolico_331a66e4.pdf (accessed on 7 September 2020).
14. Port Authority of Valencia Web Site. Available online: https://www.valenciaport.com (accessed on 16 September 2020).
15. Climeport Project. Methodology Assessment Handbook. Available online: https://www.programmemed.eu/uploads/tx_ausybibliomed/CLIMEPORT_1_Methodology_Assessment_Handbook_EN.pdf (accessed on 7 September 2020).
16. Maheras, P.; Flocas, H.A.; Patrikas, I.; Anagnostopoulou, C. A 40-year objective climatology of surface cyclones in the Mediterranean region: Spatial and temporal distribution. *Int. J. Climatol.* **2001**, *21*, 109e30. [CrossRef]
17. Lionello, P.; Malanotte-Rizzoli, P.; Boscolo, R. *Mediterranean Climate Variability*; Elsevier: Amsterdam, The Netherlands, 2006; p. 421.
18. Campins, J.; Genovés, A.; Picornell, M.A.; Jansà, A. Climatology of Mediterranean cyclones using the ERA-40 dataset. *Int. J. Climatol.* **2011**, *31*, 1596–1614. [CrossRef]
19. Ulbrich, U.; Leckebusch, G.C.; Pinto, J.G. Extra-Tropical cyclones in the present and future climate: A review. *Theor. Appl. Climatol.* **2009**, *96*, 117–131. [CrossRef]
20. Lionello, P.; Dalan, F.; Elvini, E. Cyclones in the Mediterranean region: The present and the doubled $CO_2$ climate scenarios. *Clim. Res.* **2002**, *22*, 147–159. [CrossRef]
21. Besio, G.; Mentaschi, L.; Mazzino, A. Wave energy resource assessment in the Mediterranean Sea on the basis of a 35-year hindcast. *Energy* **2016**, *94*, 50–63. [CrossRef]
22. Bargagli, A.; Carillo, A.; Pisacane, G.; Ruti, P.M.; Struglia, M.V.; Tartaglione, N. An integrated forecast system over the Mediterranean basin: Extreme surge prediction in the northern Adriatic Sea. *Mon. Weather Rev.* **2002**, *130*, 1317–1332. [CrossRef]
23. Álvarez Fanjul, E.; de Alfonso, M.; Ruiz, M.I.; Rodriguez, I.; López, J.D. Real time monitoring of Spanish coastal waters: The deep-water network. *Elsevier Oceanogr. Ser.* **2003**, *69*, 398–402.
24. Ministerio de Fomento; Puertos del Estado. Banco de Datos. Redes de Medida. Available online: https://www.puertos.es (accessed on 16 September 2020).
25. Gómez, M.; Carretero, J.C. Wave forecasting at the Spanish coasts. *J. Atmos. Ocean Sci.* **2005**, *10*, 389–405. [CrossRef]
26. Undén, P.; Rontu, L.; Jarvinen, H.; Lynch, P.; Calvo Sánchez, F.J.; Cats, G.; Cuxart, J.; Eerola, K.; Fortelius, C.; Garcia-Moya, J.A.; et al. HIRLAM-5 Scientific Documentation. Available online: http://citeseerx.ist.psu.edu/viewdoc/download?doi=10.1.1.6.3794&rep=rep1&type=pdf (accessed on 14 September 2020).
27. Group, T.W. The WAM model—A third generation ocean wave prediction model. *J. Phys. Oceanogr.* **1988**, *18*, 1775–1810. [CrossRef]
28. Gómez Lahoz, M.; Carretero Albiach, J.C. A two-way nesting procedure for the WAM model: Application to the Spanish coast. *Trans. Am. Soc. Mech. Eng.* **1997**, *119*, 20–24. [CrossRef]
29. Cornett, A.M. A global wave energy resource assessment. In Proceedings of the International Offshore and Polar Engineering Conference, Vancouver, BC, Canada, 6–11 July 2008; pp. 318–326.
30. Boronowski, S.; Wild, P.; Rowe, A.; van Kooten, G.C. Integration of wave power in Haida Gwaii. *Renew. Energy* **2010**, *35*, 2415–2421. [CrossRef]
31. Morim, J.; Cartwright, N.; Hemer, M.; Etemad-Shahidi, A.; Strauss, D. Inter- and intra-annual variability of potential power production from wave energy converters. *Energy* **2018**, *169*, 1224–1241. [CrossRef]

32. Vannuchi, V.; Cappietti, L. Wave energy resource assessment and performance estimations of the state of the art of wave energy converter in Italian hotspots. *Sustainability* **2016**, *8*, 1300. [CrossRef]
33. Veigas, M.; López, M.; Iglesias, G. Assessing the optimal location for a shoreline wave energy converter. *Appl. Energy* **2015**, *132*, 404–411. [CrossRef]
34. Robertson, B.; Hiles, C.; Luczkoa, E.; Buckhama, B. Quantifying wave power and wave energy converter array production potential. *Int. J. Mar. Energy* **2016**, *14*, 143–160. [CrossRef]
35. Rusu, E.; Onea, F. Estimation of the wave energy conversion efficiency in the Atlantic Ocean close to the European Islands. *Renew. Energy* **2016**, *85*, 687–703. [CrossRef]
36. Veigas, M.; López, M.; Romillo, P.; Carballo, R.; Castro, A.; Iglesias, G. A proposed wave farm on the Galician coast. *Energy Convers. Manag.* **2012**, *99*, 102–111. [CrossRef]
37. Carballo, R.; Sánchez, M.; Ramos, V.; Castro, A. A tool for combined WEC-site selection throughout a coastal region: RiasBaixas, NW Spain. *Appl. Energy* **2014**, *135*, 11–19. [CrossRef]
38. Behrens, S.; Hayward, J.; Hemer, M.; Osman, P. Assessing the wave energy converter potential for Australian coastal regions. *Renew. Energy* **2012**, *43*, 210–217. [CrossRef]
39. Available online: https://www.ecowavepower.com/gibraltar/ (accessed on 18 September 2020).
40. Di Lauro, E.; Lara, J.L.; Maza, M.; Losada, I.J.; Contestabile, P.; Vicinanza, D. Stability analysis of a non-conventional breakwater for wave energy conversion. *Coast. Eng.* **2019**, *145*, 36–52. [CrossRef]
41. Di Lauro, E.; Contestabile, P.; Vicinanza, D. Wave energy in Chile: A case study of the Overtopping Breakwater for Energy Conversion (OBREC). In Proceedings of the 12th European Wave and Tidal Energy Conference, Cork, Ireland, 27 August–1 September 2017.
42. Iuppa, C.; Cavallaro, L.; Musumeci, R.E.; Vicinanza, D.; Foti, E. Empirical overtopping volume statistics at an OBREC. *Coast. Eng.* **2019**, *152*. [CrossRef]
43. Castiglione, F.; Cavallaro, L.; Iuppa, C.; Contestabile, P.; Vicinanza, D.; Foti, E. Performances of a breakwater for wave energy conversion. In Proceedings of the Thirteenth European Wave and Tidal Energy Conference, Napoli, Italy, 1–6 September 2019. ISSN 2309-1983.
44. Di Lauro, E.; Contestabile, P.; Vicinanza, D. Non-Conventional overtopping breakwater for energy conversion. In *Advances in Renewable Energies Offshore, Proceedings of the 3rd International Conference on Renewable Energies Offshore (RENEW 2018), Lisbon, Portugal, 8–10 October 2018*; CRC Press: Boca Raton, FL, USA, 2018.
45. Contestabile, P.; Iuppa, C.; Di Lauro, E.; Cavallaro, L.; Andersen, T.L.; Vicinanza, D. Wave loadings acting on innovative rubble mound breakwater for overtopping wave energy conversion. *Coast. Eng.* **2017**, *122*, 60–74. [CrossRef]
46. Naty, S.; Viviano, A.; Foti, E. Wave energy exploitation system integrated in the coastal structure of a mediterranean port. *Sustainability* **2016**, *8*, 1342. [CrossRef]
47. Falcão, A.; Henriques, J. Oscillating-water-column wave energy converters and air turbines: A review. *Renew. Energy* **2016**, *85*, 1391–1424. [CrossRef]
48. Report on Greenhouse Gas Emissions in the Port of Valencia (Informe de Emisiones de Gases de Efecto Invernadero del Puerto de Valencia). 2016. Available online: https://www.valenciaport.com/wp-content/uploads/Memoria-Verificaci%C3%B3n-GEI-2016.pdf (accessed on 16 June 2020).
49. Hashemi, M.; Neill, S.P. *Fundamentals of Ocean Renewable Energy*; Academic Press: Cambridge, MA, USA, 2018; Chapter 1; ISBN 9780128104491.
50. Das Neves, L.; Samadov, Z.; Di Lauro, E.; Delecluyse, K.; Haerens, P. The integration of a hybrid Wave Energy Converter in port breakwaters. In Proceedings of the 13th European Wave and Tidal Energy Conference, Naples, Italy, 1–6 September 2019.


**Publisher's Note:** MDPI stays neutral with regard to jurisdictional claims in published maps and institutional affiliations.